University of Nevada, Reno

SEISMIC RESPONSE ASSESSMENT AND RECOMMENDATIONS FOR THE DESIGN OF SKEWED HIGHWAY BRIDGES

A dissertation submitted in partial fulfillment of the
requirements for the degree of Doctor of Philosophy in
Civil and Environmental Engineering

by

Ahmed M. Abdel-Mohti

August, 2009

UMI Number: 3369575

INFORMATION TO USERS

The quality of this reproduction is dependent upon the quality of the copy submitted. Broken or indistinct print, colored or poor quality illustrations and photographs, print bleed-through, substandard margins, and improper alignment can adversely affect reproduction.
In the unlikely event that the author did not send a complete manuscript and there are missing pages, these will be noted. Also, if unauthorized copyright material had to be removed, a note will indicate the deletion.

University of Nevada, Reno
Statewide · Worldwide

THE GRADUATE SCHOOL

We recommend that the dissertation
prepared under our supervision by

AHMED M. ABDEL-MOHTI

entitled

Seismic Response Assessment And Recommendations For The Design Of Skewed Highway Bridges

be accepted in partial fulfillment of the
requirements for the degree of

DOCTOR OF PHILOSOPHY

Gokhan Pekcan, Ph.D., Advisor

Manos Maragakis, Ph.D., Committee Member

Ian Buckle, Ph.D., Committee Member

John Anderson, Ph.D., Committee Member

Shen-Yi Luo, Ph.D., Graduate School Representative

Marsha H. Read, Ph. D., Associate Dean, Graduate School

August, 2009

Abstract

Seismic vulnerability of highway bridges remains an important problem and has received increased attention as a consequence of unprecedented damage observed during several major earthquakes. A significant number of research studies examined the performance of skewed bridges under service and seismic loads. It is noted that results of these studies are particularly sensitive to modeling assumptions in view of the interacting parameters with the skew. The present study investigates the seismic response characteristics of two-span skewed post-tensioned box-girder bridges with skew angles varying from 0 to 60 degree. To facilitate a comprehensive analytical investigation simplified modeling techniques are introduced. The accuracy of simplified beam-stick models are verified against counterpart finite element models. Effect of various parameters on the overall seismic performance was examined and implications were presented. Finally, nonlinear time history analysis was used to generate system fragility functions for a range of three-span highway bridges (straight, moderate skew, and significant skew). It is shown that simplified beam-stick models have the capability to capture the nonlinear time history response of skewed bridges. Using incremental dynamic analysis (IDA) method, fragility functions were developed and the effect of the skew on the vulnerability of highway bridges is investigated. Furthermore, analytical fragility functions of skewed bridges were compared to HAZUS fragility functions. The soil type, presence of shear keys, and aspect ratio are determined to have a significant effect on the seismic performance of skewed bridges. Also, it was concluded that highway bridges with large skew angles (> 30°) are more vulnerable to seismically induced damages compared to those with moderate skew angles ($< 30°$).

Acknowledgment

Coming from the far east of United States after I finished my Masters to the far west to Reno has been a great experience. However, the graduate student life was still the same. I am glad that I made this move since I think that I learned a lot at UNR. Also, I would like to thank all the people who helped me to achieve my goals during my graduate studies at UNR.

I would like to show my appreciation and gratitude to my academic advisors Dr. Gokhan Pekcan and Dr. Manos Maragakis. Dr. Pekcan has been a great source of advice and guidance throughout my project. I am grateful to him and I appreciate his suggestions so much. I also owe thanks to my exceptional committee members and I would like to thank Dr. Ian Buckle, Dr. John Anderson, and Dr. Shen-Yi Luo for their encouragement and support as well as valuable classes. I am thankful to Dr. Ahmad Itani, Dr. Mahmoud Zadeh, Dr. Mehdi Saiidi, and Dr. Raj Siddharthan for their contribution to my learning.

The research presented here was funded by California Department of Transportation (Caltrans) under the contract # CT59A0503.

I am grateful to my colleagues and friends Hamid Bahrami, Kevin Almer, Eric Mozon, Ashkan Vosooghi, Juian Arias, Mohamed Ayoub, Ahmed Abdel-Dayem, Ahmed Badawy, Mohamed Issa, David Hillis, and Christin Linke for their encouragements. I would like to thank Dr. Sherif Elfass for helping me to find accommodations in my first days in Reno.

I am thankful to my professors in Egypt and I would like to thank Dr. Mostafa Gad, Dr. Abdelmonam Amin, Dr. Amr Abouhashish, Dr. Mohamed Fahmy, and Dr. Samir Zien for their encouragement and advices.

I am indebted to my family my father, mother, brothers, sisters, and brothers- and sisters-in law. I owe special thanks to my mother who always cares about me.

Finally, I could not have done this without the help and dedication of my wife Haidy. I would like also to thank my newborn daughter Taleen for the nice moments that she gave to me.

Table of Contents

List of Tables

List of Figures

1. Introduction

1.1. Background

Following the 1971 San Fernando earthquake, the seismic vulnerability of highway bridges remains still to be an important problem and has received increased attention as a consequence of unprecedented damage observed during several major earthquakes. It is noted that highway bridges are required to maintain functionality following a major earthquake to provide access to hospitals, fire stations, and other important facilities. Particularly, skewed highway bridges are vulnerable to severe damage due to seismic loads. Over the past three decades a number of studies have been conducted to investigate the seismic response characteristics of skewed highway bridges, however, research findings have not been comprehensive enough to address global system response. Also, a limited number of studies have been noted in the literature that address the fragility assessment of skewed highway bridges. Therefore, the presented research aimed to be comprehensive through investigating different parameters affecting seismic response of skewed bridges for a wide range of skew angles. Also, preliminarily fragility functions for a class of highway bridges with various skew angles are developed.

1.2. Response Characteristics of Skewed Highway Bridges

Many advances have been made in developing design codes and guidelines for static and dynamic analyses of regular or straight highway bridges. However, there remains significant uncertainty with regard to the structural system response of skewed highway bridges as it is reflected by the lack of detailed procedures in current guidelines. In fact, as evidenced in by past seismic events (i.e. 1994 Northridge – Gavin Canyon Undercrossing and 1971 San Fernando – Foothill Boulevard Undercrossing), skewed highway bridges are particularly vulnerable to severe damage due to seismic loads (Figure 1-1). Even though a number of studies have been conducted over the last three decades to investigate the response characteristics of skewed highway bridges under static and dynamic loading, research findings have not been sufficiently comprehensive to address global system characteristics. Due to the fact that the current seismic design guidelines do not provide explicit procedures, a significantly large number of bridges are still at risk with consequential threat to loss of function, life safety, and economy. Many of the existing bridges may be prone to earthquake induced damage and may require substantial retrofit measures to achieve desired seismic performance and post-earthquake serviceability. Researchers and practicing design engineers need to fully understand the overall system response characteristics of skewed highway bridges for the proper detailing of system components. Therefore, it is of interest to quantify the level of ground shaking necessary to cause the occurrence of various limit states such as: (1) onset of damage, (2) life safety, and (3) partial or total collapse.

Overall response behavior of skewed highway bridges under both service and seismic loads is inherently complex. Several parameters interact with the skewness of the bridge that may lead to this complex response behavior, such as:

(1) superstructure flexibility,

(2) boundary conditions: support details at abutments and intermediate supports, shear keys, soil-foundation-structure interaction,

(3) in-span hinges and restrainers (if any) along with sub and super-structure interactions,

(4) width-to-span ratio, and

(5) mass and stiffness eccentricity.

Moreover, several seismic response issues are of importance to skew bridges as opposed to straight counterparts. These include:

(1) Bearings in skewed bridges are vulnerable to high and uneven distribution of support reactions due to atypical superstructure deformations,

(2) The potential unseating behavior in skew bridges is complex because the impact between the superstructure and abutments may cause the bridge superstructure to translate and rotate,

(3) End-diaphragms tend to increase the difference between the bearing reactions and may even produce uplift at heavily skewed abutments,

(4) The vertical component of earthquakes activates the deck inertia in the vertical direction, and the deck acts as a flexible plate. Under such conditions, there is a possibility of uplift at supports, and

(5) Increasing the skew angle introduces torsional and lateral dynamic mode coupling that leads to in-plane rotation of the superstructure. One of the adverse effects of this coupling may be elevated moments, torsions, axial and shear forces in supporting substructure components.

Degree of skew can dramatically affect the dynamic behavior of a bridge. The majority of bridges that failed during earthquakes have had significant skew, including the Foothill

Boulevard Undercrossing (San Fernando 1971) with a 60 degree skew and the I-5 Gavin Canyon Undercrossing (Northridge 1994) with a 66 degree skew. Several studies have investigated the effects of skew angle on the response of highway bridges (i.e. Maleki, 2002; Bjornsson et al., 1997; Saiidi and Orie, 1991; Maragakis, 1984). Saiidi and Orie (1991) noted the skew effects and suggested that simplified models and methods of analysis would result in sufficiently accurate predictions of seismic response for bridges with skew angles less than 15 degrees. On the other hand, Maleki (2002) concluded that slab-on-girder bridges with skew angles up to 30 degrees and spans up to 65 feet have comparable response characteristics to straight bridges, and therefore, simplified modeling techniques such as rigid deck modeling can be used in many instances. Bjornsson et al. (1997) conducted an extensive parametric study of two-span skew bridges modeled with rigid deck assumption. In this study, maximum relative abutment displacement (MRAD) was considered as a critical quantity associated with failures due to unseating. MRAD was found to be influenced strongly by the impact between the deck and the abutments. Critical skew angle as a function of span length and width was introduced to maximize the rotational impulse due to impact and was found to be between 45 and 60 degrees. It was shown that MRAD for all bridge geometries tends to increase linearly for skew angles less than 30 degrees. In general, most of the research studies agree that bridges with relatively high skew angles (typically greater than 30 degrees) exhibit complex response behavior that may require detailed seismic modeling and analysis.

In evaluating and comparing results across various seismic response studies, one must consider the underlying assumptions and idealizations implemented in the analytical treatment of the skew highway bridges. These may pertain to material modeling, inelastic (hysteretic) response characteristics of components, boundary conditions, soil-structure interaction, component geometry (i.e. idealized beam-stick versus full finite element), superstructure (i.e. rigid versus flexible), seismic mass (i.e. distributed versus lumped), etc. For instance, Meng and Lui (2000) suggested that the effects of modeling boundary conditions properly may outweigh the effects of skew angle on the overall dynamic response characteristics of a bridge. In fact, differences in assumptions can create inconsistencies in results as seen in the analysis of the Foothill Boulevard Undercrossing, which sustained severe damage during the San Fernando earthquake. A study conducted by Wakefield et al. (1991) concluded that the failure was controlled by rigid-body motion, which agreed with a previous study conducted by Maragakis (1984). But, a study conducted by Ghobarah and Tso (1974) explained that the failure was induced by flexural and torsional motion. Ghobarah and Tso (1974) assumed the deck was fixed at the abutments, while Wakefield et al. (1991) assumed free translation of the deck at the abutments.

In general, there are two main methods used to model bridge decks. The first method uses beam-stick model which simplifies the analysis greatly. The deck is modeled by a single beam-column element with the mass, length, and cross-sectional properties of the real deck. The second and more sophisticated method uses explicit model of the superstructure with 3D finite element (FE). In this method, the deck is modeled as a built-up plate, and the columns are simple beam-column elements. The beam-stick model is usually sufficient for regular bridges, but a 3D FE model is preferred for irregular bridges, such as skewed or curved bridges and bridges with significant mass and stiffness eccentricity. Neglecting deck flexibility in a rigid deck analysis can underestimate the bridge response in many ways, including the axial forces in the columns and the displacement of the deck. Therefore, the use of the rigid deck or stick model for the dynamic analysis of skew bridges with large skew angles is not recommended because the method fails to predict the major response parameters accurately in seismic analysis. By including deck flexibility in an analysis, a more realistic assessment of the structural response is expected to be obtained. Accuracy of nonlinear analysis results of bridges are affected by modeling details at abutments. To achieve a high level of accuracy, all abutment details have to be modeled explicitly. Nonlinear response characteristics of the bearing pads, abutment-soil interaction, and shear keys have to be modeled explicitly and as accurately as possible.

1.3. Seismic Fragility of Highway Bridges

It was mentioned earlier that the functionality of highway bridges after an earthquake is important. It is critical to assess the vulnerability of bridges to help in decision making; such as opening the bridge to traffic immediately after an earthquake, repair the bridge, or completely replace the bridge or some of its components. Over the last two decades, a noticeable development in the probabilistic assessment of highway bridges subjected to seismic hazard was recognized. Fragility curves have become a common tool to estimate the damage levels in bridges for a known intensity of ground motion (Yamazaki et al.,

2000; Basoz and Kiremidjian, 1997; Kircher et al., 1997; Mander and Basoz, 1999). Fragility curves reveal the relationship between the ground motion intensities and the probability of exceeding certain state of damage. The damage levels are defined by many researchers as slight, moderate, extensive, and collapse and highway bridges that experience various level of damage was reported by many researchers (Chang, 2000; JSCE, 1999; Bruneau et al., 1996; Comartin et al., 1995; Moehle, 1994; Buckle, 1994) within the context of studies of the bridges performance.

There is a number of studies addressed fragility of regular highway bridges; empirically and analytically. Empirical curves are developed using data from historical earthquakes while analytical curves are developed by conducting inelastic dynamic analysis on bridge models using large number of ground motions. Analytical methods to develop fragility curves can be considered more diverse. One of the drawbacks of the empirical fragility curves is that they require a large amount of damage data for specific class of structure. On the other hand, analytical fragility curves can be used to assess damage in bridges which did not experience any earthquake (Karim and Yamazaki, 2001). 1995 Kobe earthquake is considered as the most damaging earthquake in Japan. Yamazaki et al. (2000) and Shinozuka et al. (2000a) developed a set of empirical fragility curves based on the actual damage date due to the Kobe earthquake. Empirical fragility curves were also developed for bridges in California (Basoz and Kiremidjian, 1999). Analytical fragility curves are generated from the seismic response analysis of briges followed by verification of the results against actual earthquake data (Mander and Basoz, 1999; Hwang and Huo, 1998). Karim and Yamazaki (2001) showed that similar level of damage probability was achieved through the empirical and analytical fragility curves. A good agreement between empirical and analytical fragility curves for the 1994 Northridge and 1989 Loma Prieta earthquakes (Basoz et al., 1999 and Mander and Basoz, 1999) was observed. Developing analytical fragility curves consist of three parts: (1) simulation of ground motions, (2) modeling of bridges, and (3) developing fragility curves from response results of analysis. Different types of analysis can be used to obtain seismic response results; (1) nonlinear time history analysis (Karim and Yamazaki, 2001; Shinozuka et al., 2000b; Hwang et al., 2000a), (2) elastic spectral analysis (Hwang et al., 2000b), or (3) nonlinear static analysis (Mander and Basoz, 1999; Shinozuka et al., 2000b). Nonlinear time history analysis is considered the most accurate method among all these methods. Recently, the use of simplified nonlinear analysis methods (nonlinear static procedure) is gaining momentum, however, results of these methods don't show a constant agreement with nonlinear time history analysis in terms of predicting all levels of damage (Shinozuka et al., 2000b).

Karim and Yamazaki (2003) concluded that variation of structural parameters has a significant effect on fragility curves. The empirical fragility curves do not show information about type of the structure, structural performance, or variation of input ground motions (Mander and Basoz, 1999). Structural parameters and input motion characteristics including frequency content and duration have an effect on the vulnerability of the structure. Karim and Yamazaki (2001) found a significant effect of input ground motions on the fragility curves. Choi et al. (2004) found that continuity in prestressed concrete girder bridges improves the seismic resistance of these bridges.

Therefore, making prestressed concrete girder bridges continuous is a good practice. Hwang et al. (2001) recommended that for irregular structures, a single degree of freedom model cannot be employed and a finite element model can be more appropriate. Zhang et al. (2008) studied six classes of bridges subjected to seismic shaking and lateral spreading due to liquefaction. It was evident that bridges showed different resistance to seismic shaking and lateral spreading. Under seismic shaking, bridge load carrying capacity was improved for the case of isolation at the top of the piers and at the abutments. Seat type abutments increased the demand on the columns. Simply supported connections were found to reduce the vulnerability of columns. Most of the accomplished analytical bridge fragility studies mentioned above, considered the bridge columns only to measure the vulnerability (Mackie and Stojadinovic, 2004; Karim and Yamazaki, 2003; Shinzouka et al., 2000b). This method can be accurate for bridges when columns mainly govern the response. Nielson and DesRoches (2004) recommended the use of all major bridge components which are expected to be vulnerable. System fragility is recommended over component fragility as it is more generic and more representative of the state of damage in bridges. Nielson and DesRoches (2006) concluded that the use of any of the bridge components to assess the vulnerability of the bridge may lead to significant underestimation of the overall bridge fragility.

1.4. Modeling and Analysis Procedures

Nonlinear analysis procedures are typically used to establish performance levels and response assessment of highway bridges. The most realistic of the nonlinear procedures is nonlinear time-history analysis. However, this method of analysis requires a complex description of the analyzed system and the response is strongly sensitive to the models and the characteristics of the ground motions used in the analysis. Therefore, simplified nonlinear analysis methods have been developed based on the use of equivalent linear representation of the structural systems. The pushover analysis (hence pushover curve) constitutes an important major step in the simplified [nonlinear] method of analysis. In a practical sense it represents the capacity of the structure to resist seismic inertia forces. The pushover curve is constructed by "pushing" a computational model of the structure by monotonically increasing lateral loads in either force- or displacement-controlled manner. As the magnitude of the load increases, progressive yielding of the model occurs, which is accompanied by a change in the dynamic properties of the structure. Ideally, the pattern of loads (or displacements) should be consistent with the expected distribution of the inertia forces (or deformations) in the yielding structure.

Following the major earthquakes in the recent past, it has been widely recognized that changes are needed in the existing seismic design methodology implemented in the codes. Among several methods, the pushover analysis is generally recommended for simple or regular structures while the elastic and/or nonlinear time-history analysis is recommended for relatively complex structures with irregularities leading to mode-coupling and higher mode effects on the overall seismic response. Therefore, the inherent assumption of pushover analysis that the overall seismic response of the structure is controlled by the fundamental mode introduces severe limitations for its application to complex structural systems, such as highway bridges. Thus it is realized to be more rational to adopt both the pushover analysis and elastic or nonlinear time-history analysis, where the former is used for simple or regular structures and the latter is used for complex structures. Moreover, it should also be recognized that simplified pushover procedures have been predominantly developed and tested for building structural systems and that their implementation for bridge structures may not be as accurate or straightforward.

Several pushover methods for the seismic response evaluation of structures have been developed, which can potentially take into account the influence of higher modes. Three methods that are most widely referred are:

(1) modal pushover analysis (MPA) (Chopra and Goel, 2002; 2004);

(2) modal adaptive non-linear static procedure (MANSP) (Reinhorn, 1997; De Rue, 1998); and

(3) incremental response spectral analysis (IRSA) (Aydinoglu, 2004).

As an improvement over these methods, adaptive pushover procedures have also been developed and proposed by Bracci et al. (1997), Gupta and Kunnath (2000), Elnashai (2001) and Aydinoglu (2003). The applied methods have been mainly developed and tested for buildings. Fewer results are available regarding their use for the analysis of bridges (i.e. Kappos et al; 2004). Moreover, modal analysis procedures in the inelastic domain may not be appropriate for complex structural systems. However, in multimode pushover-based methods, the modal analysis as well as the mode combinations is unavoidable. The non-adaptive methods listed above address this issue using piecewise elastic (spectral) modal analysis with constant structural and dynamic properties between the occurrences of the two successive plastic hinges. Specifically, the MPA method takes into account the influence of higher modes. All of the important modes, identified in the initial elastic state, are used to determine the distribution of forces for the pushover analyses. Each mode is considered separately. Therefore, the number of analyses is equal to the number of the important modes in the elastic range. When all of the analyses are completed, the results are combined using the Complete Quadratic Combination (CQC) rule. The method assumes that the mode shapes are not changing during the response, whereas, the MANSP takes into account the influence of higher modes as well as the possibility that the deflection line could change significantly during the response. To obtain the capacity diagram, one pushover analysis with variable distribution of forces should be performed. The distribution depends on the variable dynamic properties of the structure. The forces are calculated combining the modal inertial forces (in each node forces related to different important modes are combined using SRSS). Each time the new hinge occurs in the structure, the structural system as well as its dynamic properties changes. After each occurrence of a new hinge, a new distribution of the forces for pushover is calculated based on the new dynamic properties of the structure.

The IRSA method also takes into account the influence of all important modes as well as the variable distribution of inertial forces. However, the variable distribution of modal forces is not shown explicitly, since this displacement controlled method works directly with modal displacements that are compatible with those modal forces. The method takes into account the changes of the dynamic properties of the structure each time a new hinge occurs. Each time a new hinge occurs in the structure, the spectral elastic modal analysis is performed. The slope of the capacity diagram is determined based on the instantaneous frequency of the structure. Monitoring some relevant chosen quantity, the level of load corresponding to the occurrence of the new hinge is determined. The frequency of the new structural system is changed, and a new slope of the capacity diagram is calculated. Several capacity diagrams, corresponding to each important mode, are constructed separately. During each analysis step the response quantities are combined using the standard modal combination rule (CQC or SRSS).

The response spectrum method is preferred tool since the maximum response can be calculated directly from the response spectrum. The response spectrum method is restricted only to elastic analysis (EM, 1999). The response spectrum method is one of the methods which is implemented in many design guidelines including Caltrans seismic design criteria. To generate response spectra for actual earthquake time histories, Duhamel's integral is used and maximum values are reported and plotted against periods or frequencies with respect to the damping ratio.

Selection of method of analysis is on basis of type of the bridge and its characteristics such as; regularity, number of spans, and geometry. There are three methods for conducting spectral analysis which are (1) Uniform Load Method (U), (2) Single Load Method (S), and (3) Multi-mode Method (M). Table 1-1 shows a recommendation for selection of type of analysis on basis of bridge characteristics. As moving from method-1 to method-3, the accuracy increases.

Not all bridges can be analyzed accurately using single mode method. Bridges can vibrate in other mode shapes with considerable mass participation factor. The importance of higher mode shapes can be determined by their frequency of vibration and direction of inertial forces for each mode. Level of mass participation is influenced by modes energy contents. Therefore, modes with highest energy content are more significant.

Modal coupling and higher mode effect is more pronounced in bridges with irregularity. Herein, effect of higher modes need to be incorporated. Many of these analysis methods can be performed using hand calculations or computer programs. As the bridge becomes complex, the hand calculations consumes more time hence using a computer based solution becomes better option.

In Table 1-1, bridges are classified as regular and irregular. A regular bridge is defined as it does not have a significant change in mass, stiffness, or geometry along its length. On the other hand, an irregular bridge does not follow the previous definition of a regular bridge. Mode coupling are more likely to occur in irregular bridges.

A single mode method (Figure 1-2) can be used safely in analysis of regular bridges. A hand calculation can be performed to apply this method in analysis of bridges. The use of this method gives accurate results in most of cases. It should be noted that the main assumption of this method is that a single mode of vibration controls the behavior of the bridge. In other words, throughout the duration of an earthquake the bridge behaves in a single mode of vibration. Therefore, approximations of this method can lead to uncertainties in results of bridges which have some irregularity but analyzed as regular bridges (Buckle et al., 1987). The single mode method is performed in the longitudinal and transverse direction followed by combing the results of these analyses. As a last step, the design forces and displacements will be calculated.

As discussed earlier, multi-mode spectral analysis method (Figure 1-3) is the most accurate for irregular bridges since a large number of mode shapes can be considered in this method. Irregularity can be in mass, stiffness, or geometry. In this method, the response is not limited to or governed by a single mode. A wide range of mode shapes can contribute to the total response of the bridge. A computer based program can be used to perform this type of analysis since it is impractical to conduct this analysis manually. The number of modes included in the analysis must ensure 90 percent mass participation. Therefore, the mass and stiffness of the system should be modeled accurately (Buckle et al., 1987).

The maximum responses evaluated for individual modes are combined in order to determine the system response. It should be noted that maximum response in each mode may not occur simultaneously. Usually, Square Root of the Sum of the Squares (SRSS) method or Complete Quadratic Combination (CQC) is used to combine modal responses. SRSS are accurate for structures with well separated mode shapes. Whereas, CQC is more suitable, accurate, and graceful for bridges which has mode coupling. The CQC method accounts for the statistical correlation between the various modal responses. It should be noted that the use of SRSS method to analyze bridge with mode coupling can lead potentially to overestimating the response. The use of SRSS method should be limited to structures with well separated mode shapes (Buckle et al., 1987) because it does not consider the effect of the cross-modal terms. SAP2000 incorporates both of SRSS and CQC methods to combine modal responses.

1.5. Objectives, Scope, and Organization

The objectives of the present study are:

(1) assess analytically the seismic response characteristics of skewed highway bridges

(2) develop preliminary fragility curves for skewed highway bridges

For the first part of the study, a series of benchmark bridge models were selected and included in this analytical investigation. It should be noted that the selection of the benchmark bridges was intentionally constrained to California bridges. The benchmark bridge is a post-tensioned (PT) box girder two-span bridge. Various parameters and configurations that may interact with the skewness of highway bridges are also investigated. For each set of analyses, variation of various response parameters versus the skew angle and aspect ratio was monitored; these parameters were namely; the deck displacement, column axial force, shear force, and moment, and abutment-soil structure interaction and shear keys response. Accordingly, the goal of this computational study is to investigate:

(1) accuracy of simplified beam-stick models versus more complex models used for modal and time history analyses in capturing the response characteristics of highway bridges with various skew angles and to provide guidelines

(2) effect of changing skew angle on the bridge response

(3) effect of ground motion intensity

(4) effect of soil type

(5) effect of abutment support conditions (with and without shear keys)

(6) effect of support (foundation) conditions

(7) effect of aspect ratio (width to span ratio)

To achieve this goal, simplified beam-stick (BS) and finite element (FE) models of a two-span continuous concrete box girder bridge were developed with six different skew angles, namely 0, 20, 30, 45, 52, and 60 degrees. Models included nonlinearities in the bent columns and abutment details; i.e. shear keys, support bearings, and abutment-soil interaction. The analysis types include;

(1) modal analysis to establish structural dynamic characteristics and variations,

(2) nonlinear time history analysis to investigate variations in various response quantities such as deformations, column axial force, shear force and moments.

(3) Multi-mode spectral analysis.

For nonlinear time history analysis, the two horizontal ground motions components were applied in the longitudinal and transverse directions of the bridge models. It is important to note that the effect of direction of the two horizontal components of the strong motions relative to the longitudinal and transverse direction did not have any significant effect on the overall response (Figure 1-4). The two cases are designated as ST and SL. For the ST, the stronger component was applied in the transverse direction while the opposite was true for the SL case.

Chapter 1 of this document discusses previous findings on seismic behavior of highway bridges. Chapter 2 presents modeling of bridges and parametric study for nonlinear modeling. Modal analysis was performed on each skew and model type for models with 0.3 aspect ratio and fixed foundation in order to measure the accuracy of the models in chapter 2. The aspect ratio is the ratio of the width of the bridge to its span length. Relative accuracy of simplified beam-stick models are verified against counterpart finite element models in chapter 2. Simplified beam-stick models are used for time history analyses performed on each skew. Also, both SAP2000 and DRAIN3DX programs were evaluated and the latter was used for nonlinear time history analyses of the bridges for the first part of the study in chapter 3. Summary and recommendations were made in light of the findings of this analytical study in chapters 4 and 6. Design tools were developed on basis of multi-mode spectral analysis and presented in chapter 4.

For the second part of the study, in order to assess the vulnerability of skewed highway bridges and develop a tool to determine the functionality of skewed bridges after an earthquake, fragility analysis was conducted in chapter 5. There is a limited number of studies addressing explicitly the effect of the skew angle on the vulnerability of highway bridges found in the literature. A Federal Highway Administration (FHWA) RC box-girder three-span bridge with a large aspect ratio was selected and 45 ground motions were used. The PGA's were scaled ranging from 0.1g to 1.0g. The fragility curves for bridges with three different skew angles (0, 30, and 60) were generated and damage states were evaluated.

1.6.   What is Particularly New in This Dissertation?

What is new in this dissertation is that the study is comprehensive to cover a large number of parameters and wide range of skew angles with accurate modeling techniques. What is also interesting is that fragility functions are developed for skewed highway bridges.

Improved beam stick models were developed, calibrated against more complex finite element models, and used for nonlinear analyses. Parametric studies were conducted to obtain consistent modeling assumptions for various structural details and analysis techniques. This study presented also recommendation for modeling of bridges. The effect of a wide range of parameters which are expected to affect the seismic performance of highway bridges with skew is investigated.

It was important to analyze skewed bridges using a simple tool such as response spectrum analysis. Multi-mode response spectra analyses were conducted and compared to the results of nonlinear time history analysis. A set of design equations are proposed for skewed highway bridges. Furthermore, three sets of fragility curves for three skew angles (0, 30, and 60) for highway bridges with 3-spans are generated to examine the effect of skew on the vulnerability of highway bridges and are compared to HAZUS functions.

Table 1-1. Analysis Method Recommendations (Buckle et al., 1987)

Seismic Performance Category	Regular Bridges With 2 or more spans	Irregular Bridges With 2 or more spans
A	-----	-----
B	Single mode	Single mode
C	Single mode	Multi-mode
D	Single mode	Multi-mode

Figure 1-1. I-5 Collapse at Gavin Canyon under 1994 Northridge Earthquake (http://nisee.berkeley.edu/elibrary/Image/NR788)

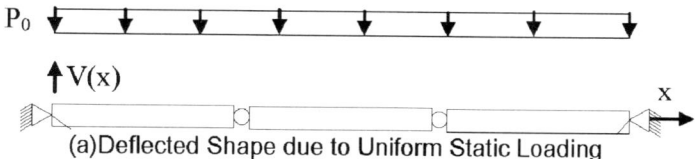

P_0

$V(x)$

x

(a)Deflected Shape due to Uniform Static Loading

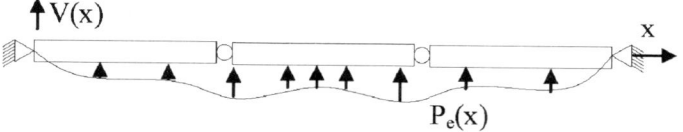

$V(x)$

x

$P_e(x)$

(b)Characteristic static Loading Applied to the Bridge System

Figure 1-2. Single Mode Method of Analysis (Buckle et al., 1987)

m_1 ϕ_{11} ϕ_{12} ϕ_{13}

m_2 ϕ_{21} ϕ_{22} ϕ_{23}

m_3 ϕ_{31} ϕ_{32} ϕ_{33}

First Mode   Second Mode   Third Mode

(a)Compute mode shapes and natural periods

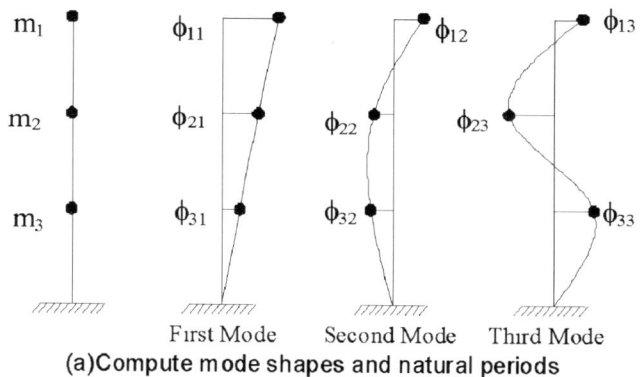

Spectral Acceleration

S_{a3}

S_{a2}

S_{a1}

Period

T_3 T_2 T_1

(b)Obtain Spectral acceleration for all modes

Figure 1-3. Illustration of Multi-modal Spectral Analysis (EM, 1999)

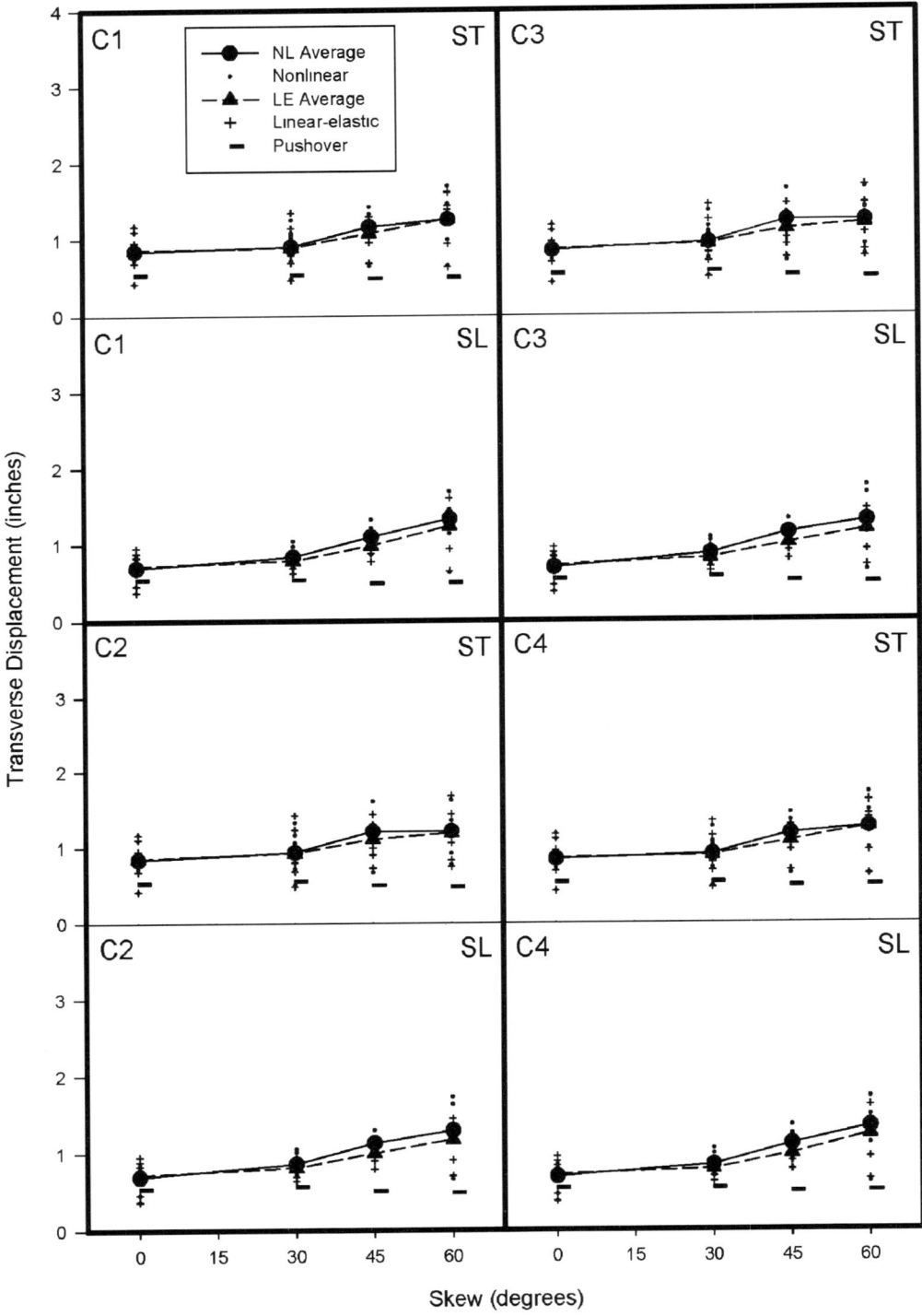

Figure 1-4. Effect of Ground Motion Direction on the Transverse Displacement at the Top of the Columns (Schroeder, 2006)

2. Modeling of Bridges and Parametric Studies

2.1. Introduction

This section presents details of the two sets of bridge models with various skew angles. The first set of models is based on a Caltrans benchmark bridge and it covers a range of skew angle from 0 to 60 degree. A set of nonlinear time history analyses is conducted using these models to study the effect of a range of parameters on the seismic performance of skewed bridges. The second set of models is based on a FHWA benchmark bridge with skew angles (0, 30, and 60 degrees) to establish preliminary fragility curves. It is noted that nonlinear time history analysis of a complete three-dimensional finite element model of a bridge requires significantly longer time than that of a simplified model. Therefore, simplified beam-stick models have been developed and used. For the first set of analyses, SAP2000 was employed to develop detailed three-dimensional (3D) finite element (FE) and improved beam-stick models (BS). In addition, DRAIN3DX was employed to develop improved beam-stick models (BS). For the second set of analysis, single spine models were developed and verified against counterpart finite element (FE) model (chapter 5).

In all of the models, the superstructure was assumed to behave linearly elastic and all of the nonlinearity was assumed to take place in the substructure elements including bents, shear keys, bearings, and abutment-soil interaction. This chapter presents the details of the FE and BS computational models in both SAP2000 and DRAIN3DX and how the nonlinearity was accounted for. Subsequently, structural dynamic properties of various models are compared and the accuracy of the improved simplified models from both SAP2000 and DRAIN3DX in presenting the modal characteristics is discussed in comparison to those obtained from the counterpart FE models for the first set of analysis. Whereas, for the second set of analysis (targeting fragility) accuracy of single spine model was verified against more complex FE model using SAP2000.

2.2. Selection of Benchmark Bridges

Two benchmark bridges were selected to investigate the dynamic behavior of skewed bridges and to develop fragility curves, respectively. Figure 2-1 through Figure 2-4 show general views and cross-sectional views of the selected bridges.

For the first set of analysis, Table 2-1 summarizes the data compiled from the twelve (12) Caltrans highway bridges considered. These twelve bridges are 42-0427L/R, 42-0431L/R, 39-0146, 39-0149, 19-0192R, 42-299L/R, 38-0075, 19-0197R, 17-0104, 17-0105, 53-2790R/L, and 390-0042. The properties, which were of interest, are span length, depth of superstructure, width of the bridge, width-to-span ratio, skew angle, number of spans, and number of columns per bent. These properties were summarized and tabulated as per Table 2-1. Also, the average of these properties was determined and was a key factor in selecting the benchmark bridge. The bridge that can be represented by the average properties, bridge 42-0427L/R, was selected for further analytical investigation. Figure 2-1 shows the elevation view and plan view of the selected bridge. Figure 2-4 shows the cross section details of the bridge. In addition since the skew angle is one of the critical parameters in this study, the selected bridge has the largest skew angle amongst the 12 bridges. The selected benchmark bridge has an aspect ratio of 0.3. The aspect ratio is defined as the ratio of bridge width including the overhang to the span length of the bridge. The bridge is a continuous post-tensioned two-span box-girder bridge with 268 ft total length, spans of 134 ft, and 52 degree skew (Figure 2-1). The cap beams of bents are integral with superstructure and each bent has two circular columns. The abutments of the bridge are seat-type abutments. The selected bridge was altered to further model bridges with larger aspect ratios of 0.54 and 1.1 later in this study. Figure 2-5 and Figure 2-6 show the cross section of the bridges with larger aspect ratios. These ratios were selected to study the effect of aspect ratio on the seismic performance of the skewed bridges.

For the second set of analysis, a highway bridge was chosen from Federal Highway Administration's (FHWA) Seismic Design of Bridges Series (Design Example No.4). The bridge was studied previously by Schroeder (2006). The bridge is a continuous three-span box-girder bridge with 320 ft total length, spans of 100, 120, and 100 ft, and 30 degree skew (Figure 2-2; FHWA, 1996). The bridge has an aspect ratio of 0.43. The cap beams of bents are integral with superstructure and each bent has two circular columns. The abutments of the bridge are seat-type abutments with a gap of 6 in between the deck and the abutment back-wall. The abutment seat width is 42 in. measured normal to the center line of the bearing from the edge of the bridge deck to the edge of the abutment (Figure 2-7).

2.3. Properties of Benchmark Bridges

The superstructures of the selected bridges are assumed to remain linear-elastic.
Therefore, properties of the deck were calculated based on uncracked properties. All
nonlinearities are assumed to take place at the abutments and the columns. Bent cap beam
is treated as rigid element. Column properties were calculated based on cracked
properties as most of nonlinearities takes place in columns. These properties are area (A),
torsion (J), and the moment of inertia $(I_y \ and \ I_z)$ about the weak and strong bending axis,
respectively. Appendix I presents the properties calculations. Table 2-2 through Table 2-4
show a list of properties of superstructure, bent cap beam, and bent columns for bridges
with different aspect ratios for Caltrans bridge. It should be mentioned that larger aspect
ratios required larger number of columns per bent. Originally, number of columns per
bent for the original aspect ratio of 0.3 is two. For bridges with aspect ratio of 0.54, the
number of columns increased to three columns while for that with aspect ratio of 1.1, the
number increased to four. However, the span length of 134 ft was kept to be the same for
all bridges. Table 2-5 shows section properties of FHWA bridge.

2.4. Finite Element (FE) Models

The Caltrans benchmark bridge, having 52 degree skew angle and aspect ratio of 0.3, was
altered to produce four other bridge models with skew angles of 0, 20, 30, 45, and 60
degree. The models are suitable for nonlinear static and dynamic analyses in both
SAP2000 (CSI, 2005) and DRAIN3DX (Prakash et al., 1994). SAP2000 allows the
modeling of the complete three-dimensional geometry of the bridge with a finite element
(FE) mesh Figure 2-8; including the super- and sub-structure (with post-tension,

diaphragms, bent and foundation details, etc.), and boundary conditions (with shear-keys, bearings, hinges, abutment-soil interaction). Although many features for nonlinear modeling and analyses are available in SAP2000, modeling of shear-keys and abutment-soil interaction for nonlinear time-history analyses was challenging due to the absence of an element with suitable hysteretic properties. On the other hand, DRAIN3DX allows detailed and complete nonlinear modeling of the bridge including the shear-keys, even though the super-structure geometry has to be approximated by a beam-stick model.

FE model of the FHWA benchmark bridge (Figure 2-9), having 30 degree skew was that developed by Schroeder (2006). Additional details and improvements were implemented particulary for the modeling of abutment-soil and shear key responses. A single spine model is developed and calibrated as per chapter 3 in SAP2000 for fragility analysis. Figure 2-10 and Figure 2-11 show the beam stick models for both bridges.

SAP2000 was employed to develop the complete FE models using shell and solid elements which are considered as the most accurate type of models in this study. The models for the first set of analysis have different skew angles; namely 0, 20, 30, 45, 52 (as is), and 60 degrees. The benchmark bridge is a two-span bridge with a two-column-interior bent cap and end diaphragms at abutments. For each skew, a finite element mesh was used to model the deck, soffit, girders, and diaphragms. Figure 2-12 shows an overview of details of bent cap beam of the selected bridge. The internal bent cap and end diaphragms at the abutment were modeled explicitly as part of the superstructure. The nodes making up each diaphragm were constrained so that the joints move together as a planner diaphragm that is rigid against membrane deformation. It should be mentioned that each diaphragm and bent is aligned along the skew angle being modeled.

The bent columns and footings were modeled using 3D beam-column elements. The properties of these elements are summarized in Table 2-2 through Table 2-5. Nonlinearity in bridges is assumed to take place in columns, bearing pad springs, abutment-soil interaction springs, and shear key springs. Nonlinearity in columns was localized at top and bottom of columns. The Fiber P-M2-M3 hinge was utilized to model plasticity in columns of Caltrans bridge. Whereas, uncouple axial and moment hinges were used for FHWA bridge. For the fiber hinge, the parametric study presented in section 2.8 resulted in the selection of number of fibers, length of plastic hinge, and location of plastic hinge with respect to the height of the column to ensure accurate model details. SAP2000 manual has the following definition of PMM fiber hinge "The Fiber P-M2-M3 (Fiber PMM) hinge models the axial behavior of a number of representative axial "fibers" distributed across the cross section of the frame element. Each fiber has a location, a tributary area, and a stress-strain curve. The axial stresses are integrated over the section to compute the values of P, M2 and M3. Likewise, the axial deformation U1 and the rotations R2 and R3 are used to compute the axial strains in each fiber. User defined fiber hinges are possible by explicitly specifying the location, area, material and its stress-strain curve for each fiber, or the program can automatically create fiber hinges for circular and rectangular frame sections.

The Fiber PMM hinge is more "natural" than the coupled PMM hinge, since it automatically accounts for the interaction, changing moment-rotation curve, and plastic axial strain. However, it is also more computationally intensive, requiring more computer storage and execution time. It is recommended to experiment with the number of fibers needed to get an optimum balance between accuracy and computational efficiency. Strength loss in a fiber hinge is determined by the strength loss in the underlying stress-strain curves. Because all the fibers in a cross section do not usually fail at the same time, the overall hinges tend to exhibit more gradual strength loss than hinges with directly specified moment-rotation curves. This is especially true if reasonable hinge lengths are used. For this reason, the program does not automatically restrict the negative drop-off slopes of fiber hinges. However, it is recommended to pay close attention to the modeling of strength loss, and modify the stress-strain curves if necessary". Therefore, user-defined stress-strain curves for both concrete and steel was developed and inserted in SAP2000. Furthermore, a parametric study was conducted to find the proper properties of PMM fiber element and is presented later.

For Caltrans bridge, the footing was assumed to present either a pinned or a fixed condition. The external shear-keys, bearings, and abutment-soil interaction were modeled using nonlinear springs. There are four bearings beneath each of the four girders. The properties of 17.7 x 17.7 x 1.97 in. electrometric bearings were determined using:

$$k_o = \frac{G \times A}{h} \qquad\qquad (2\text{-}1)$$

$$F_y = k_o \times \Delta_y \qquad\qquad\qquad\qquad \textbf{(2-2)}$$

The bearing was designed based on a shear modulus of elasticity (G) of 150 psi, and bearing dimensions of 2 in. in height, 314 in.^{2} of area. The initial stiffness (k_0) was calculated to be 23.9 kip/in., and yield force (F_y) of 47 kips. Failure strain of bearings according to Caltrans specifications is 150%. It was assumed that the yielding occurs at 100% strain. Therefore, the yield deformation (Δ_y) is equal to 1.97 in. The bearing was assumed to have an elastic-perfectly plastic hysteresis (Figure 2-13a). The plastic (Wen) link element available in SAP2000 was used to model the hysteresis of bearing pads. The bearing links were assigned in the longitudinal direction of the bridge and the transverse direction as well.

The modeling of abutment-soil interaction is complex. The hysteresis of abutment-soil interaction is elastic-perfectly plastic and compression only activated. Therefore a single link element in SAP2000 is not sufficient to model the complete hysteresis. Figure 2-14 represents the backbone curve provided by Caltrans as the force-displacement relationship.

For the Caltrans bridge, there is no gap between the deck and the abutment back-wall. The soil will not be activated unless it becomes in full contact with the abutment. Therefore, the induced gap following the first "yielding" cycle, which varies in width, needs to be overcome. This mechanism seems to be complex and difficult to model.

Nevertheless, combination of elements can be used in SAP2000 to achieve the behavior and is presented in detail later.

Figure 2-15 shows the developed hysteresis of the external shear keys for Caltrans and FHWA bridges. Hysteretic properties of shear keys were determined after the experimental study reported by Megally et al. (2002). Shear keys are external and four shear keys were introduced for the bridge, one at each corner of the bridge. The shear keys were aligned along with the absolute transverse direction to comply with details of bridge drawing provided by Caltrans and to closely simulate the behavior of external shear keys. In SAP2000, there is no single element that can be used to model behavior of shear keys thereby a combination of elements is developed in SAP2000 and used.

2.5. Improved Simplified Beam-Stick (BS) Models Using SAP2000

As mentioned earlier, in order to ensure accuracy and consistency throughout this study, it was important to develop these simplified models as accurately as possible. The stick models were developed based on a refined stick model proposed by Meng and Lui (2002). Appendix II presents the details of the calculations of the properties of beam stick models. The refined stick modeling approach was introduced particularly for the modeling of highway bridges with relatively large skew angles and to capture dynamic response characteristics (coupling) more accurately. Accordingly, in that refined modeling approach, the super-structure is modeled with two lines of girder elements. Particularly in this study, this approach is modified by introducing three lines of girder elements or more to further improve the accuracy and to better represent the distributed nature of the seismic mass and geometry of the superstructure (Figure 2-10). The properties of each beam stick were calculated based on the following equations:

$$A_s = \frac{A}{n} \tag{2-3}$$

$$I_{sy} = \frac{W_m}{W_T} \times I_y \tag{2-4}$$

For the actual deck of the bridge:

$$I_{mz} = \sum_{i=1}^{2} [\frac{M_i}{12}(L^2 + \frac{W_i^2}{\cos^2 \theta})] + \sum_{j=1}^{n_w} [M_j (\frac{L^2}{12} + \frac{d_j^2}{\cos^2 \theta})] \tag{2-5}$$

For the model:

$$I_z = nI_{sz} + (n-1)A_s S^2 \tag{2-6}$$

Where A_s = cross-sectional area of each stick; A = cross-sectional area of deck; I_{sy} = moment of inertia of each stick about the Y-axis; W_m = tributary width of each stick; W_T = total width of the deck; I_y = moment of inertia of deck cross section about Y-axis; I_{mz} = mass moment of inertia of the actual bridge deck; M_i = mass of top or bottom flange of the bridge deck; L = span; W_i = width of top or bottom flange of the bridge deck; M_j = mass of the jth web; d_j = perpendicular distance of the jth web with respect to the centerline of the deck; n_w = number of webs; I_z = moment of inertia of stick model about Z-axis; n = number of stick beams; I_{sz} = moment of inertia of each stick about the Z-axis; and S = spacing between stick beams.

Figure 2-16 through Figure 2-18 present the bent elevation of the bridge models with aspect ratios of 0.3, 0.54, and 1.1, respectively. It should be noted that the effect of skew was considered thoroughly in the modeling processes for SAP2000 models. Table 2-6 through Table 2-8 present the properties of the simplified models for all aspect ratios.

The interior bent cap beam was modeled using 3-D frame element as rigid whereas the end diaphragms were explicitly modeled using 3-D frame elements. All nonlinear link elements were attached to the end diaphragms, at abutment locations, at their proper stations similar to the FE models. Locating links at their proper locations realistically is expected to enhance the accuracy of the beam stick models. Properties of nonlinear link elements, columns, and column PMM hinges were the same as those of FE models. This level of accuracy was maintained in order to rely on the improved beam stick models for nonlinear analyses.

2.6. Improved Simplified Beam-Stick (BS) Models in DRAIN3DX

The beam stick models are developed based on a refined stick model proposed by Meng and Lui (2002) as detailed in Appendix II. Similar to the simplified model developed using SAP2000, the superstructure was modeled using three 3-D frame elements having the properties of the deck. It should be noted that the effect of skew was considered thoroughly in the modeling processes for DRAIN3DX models as well. Table 2-6 through Table 2-8 present the properties of the simplified models.

The interior bent cap beam was modeled using 3-D frame element as rigid whereas the end diaphragms were explicitly modeled using 3-D frame elements. All nonlinear springs were attached to the end diaphragms, at the abutment, at their proper locations. Modeling of the abutment-soil interaction and shear key response was more direct in DRAIN3DX compared to SAP2000. Hysteresis of bearing pads was explicitly modeled using element 01 (Prakash et al., 1994). Element 01 is a simple inelastic bar element that can be used for truss bars, simple columns, and nonlinear supports springs. It is permitted to align the element in any direction, however, it can carry axial load only. In this element, inelastic behavior can be modeled in two ways:

(1) Yielding in both tension and compression and

(2) Yielding in tension but elastic buckling in compression.

Also, strain hardening can be accounted for by collapsing each element into two elements; one that has elastic properties while the other has the inelastic properties. On the other hand, element 09 was employed to accurately model the hysteretic behavior of both shear key and abutment-soil interaction. This element is either compression-only or tension-only element. Both shear key and abutment springs are assumed to resist compression only which simulates the actual behavior. An initial gap can be assigned to this element and the assigned stiffness will not be active unless the developed gap (if any) is overcome at each time step during time history analysis. Figure 2-19 shows the typical hysteresis of element 09. Element 15 (distributed plasticity) was employed to model the nonlinearity in columns. This element can be used to model steel, reinforced concrete, composite steel-concrete beams and column. The element is divided into segments. Each cross section is made up of various fiber elements that may correspond to the longitudinal steel, concrete; etc. Each fiber element, hence, is assigned its associated nonlinear stress-strain properties.

Figure 2-20 (a and b) shows the nonlinear stress-strain curves used for the steel and concrete for the modeling of the bent columns. These were also used to establish the moment curvature characteristics shown Figure 2-20c. The ultimate moment was found to be about 5460 k.ft. The cross-sectional properties were assumed to be the same throughout the individual segment. It accounts for yielding of steel, strain hardening, crushing of concrete, and tension stiffening. The element accounts for the formation of plastic hinges over the cross section and along the length of the member. Localized plastic hinges can be modeled as a special case if needed. In this study, distributed plasticity (element 15) was utilized for modeling column plasticity. The behavior of this element is subjected to several assumptions:

(1) Plane sections remain plane.

(2) Elastic behavior under torsion and shear; these stress are not expected to cause failure in the models under study.

(3) Uniform properties over each cross section.

It is recommended to conduct a parametric study to arrive at the proper number of fibers in a cross section and segments over the length of the element. This recommendation along with recommendation of SAP2000 manual was found in line with the original project scope and was the motivation to conduct a parametric study to determine:

(1) the proper plastic hinge length,

(2) the proper plastic hinge location,

(3) the proper number of fiber in cross section, and

(4) the proper number of slices over the length of the element

2.7. Single Spine Models for Fragility Analysis Using SAP2000

Developing reliable and relatively simple but representative models is the first step in fragility analysis. As was mentioned earlier, the FHWA bridge is a three-span RC box-girder bridge with seat type abutment and with 6 in. gap between the deck and the abutment back-wall. Four external shear keys located at each corner of the bridge. The benchmark bridge has a skew angle of 30 degree and it was altered to produce bridge with 0, [30 (as is)], and 60 degree skew.

Single spine models were developed and calibrated against 3D FE models of the benchmark bridge. In the models, the superstructure was assumed to be linear-elastic, and all the nonlinearity was assumed to take place in the substructure elements, external shear keys, bearings, and abutment-soil springs. Table 2-5 shows the properties of the single spine model. Figure 2-11 shows overview of single spine model. The bridge deck was represented by a single beam element having the equivalent properties of the entire deck. It should be mentioned that the bents were modeled explicitly. The bent cap was modeled using a 3D frame element with a high moment of inertia to facilitate the force distribution to the columns. Columns and footings were modeled using 3-D frame elements. The nonlinearity is assumed to take place in the form of localized plastic hinges at the top and bottom of columns. Figure 2-21 presents the moment-rotation properties of lumped moment hinges. It should be noted that coupling between moment about the two principle axes was not considered in modeling. Footing was assumed to present fixed conditions. At abutment locations, bearings, abutment-soil, rotational, and shear key springs were assigned to the deck element. Figure 2-13 shows the properties of the nonlinear lumped bearing pads. The plastic (Wen) link element available in SAP2000 was used to model the hysteresis of bearing pads. The bearing links were assigned in the longitudinal and transverse directions of the bridge. As was mentioned earlier, modeling of the abutment-soil interaction was complex since the hysteresis of abutment-soil interaction is compression only activated. A combination of nonlinear link elements in SAP2000 can simulate the behavior as will be discussed later and an initial gap of 6 in can be assigned. The soil will not be activated unless the initial gap is overcome and it becomes with full contact with the abutment. Figure 2-14 represents the backbone curve provided by Caltrans of the force displacement relationship of the soil. The rotational spring was modeled using the multilinear elastic link element. Introducing the rotational spring is to account for the rotational stiffness of the bridge at the abutment locations. Excluding the rotational springs may lead to unrealistic forces at the abutments. The hysteretic properties of shear keys were determined after the experimental study reported by Megally et al. (2002). Figure 2-15b shows the backbone of force-displacement relationship of the shear keys. The shear keys are external and four shear keys were assigned, one at each corner of the bridge. The shear keys were aligned along the absolute transverse direction. A combination of nonlinear link element was used in SAP2000 to simulate the hysteretic behavior of shear keys.

2.8. Parametric Studies and Model Calibration

Nonlinear models of the benchmark bridge was further refined and updated following a detailed parametric study presented in what follows. Significant effort was necessary to arrive at consistent modeling assumptions for various structural details and analysis techniques. Furthermore, due attention was given to ensure that the numerical models were general enough to capture the global bridge response and at the same time detailed enough to allow accurate estimation of component-level seismic response both in linear and nonlinear range.

Therefore, the main objective of this phase of the study was to establish guidelines for the modeling of nonlinear hysteretic response of bent columns, abutment-soil interaction, and shear keys to enable accurate and reliable system response prediction and assessment. In order to successfully complete this task and undertake the comprehensive parametric study, various modeling approaches for both SAP2000 and DRAIN3DX have been investigated and proposed model details have been verified and calibrated using some of the available experimental data.

2.8.1. Bent Columns with Circular Cross-Section

In SAP2000, it is possible to assign any number of plastic hinges at any location along the clear length of the column, but obviously that increases the computational time. It should be noted that lumped plastic hinges can be assigned only to frame elements. The plastic deformations are integrated over the length of the plastic hinges which is defined by the user (CSI, 2005). There are many available types of hinge definitions which can be used to model the nonlinearity on the basis of expected major type of loading such as: shear, axial, and bending or a combination. As per SAP2000 manual, PMM fiber hinge is more natural than other types of fibers when interaction between axial load and bi-axial moment is investigated. The column cross section has to be modeled using a sufficient number of fibers and accurate properties. Therefore, it is important to investigate and determine proper number of fibers to capture the most accurate nonlinear response and initial yielding with reasonable execution time.

In DRAIN3DX, there are two distinct ways to model nonlinearity in the columns by either using Element 08 or Element 15. For both of these elements, fibers should be defined in a similar fashion as in SAP2000. Element 15 is a fiber element for which plasticity is distributed throughout the length of the member. Fibers used in defining this element are PMM fibers. The element behaves elastically under both shear and torsional loading. Also for special cases, this element can be used to model lumped plasticity located at predefined zones (Prakash et al., 1994). Element number 08 is used to model beam-column elements as elastic with lumped plasticity at predefined locations. Nonetheless, the plasticity is distributed along the assigned hinge length. The fibers defined for this element are PMM fibers as well. For lumped plasticity elements in both DRAIN3DX and SAP2000, lumped shear hinges can be also defined in these elements.

For both DRAIN3DX and SAP2000, it is recommended to investigate these elements and calibrate prior to their implementation. There are many parameters that can affect the results in case of using these types of elements in both DRAIN3DX and SAP2000. These parameters are number of fibers, distribution of fibers on the cross-section, plastic hinge length, and location of plastic hinges. In order to establish a reasonably accurate guideline for the use of these elements in modeling the skewed highway bridges, the following parametric study was conducted.

This study is being undertaken in two phases: first phase involves the determination of the proper number of fibers for circular columns under flexure and their distributions.

2.8.1.1. First Phase

This part of the study was done using SAP2000 to simulate monotonic and cyclic response of various columns. The results are presented in comparison to the experimentally observed responses. Three sets of experimental data were chosen from published literature, namely by, Esmaeily and Xiao (2002); Murat and Baingo (1999); and Cheok and Stone (1990). Properties of the columns used in these studies are summarized in

Table 2-9. These columns have different concrete compressive strength and were tested under different loading conditions, but all of them failed in flexure. Table 2-10 presents the properties of steel and concrete of these columns. Column Four (Esmaeily and Xiao, 2002) was tested under monotonically increasing lateral loads up to complete failure without axial loading. Column RC-2 (Sucuoglu and Baingo, 1999) was first axially loaded incrementally up to axial load of 1000 kN and loaded laterally in cyclic displacement. Three cycles in the elastic range were applied followed by complete nine cycles in the inelastic range. Column N6 (Cheok and Stone, 1990) was first loaded incrementally up to axial load of 68.7 kN and loaded laterally in cyclic displacement. This lateral loading protocol is shown in Figure 2-22 in which ductility ratio is defined as the ratio of applied displacement to yielding displacement.

All of the aforementioned specimens were modeled using SAP2000. In SAP2000, the same material properties and section properties were used. The PMM fiber hinge was assigned at the base of columns with an assumed hinge length of 10% of column length. A sufficient number of fibers are important to represent the cross section accurately. Various number of fibers distributed along the cross section were examined (Figure 2-23). It was assumed that the reinforcement can be modeled by eight lumped steel fibers regardless of the distribution of reinforcement. Four of them located at the absolute corners and the other four located at 45° (Figure 2-23). Properties of concrete and steel fibers were defined by their respective stress-strain curves. Pushover analysis was performed to simulate the different types of loading applied on specimens as reported by the authors.

The results of the pushover analyses were compared against the experimental response (Figure 2-24 through Figure 2-26) and the percent errors were calculated (Table 2-11). It can be observed from the presented results that the use of 20 fibers and 32 fibers did not capture the initial yielding of specimens due to pushover. On the other hand, by introducing concrete fibers surrounding the location of reinforcement, the predicted initial yielding was reasonably accurate. This is attributed to that the section behavior in flexure is controlled by the outer fibers and not the inner ones. Therefore, a single ring is enough for the core of the cross section. It can be seen in Table 2-11, the percent of error did not exceed 3% when fibers at reinforcement location were introduced. Accordingly, it is recommended that 32@reinf. fiber distribution (Figure 2-23c) is used in order to capture accurately the initial yielding of specimens. Also, the same fiber distribution can be used for columns with circular sections experiencing flexural failure. The presented distribution of fibers can be used for columns with different dimensions, however, the number of wedges will change to maintain the area of fibers to be the same as those used in the present study.

2.8.1.2. Second Phase

This study was performed on a complete bent in order to simulate the actual case of bridge at which plastic hinges are formed following a mechanism. To choose reasonable dimensions for the bent understudy, a previously examined column, which dominantly failed in flexure, was selected and considered as part of the bent. The bent consisted of two typical columns and a rigid beam in accordance with Caltrans bridges. The column is designated as Column Four from study done by Esmaeily and Xiao (2002). A pushover analysis is performed for the current parametric study.

Firstly, pushover experimental results of the individual column were compared to results of pushover analysis on the column using DRAIN3DX. The objective was to find the suitable number of fibers using element 08 to achieve close results to the experimental. A range of number of fibers was tested to find the proper number of fibers (segment) which can be used. 2, 5, 10, and 15 fibers were experimented. Figure 2-27 presents the applied load versus drift of the columns having different number of fibers. The drift is defined as the ratio of lateral displacement to the height of the column (H). The two fiber columns results were found to extremely overshoots. There was not much difference between the results of 10 fiber column and that of 15 fiber column. It was found that using 10 fibers will lead to reasonable accuracy in comparison to the experimental results.

Subsequently, two columns with 10 segments were used as a part of the complete bent in DRAIN3DX. A pushover analysis was done on the complete bent and results were further utilized to calibrate lumped plasticity bents developed using SAP2000. In order to achieve the objectives of the current study, a large number of bents with different aspect ratios were modeled and locations of plastic hinges were varied as well. The aspect ratio is defined as the ratio of height of bent (H) to column diameter (D). Aspect ratios (3.5, 5, and 7) were chosen to achieve flexural failure of bents. For each aspect ratio, two plastic hinged were assigned at top and bottom of both columns. The assumed plastic hinge lengths 0.05H, 0.1H, 0.15H, 0.25D, and 0.5D were examined individually. For each plastic hinge length, locations of plastic hinge were examined by varying locations at ends, 0.05H, and 0.1H. Thereafter, 45 bents were modeled using SAP2000 and compared to uniformly distributed plasticity done using DRAIN3DX. Figure 2-28 presents notations used for location of plastic hinges and their length. Figure 2-29 through Figure 2-43 show results of the study. Also, Table 2-12 through Table 2-14 present a comparison of SAP2000 results against more accurate results determined using DRAIN3DX. From the figures and the tables, two lumped plastic hinges at ends of columns with plastic hinge length of 0.1H can be assigned to columns to provide reasonably accurate results. This conclusion was confirmed by looking at the error in predicting the forces and displacements.

2.8.2. Abutment-Soil Interaction

The skewed-bridge abutments, foundations and surrounding soils constitute a strongly coupled system and the dynamic behavior of a skewed bridge structure and the abutment-embankment soil has been identified as having the first order influence on dynamic response of the bridge (Shamsabadi, et al., 2004). It was suggested based on past experiences that even though the bridge structure during a seismic event could remain linear, the nonlinear behavior of the abutment embankment soil can result in significant nonlinear behavior of the bridge structure. Therefore, it was recommended for the realistic system response prediction, the abutment-soil interaction should be included in the bridge response assessments. However, it is not possible in SAP2000 to model the complete hysteresis response characteristics due to abutment-soil interaction using one single element.

In this study, nonlinear abutment-soil response is modeled using a series combination of springs as shown in Figure 2-44. The nonlinear gap element ensures no pressure is applied to the abutment wall during a reversal whereas the nonlinear elasto-plastic spring constitutes the soil properties (with optional post yield stiffness). Further refinement to this modeling approach is possible by introducing dashpot elements to model damping associated with soil response. This model was developed for the p-y relation for passive pressure at abutment backwall that was provided by Caltrans as shown in Figure 2-45. As can be seen in Figure 2-46, various other types of abutment soil response can also be modeled.

2.8.3. Shear Key Response

The modeling of the nonlinear shear-key response is critical for the reliable and accurate assessment of the overall seismic response of skewed highway bridges. Megally, et al. (2002) studied experimentally cyclic pushover response of both internal and external shear-keys. Capacity determination of external shear-keys presented in this study has been adopted and nonlinear external shear-key response is modeled using a combination of link elements as shown in Figure 2-47 to simulate the entire hysteretic response using SAP2000.

Experimental cyclic pushover response of a shear-key is compared to that of analytical prediction in Figure 2-48 (Megally et al., 2002). Furthermore, response prediction from a nonlinear time history analyses is shown in Figure 2-49.

2.9. Analytical Matrix

Table 2-15 shows the analytical matrix which was used to examine Caltrans bridges. The entire analyses presented in this study follow the analytical matrix in order to investigate the effect of a wide range of parameters. These parameters are the effect of ground motion intensity, soil conditions, abutment support and base support conditions, and aspect ratio interacting with skew angle, on the seismic response of skewed highway bridges (chapter 3). Development of fragility curves for FHWA bridges is presented in chapter 5. Prior to conducting the analyses using DRAIN3DX, modal analyses and preliminary nonlinear time history analysis was conducted to compare the accuracy of the simplified models in DRAIN3DX to those in SAP2000 and finite element models in SAP2000.

2.10. Modal Analysis

The modal analyses were conducted for a large range of skew angles namely; 0, 20, 30, 45, 52, and 60 degree of bridges with the aspect ratio of 0.3. The modal analysis was conducted for FE models (shell and solid) and BS models (using SAP2000 and DRAIN3DX). The modal analysis was conducted for fixed foundation case and included the shear keys and abutment-soil interaction elements. Structural dynamic characteristics of bridges are expected to be captured by the FE models more accurately. Hence, results of modal analysis of BS models were compared against the counterpart FE models. The objective of the modal analysis was to:

(1) Measure the accuracy of the BS models against more complex FE models and

(2) The change in modal characteristics by varying the skew angle.

Table 2-16 through Table 2-21 present the modal analysis results of the bridge models. For the sake of comparison, the first ten modes of vibration are considered and tabulated. The mass participation factors were used to describe the mode shapes as shown in the tables. Figure 2-50 and Figure 2-51 show a comparison of mode shape of the first and second modes of the BS model, FE-shell model, and FE-solid model for 52 degree skew. In comparison, transitional and rotational modes are presented. The torsional mode will be that mode at which rotation about x (Rx) is dominant. Also, the first mode of vibration is referred to as the dominant mode.

The following observations are made for straight (0° skew) bridge models:

- The first mode of vibration (dominant mode) was predicted accurately by the three models. In this mode, the bridge translates longitudinally (Ux) accompanied by rotation about the transverse axis (Ry).

- For the second mode, the three models agree well in predicting translation in longitudinal direction accompanied by in-plane rotation (Rz).

- For the third mode, the three models agree well in predicting translation in transverse direction accompanied by torsion (Rx).

- For the fourth mode, the three models agree well in predicting translation in vertical direction (Uz) accompanied by torsion (Rx).

- For the fifth mode, the FE model predicted a large in-plane rotation while BS models predicted almost no rotation.

- For the sixth mode, the three models agree in predicting the torsional mode.

- For the seventh mode, the FE model predicted rotation about the transverse axis while BS model did not.

- For the eighth mode, the FE model predicted in-plane rotation while BS models predicted vertical translation only.

- For the ninth mode, the FE model predicted the vertical translation expected by the eighth mode in BS models while BS model predicted in-plane rotation only.

- For the tenth mode, the FE model predicted translation in the transverse direction accompanied by torsion while BS model did not.

In view of the participation factors and periods of vibration, BS models accurately predicted the dominant mode of vibration. BS models predicted the time periods of all modes in good agreement with the FE models. The mode of vibration was consistently and accurately predicted up to the sixth mode. Beyond this mode, mode shifting occurred. Nonetheless, some agreement was reported in predicting higher modes and mode coupling.

The following observations are made for straight (20° skew) bridge models:

- The first mode of vibration (dominant mode) was predicted accurately by the three models. In this mode, the bridge translates longitudinally (Ux) accompanied by rotation about the transverse axis (Ry).

- For the second mode, the three models agree in predicting translation in longitudinal direction accompanied rotation about the transverse axis (Ry).

- For the third mode, the three models agree well in predicting translation in transverse direction accompanied by torsion (Rx).

- For the fourth mode, the three models agree well in predicting translation in vertical direction (Uz) accompanied by torsion (Rx).

- For the fifth mode, the FE and BS models predicted in-plane rotation.

- For the sixth mode, the FE model predicted a torsional mode while BS models predicted translation in the longitudinal direction.

- For the seventh mode, the FE and BS models agree in predicting no rotation but FE model predicted a translation in the vertical axis.

- For the eighth mode, the FE model predicted a torsional mode while BS models predicted vertical translation only.

- For the ninth mode, the FE model predicted the vertical translation expected by mode eighth in BS models while BS model predicted in-plane rotation only.

- For the tenth mode, the FE and BS models predicted torsional mode.

BS models predicted the time periods of all modes in good agreement with the FE models. The mode of vibration was consistently predicted between FE and BS models up to the fifth mode. Beyond this mode, mode shifting occurred. Nonetheless, some agreement was reported in predicting higher modes and mode coupling.

The following observations are made for straight (30° skew) bridge models:

- The first mode of vibration (dominant mode) was predicted accurately by the three models. In this mode, the model translates longitudinally (Ux) accompanied by rotation about the transverse axis (Ry).

- For the second mode, the three models agree in predicting translations in longitudinal and transverse direction accompanied by rotation about the transverse axis (Ry).

- For the third mode, the three models agree well in predicting translation in transverse direction accompanied by torsion (Rx).

- For the fourth mode, the three models predicted translation in vertical direction (Uz) accompanied by torsion (Rx).

- For the fifth mode, the FE model predicted in-plane rotation while BS models predicted vertical translation.

- For the sixth mode, the FE and BS models predicted torsional mode.

- For the seventh mode, the FE and BS models agree in predicting no rotation but FE model predicted a translation in the vertical axis.

- For the eighth mode, the FE model predicted a torsional mode while disagreement was demonstrated by BS models.

- For the ninth mode, the FE model predicted coupled translation in both of longitudinal and transverse direction while BS model predicted in-plane rotation only.

- For the tenth mode, the FE model predicted a vertical translation.

BS models predicted the time periods of all modes in good agreement with the FE models. The mode of vibration was consistency predicted between FE and BS models up to the fifth mode. Beyond this mode, mode shifting occurred. Nonetheless, some agreement was reported in predicting higher modes and mode coupling.

The following observations are made for straight (45° skew) bridge models:

- The first mode of vibration (dominant mode) was predicted accurately by the three models. In this mode, the model translates longitudinally (Ux) accompanied by rotation about the transverse axis (Ry).

- For the second mode, the three models agree in predicting translation in transverse direction accompanied by torsion.

- For the third mode, the three models agree well in predicting translation in transverse direction accompanied by torsion (Rx). However, BS models marginally under-predicted the mass participation factor.

- For the fourth mode, the FE model predicted vertical translation accompanied by torsion while BS models predicted vertical translation only.

- For the fifth mode, the FE model predicted translation in the transverse direction and BS-SAP2000 model predicted large vertical translation while BS-DRAIN3DX model agrees with the FE model result.

- For the sixth mode, a disagreement between results of FE and BS models can be observed.

- For the seventh mode, the FE model predicted a vertical translation while BS model did not predict any translation.

- For the eighth mode, the FE model predicted a mode having longitudinal and transverse translation and torsion while BS model results did not agree.

- For the ninth mode, the FE model predicted no rotation while the BS models predicted a large in-plane rotation.

- For the tenth mode, the FE and BS models results agree together.

The modal analysis of FE model results showed significant coupling throughout the modes. Also, BS models did not predict well all the modes of vibration and the coupling when compared to FE models. However, BS models predicted all the necessary modes reasonably.

The following observations are made for straight (52° skew) bridge models:

- For 52 skew, the first mode of vibration (dominant mode) was predicted accurately by the three models. In this mode, the model translates longitudinally (Ux) accompanied by rotation about the transverse axis (Ry).

- For the second mode, the three models agree in predicting translations in transverse direction accompanied by torsion.

- For the third mode, the three models agree well in predicting translations in longitudinal and transverse direction accompanied by rotation about transverse axis (RY).

- For the fourth mode, the three models predicted translation in vertical direction (Uz) accompanied by torsion (Rx), but BS models predicted lower participations.

- For the fifth mode, the FE model predicted in-plane rotation while BS-SAP2000 model predicted vertical translation and BS-DRAIN3DX agrees with the FE model results.

- For the sixth mode, no agreement can be observed.

- For the seventh mode, the FE- and BS-DRAIN3DX models agree in predicting vertical translation.

- For the eighth mode, the FE model predicted a torsional mode accompanied by transverse translation while disagreement was demonstrated by BS models.

- For the ninth mode, the BS models predicted in-plane rotation.

- For the tenth mode, small participation factors are reported.

BS models predicted the time periods of all modes in a good agreement with FE models. The mode of vibration was consistency predicted between FE and BS models up to the fifth mode. Beyond this mode, mode shifting occurred. However, BS models predicted all the necessary modes reasonably.

The following observations are made for straight (60° skew) bridge models:

- For 60 skew, the first mode of vibration (dominant mode) was predicted accurately by the three models. In this mode, the model translates longitudinally (Ux) accompanied by rotation about the transverse axis (Ry).

- For the second mode, the three models agree in predicting translations in transverse direction accompanied by torsion.

- For the third mode, the three models agree well in predicting translation in transverse direction accompanied by rotation about transverse axis (Ry).

- For the fourth mode, the FE model predicted translation in vertical direction (Uz) accompanied by torsion (Rx), but BS models results disagree with FE model's.

- For the fifth mode, the FE- and BS-DRAIN3DX model predicted transverse translation.

- For the sixth mode, no agreement can be observed.

- For the seventh mode, the FE- and BS-DRAIN3DX models agree in predicting vertical translation.

- For the eighth mode, the FE model predicted a torsional mode accompanied by transverse translation while disagreement was demonstrated by BS models.

- For the ninth mode, the BS models predicted in-plane rotation.

- For the tenth mode, small participation factors are reported.

BS models predicted the time periods of all modes in a good agreement with FE models. The mode of vibration was consistency predicted between FE and BS models up to the fourth mode. Beyond this mode, mode shifting occurred. BS models predicted all the necessary modes reasonably.

It can be concluded that BS models fairly predicted the periods of all modes for all skew angles while the mass participation factors of the dominant mode of vibration was predicted accurately. BS models developed using DRAIN3DX are more accurate in predicting modal characteristics compared to those developed using SAP2000. BS models can be used to predict modal characteristics even for higher modes. BS models predicted mode shapes in agreement with FE models up to the fifth mode followed by mode shifting beyond the fifth mode for all skews. As skew angle increases, the second mode shifts from longitudinal mode to transverse mode. Also, the torsional mode moves from the third mode to the second mode.

2.11. The Effect of Post-tensioning

Figure 2-10 shows a 3-D view of the improved beam stick model showing the nodes at which the response was recorded (M and Q) for the preliminarily model verification study. In SAP2000 and DRAIN3DX, post-tensioning is modeled differently. In SAP2000, for both FE and BS models; tendons were modeled explicitly using the available "tendon elements". These elements are connected automatically (by the program) to the superstructure while it is the responsibility of the user to define the complete profile of tendon, its cross-section, and the applied forces and distribution of tendons as well (CSI, 2005). On the other hand, in DRAIN3DX, the equivalent loads due to prestressing was determined, calibrated, and applied at their predefined locations along the stick girders to model the effect of prestressing. There is no other direct way to model the tendons in DRAIN3DX.

In order to investigate the effect of post-tensioning on dynamic performance of benchmark bridge, nonlinear time history analyses were conducted with and without post-tensioning also accounting for the self-weight of the bridge. Figure 2-52 represents the prestressing profile in the box girders of the bridge. The bridge understudy has a total post-tensioning force of 8,197 kips. Figure 2-53 through Figure 2-55 present the comparison of these results. The comparison was presented in terms of the longitudinal and transverse deck displacements at the bent and abutment link hysteresis. It was concluded that the prestressing did not have any significant effect on the overall seismic response of the bridge structure and will be neglected for the rest of the study.

2.12. Preliminary Time-History Analyses: Model Verification

1940 El Centro S00E ground motion acceleration record (scaled to PGA= 0.6g) was chosen in order to conduct a series of preliminary time-history analyses to compare response obtained from each of the three different models of the benchmark bridge (FE-, BS-SAP2000, and BS-DRAIN3DX). A nonlinear static analysis including both dead load and post-tensioning preceded the time history analyses. The response quantities at nodes M and Q in (Figure 2-10) were monitored in order to compare the models. Figure 2-56 through Figure 2-60 show comparison of response of the models due to aforementioned ground motion. The comparison was in terms of longitudinal and transverse deck displacement at bent and through the deck and the abutment spring hysteresis. A very good agreement between the various models can be seen. Nonetheless, there is a difference in the force-displacement relationship of the abutment-soil elements between SAP2000 models and DRAIN3DX model. This is attributed to the inevitable difference in modeling approach (use of nonlinear elements) as explained earlier which, however, does not affect significantly the global seismic response of the bridge.

2.13. Recommendations for Modeling

Modeling of highway bridges with skew and incorporating inelastic response characteristics of various components accurately is an important but complex task. Also, there are several parameters interacting with the skew angle affecting the seismic response of skewed highway bridges such as; superstructure flexibility, boundary conditions, and width-to-span ratio. Therefore it is crucial to achieve accurate modeling techniques, especially, for bridges whose behavior is not well understood such as; skewed and curved bridges. Modeling of these bridges in 2-D is in general not sufficient to study overall system response behavior and 3-D model is needed. It is important to ensure that the models are general and detailed enough to capture the global bridge response to allow accurate estimation of component-level seismic response in linear and nonlinear ranges. Since the bridge deck is assumed to remain linear elastic, the focus is on modeling nonlinear elements in the bridge such as, nonlinearity in columns, bearing pads, abutment-soil interaction, and shear keys. Also, some of the limitations in modeling nonlinearity in SAP2000 is presented along with ways to overcome these limitations. It should be noted that the discussion here is not limited to specific software, however, it can be applicable to any analytical tool of interest. In light of the modeling techniques and parametric studies performed in this chapter, the following recommendations are made:

1- It is a practical conjecture to assume that the bridge deck remains linear elastic. However, nonlinearity should be modeled in elements whose inelastic behavior is anticipated to affect the global response such as; columns (lumped plastic hinges or distributed plasticity), bearing pads, abutment-soil interaction, and shear keys.

2- The fiber PMM hinge is the most natural type of hinge which can be used to model the nonlinearity of columns. However, it requires more computer storage and execution time. Also, using the distributed plasticity elements to model columns is the most accurate. It is preferable to model bridges with irregularity such as; skewed bridges and curved bridges using the fiber PMM hinge and the distributed plasticity elements. The number of segments used to divide the distributed plasticity element is important and it increases the execution time as well. DRAIN3DX has the distributed plasticity element namely, fiber beam-column element (element 15). It is recommended to divide the fiber column into sufficient number of segments to obtain two segments at top and bottom of the column with length of 10% of the column height or less in order to obtain accurate results.

3- Most of the other analysis programs do not provide distributed plasticity element. At this time, lumped plasticity hinges may be used. As mentioned earlier, the fiber hinge is the most accurate and natural type of hinges and it is available in SAP2000 and DRAIN3DX. The fiber hinge consists of a number of fibers distributed across the cross section of the frame element. These fibers are typically concrete and steel in case of bridge columns. Each fiber has a location and a tributary area. The number and distribution of fibers play also a role in increasing the execution time. Therefore, the minimum number of fibers establishing the accuracy should be assigned. 32@reinf. fiber distribution (Figure 2-23c) is used in this study. The fiber distribution consists of 24 concrete fibers and 8 lumped steel fibers. Approximately, 12 concrete fibers can be used to represent the concrete core and another 12 fibers as close to the concrete cover in addition to 8 steel lumped fibers. The same distribution of fibers described above can be used for columns with different dimensions; however, a sufficient number of fibers has to be used to obtain section properties (area and moment of inertia) within 5% of those of the actual column cross section (Aviram et al., 2008).

4- It is possible to assign any number of lumped plastic hinges throughout the length of the column frame element, but it will increase the execution time. Therefore, it is important to investigate the location of the plastic hinges and minimize the number of the hinges to achieve an acceptable level of accuracy. Many options were investigated and calibrated against available experimental results of flexural dominant columns. From the results of this study, it is recommended to insert the plastic hinges at the ends of doubly curvature columns.

5- The plastic hinge length is one of the important parameters which affect the accuracy of the nonlinear analysis. The plastic deformations are integrated over the length of the plastic hinges. Therefore, it is important to assign the proper plastic hinge length to achieve accurate nonlinear response for the component and the system in general. A plastic hinge length of 10% of the column height is recommended.

6- Bearing pads are considered sacrificial elements and they fail when they reach the shear capacity (shear strain). Elasto-plastic spring can be used to model the bearing pad hysteresis. Examples for this type of element are the plastic (Wen) link in SAP2000 and inelastic truss bar element "element 01" in DRAIN3DX.

7- Based on past research such as Shamsabadi, et al. (2004), it was recommended to include the nonlinear behavior of abutment-soil interactions. The nonlinear abutment-soil interaction response is complex since it is compression activated. There is not a single spring can simulate the nonlinear response of abutment-soil interaction solely in most of the available softwares such as; SAP2000. Based on a parametric study, it is recommended to use a combination of gap element and elasto-plastic spring connected in series to model abutment-soil interaction hysteresis accurately (Figure 2-44). Furthermore, initial gap opening can be easily incorporated to this model.

8- Shear keys [internal or external] are one of the elements which affect the response of the system significantly. Therefore, it is important to model the nonlinear hysteresis of the shear keys accurately. Shear keys are also compression activated elements therefore a combination of springs are needed to achieve the response since a single element with the needed hysteresis is not available in most of the softwares. It is recommended to use a combination of gap, multilinear elastic, and multilinear plastic springs to model the hysteresis of the internal or external shear keys (Figure 2-47).

9- As was mentioned earlier, most of the available softwares do not have a single element to model the hysteresis of abutment-soil interaction and shear keys. However, DRAIN3DX has this capability, hysteresis of abutment-soil interaction and internal and external shear keys can be modeled accurately using element 09 in DRAIN3DX.

Table 2-1. Summary of Properties of 12 Caltrans Bridges

Bridge ID	Span length, ft	Depth, ft	Width of bridge, ft	Width/Span ratio*	Skew angle,°	Number of spans	No. of columns/bent
42-0427L/R	134.00	5.50	32.00	0.24	52	2	2
42-0431L/R	121.00	5.10	34.00	0.28	47	2	2
39-0146	128.00	5.50	59.00	0.46	32	2	3
39-0149	145.00	6.00	47.00	0.32	40	2	3
19-0192R	164.00	5.00	35.00	0.21	40	2	2
42-299L/R	157.00	6.80	46.00	0.29	33	2	2
38-0075	133.00	5.25	41.00	0.31	41	3	2
19-0197R	115.00	4.60	35.50	0.31	30	1	0
17-0104	164.00	7.50	28.30	0.17	45	1	0
17-0105	66.00	4.00	32.00	0.48	35	3	2
53-2790R/L	260.00	11.50	60.00	0.23	45	3	3
390-0042	59.00	3.50	40.50	0.69	34	3	2
Average	137.17	5.85	40.86	0.33			

*excluding overhang length

Table 2-2. Section Properties (Aspect Ratio of 0.3)

	Superstructure	Bent Cap Beam	Bent Column
Area, ft2	66	33	12.5
J, ft4	91	100000	12.6
Iy, ft4	297	100000	8.3
Iz, ft4	7819	100000	8.3

Table 2-3. Section Properties (Aspect Ratio of 0.54)

	Superstructure	Bent Cap Beam	Bent Column
Area, ft2	122	33	12.5
J, ft4	462	100000	12.6
Iy, ft4	572	100000	8.3
Iz, ft4	48199	100000	8.3

Table 2-4. Section Properties (Aspect Ratio of 1.1)

	Superstructure	Bent Cap Beam	Bent Column
Area, ft2	253	33	12.5
J, ft4	3603	100000	12.6
Iy, ft4	1212	100000	8.3
Iz, ft4	426482	100000	8.3

Table 2-5. Section Properties (FHWA Bridge Example 4)

	Superstructure	Bent Cap Beam	Bent Column
Area, ft2	72.76	27.02	12.59
J, ft4	1177	100000	25.49
Iy, ft4	400.84	100000	12.74
Iz, ft4	9695.39	100000	12.74

Table 2-6. Summary of Section Properties of Caltrans Bridge (Aspect Ratio of 0.3)

Deck	Bent Cap	Columns
$A_s = 3.17E3\ in.^2$ $I_{sy(middle)} = 1.593E6\ in.^4$ $I_{sy(edge)} = 2.282E6\ in.^4$ $I_{sz} = 23.845E6\ in.^4\ (typ.)$ $J_{s(edge)} = 947.657E3\ in.^4$	$A_s = 4.752E3\ in.^2$ $I_{ex} = 1.050E6\ in.^4$ $I_{ez} = 1.073E6\ in.^4$	$A_s = 1.8E3\ in.^2$ $I_e = 171.472E3\ in.^4$

Table 2-7. Summary of Section Properties of Caltrans Bridge (Aspect Ratio of 0.54)

Deck	Bent Cap	Columns
$A_s = 5.88E3\ in.^2$ $I_{sy(middle)} = 3.822E6\ in.^4$ $I_{sy(edge)} = 4.027E6\ in.^4$ $I_{sz} = 2.407E7\ in.^4\ (typ.)$ $J_{s(edge)} = 4.786E6\ in.^4$	$A_s = 4.752E3\ in.^2$ $I_{ex} = 1.050E6\ in.^4$ $I_{ez} = 1.073E6\ in.^4$	$A_s = 1.8E3\ in.^2$ $I_e = 171.472E3\ in.^4$

Table 2-8. Summary of Section Properties of Caltrans Bridge (Aspect Ratio of 1.1)

Deck	Bent Cap	Columns
$A_{se} = 7.189E3\ in.^2$ $A_{sm} = 1.106E4\ in.^2$ $I_{sy(middle)} = 6.9E6\ in.^4$ $I_{sy(edge)} = 7.4E6\ in.^4$ $I_{sz} = 1.902E7\ in.^4\ (typ.)$ $J_{s(edge)} = 1.868E7\ in.^4$	$A_s = 4.752E3\ in.^2$ $I_{ex} = 1.050E6\ in.^4$ $I_{ez} = 1.073E6\ in.^4$	$A_s = 1.8E3\ in.^2$ $I_e = 171.472E3\ in.^4$

Table 2-9. Properties of columns

Authors	Column Designation	Diameter, in.	Height, in.	Concrete Compressive Strength, *psi*	Reinforcing Steel
Esmaeily and Xiao, 2002	Four	16	81	7300	12#4
Murat and Baingo, 1999	RC-2	10	72.6	9427	8#5
Cheok and Stone, 1990	N6	10	72.1	3367	25D6

Table 2-10. Properties of steel and concrete

Authors	Column Designation	f_c', psi	ε_c	f_y, psi	ε_y	f_u, psi	ε_u
Esmaeily and Xiao, 2002	Four	7300	0.0029	60000	0.0021	84000	0.07
Murat and Baingo, 1999	RC-2	9427	0.002	60000	0.0021	90000	0.08
Cheok and Stone, 1990	N6	3367	0.0022	60000	0.0021	90000	0.08

Table 2-11. Summary comparison of analytical and experimental results

1. Specimen Four (Esmaeily and Xiao, 2002)

	$\phi^{(1)}$, (1/M)	Error, %	$My^{(2)}$, kN.m	Error, %
Experimental	0.020		127.0	
20 Fibers	0.014	30.0	125.5	1.2
32 Fibers	0.014	30.0	129.5	2.0
32@reinf.	0.015	25.0	130.1	2.4

2. Specimen RC-2 (Sucuoglu and Baingo, 1999)

	$\Delta y^{(3)}$, mm	Error, %	My, kN.m	Error, %
Experimental	18.5		97	
20 Fibers	33.6	81.6	114	17.8
32 Fibers	33.0	78.4	112	15.7
32@reinf.	26.0	42.7	98	1.2

3. Specimen N6 (Cheok and Stone, 1990)

	Δy, mm	Error, %	$Py^{(4)}$, kips	Error, %
Experimental	0.66		6.20	
20 Fibers	0.60	9.1	6.30	1.60
32 Fibers	0.72	9.1	6.28	1.29
32@reinf.	0.61	7.6	6.16	0.60

(1) Curvature
(2) Yield moment
(3) Yield displacement
(4) Lateral force at first yield

Table 2-12. Aspect ratio of 3.5

Hinge Location Response	at ends				at 0.05 H				at 0.1 H			
	Yield Force, kips	Yield Displacement, in	error, % Force	error, % Displacement	Yield Force, kips	Yield Displacement, in	error, % Force	error, % Displacement	Yield Force, kips	Yield Displacement, in	error, % Force	error, % Displacement
Lp= 0.05H	85.84	0.16	2.19	20.00	106.8	0.21	27.14	5.00	112.13	0.18	33.49	10.00
Lp= 0.1H	86.8	0.25	3.33	25.00	89.919	0.21	7.05	5.00	98.67	0.21	17.46	5.00
Lp= 0.15H	79.68	0.26	5.14	30.00	83.42	0.23	0.69	15.00	100.6	0.25	19.76	25.00
Lp= 0.25D	84.5	0.18	0.60	10.00	91.8	0.18	9.29	10.00	110.3	0.21	31.31	5.00
Lp= 0.5D	79.55	0.25	5.30	25.00	86.07	0.24	2.46	20.00	99.9	0.25	18.93	25.00
Experimental	84	0.2										

Table 2-13. Aspect ratio of 5.0

Hinge Location Response	at ends				at 0.05 H				at 0.1 H			
	Yield Force, kips	Yield Displacement, in	error, % Force	error, % Displacement	Yield Force, kips	Yield Displacement, in	error, % Force	error, % Displacement	Yield Force, kips	Yield Displacement, in	error, % Force	error, % Displacement
Lp= 0.05H	59.87	0.34	1.47	20.93	65.8	0.32	11.53	25.58	69.98	0.3	18.61	30.23
Lp= 0.1H	54.4	0.41	7.80	4.65	62.67	0.43	6.22	0.00	67.88	0.38	15.05	11.63
Lp= 0.15H	49.21	0.43	16.59	0.00	56.81	0.43	3.71	0.00	62.4	0.43	5.76	0.00
Lp= 0.25D	59.87	0.32	1.47	25.58	64.83	0.33	9.88	23.26	70.12	0.3	18.85	30.23
Lp= 0.5D	54.45	0.39	7.71	9.30	57.98	0.35	1.73	18.60	68.25	0.39	15.68	9.30
Experimental	59	0.43										

Table 2-14. Aspect ratio of 7.0

Hinge Location Response	at ends				at 0.05 H				at 0.1 H			
	Yield Force, kips	Yield Displacement, in	error, % Force	error, % Displacement	Yield Force, kips	Yield Displacement, in	error, % Force	error, % Displacement	Yield Force, kips	Yield Displacement, in	error, % Force	error, % Displacement
Lp= 0.05H	39.97	0.53	2.51	26.39	42.9	0.5	4.63	30.56	46.74	0.49	14.00	31.94
Lp= 0.1H	37.46	0.68	8.63	5.56	39.89	0.613	2.71	14.86	41.03	0.53	0.07	26.39
Lp= 0.15H	36.54	0.85	10.88	18.06	40.69	0.83	0.76	15.28	40.36	0.65	1.56	9.72
Lp= 0.25D	42.92	0.53	4.68	26.39	44.73	0.463	9.10	35.69	51.69	0.51	26.07	29.17
Lp= 0.5D	38.16	0.59	6.93	18.06	43.13	0.6	5.20	1.67	45.55	0.55	11.10	23.61
Experimental	41	0.72										

Table 2-15. Analytical Matrix

Analysis Type	Parameters		
Modal Analysis	W/L= 0.3		
	Skews: (0, 20, 30, 45, 52, 60)		
	With Shear Keys		
	Fixed Foundation		
	Two column bent		
Time History Analysis	W/L= 0.3	W/L= 0.54	W/L= 1.1
	Skews: (0, 20, 30, 45, 52, 60)		
	With Shear Keys	Without Shear Keys	
	Pinned Foundation	Fixed Foundation	
	Two column bent	Three column bent	Four column bent
	Two levels of intensity of ground motions (Mw – PGA pair)		
	Two different soil conditions (B and D)		

Table 2-16. Modal Analysis of Bridge Model with no skew

Mode	FE-SAP2000 Period, sec	Mass Participation, %		BS-SAP2000 Period, sec	Mass Participation, %		BS-DRAIN3DX Period, sec	Mass Participation, %	
1	0.588	UX	93.56	0.562	UX	64.55	0.558	UX	61.79
		UY	0		UY	0		UY	0
		UZ	0		UZ	0		UZ	0
		RY	24.47		RY	43.67		RY	N/A
2	0.507	UX	4.80	0.451	UX	33.94	0.448	UX	36.06
		UY	0		UY	0		UY	0
		UZ	0		UZ	0		UZ	0
		RZ	6.06		RZ	11.23		RZ	N/A
3	0.363	UX	0	0.380	UX	0	0.355	UX	0
		UY	86.05		UY	83.19		UY	80.04
		UZ	0		UZ	0		UZ	0
		RX	54.34		RX	75.62		RX	N/A
4	0.344	UX	0	0.334	UX	0	0.339	UX	0
		UY	0		UY	0		UY	0
		UZ	69.05		UZ	64.21		UZ	59.82
		RX	21.19		RX	4.04		RX	N/A
5	0.165	UX	0	0.265	UX	0	0.315	UX	0
		UY	0		UY	0		UY	0
		UZ	0		UZ	0		UZ	0
		RZ	83.98		RZ	0		RZ	0
6	0.138	UX	0	0.231	UX	0	0.287	UX	0.22
		UY	0		UY	0		UY	0
		UZ	0		UZ	0		UZ	0
		RX	5.81		RX	2.30		RX	N/A
7	0.136	UX	0	0.198	UX	0	0.267	UX	0
		UY	0		UY	0		UY	0
		UZ	0		UZ	0		UZ	0
		RY	12.25		RY	0		RY	0
8	0.132	UX	0	0.189	UX	0	0.263	UX	0
		UY	0		UY	0		UY	0
		UZ	0		UZ	0.28		UZ	4.12
		RZ	1.63		RZ	0		RZ	N/A
9	0.115	UX	0	0.172	UX	0	0.154	UX	0
		UY	0		UY	0		UY	0
		UZ	1.25		UZ	0		UZ	0
		RZ	0		RZ	87.61		RZ	N/A
10	0.105	UX	0	0.129	UX	0	0.130	UX	0
		UY	11.96		UY	0		UY	0
		UZ	0		UZ	0		UZ	0
		RX	7.20		RX	0		RY	N/A

Table 2-17. Modal Analysis of Bridge Model with 20° skew

Mode	FE-SAP2000			BS-SAP2000			BS- DRAIN3DX		
	Period, sec	Mass Participation, %		Period, sec	Mass Participation, %		Period, sec	Mass Participation, %	
1	0.598	UX	95.20	0.547	UX	87.19	0.541	UX	84.29
		UY	0		UY	0		UY	0
		UZ	0		UZ	0		UZ	0
		RY	19.29		RY	29.66		RY	N/A
2	0.479	UX	02.03	0.420	UX	11.22	0.416	UX	13.41
		UY	0		UY	1.02		UY	0
		UZ	0		UZ	0.98		UZ	0
		RY	40.79		RY	25.63		RY	N/A
3	0.364	UX	0	0.378	UX	0	0.356	UX	0
		UY	84.73		UY	83.09		UY	79.94
		UZ	0		UZ	0		UZ	0
		RX	53.77		RX	74.78		RX	N/A
4	0.339	UX	0	0.309	UX	0	0.316	UX	0
		UY	0		UY	0		UY	0
		UZ	68.03		UZ	64.33		UZ	2.02
		RX	20.79		RX	4.34		RX	N/A
5	0.164	UX	0	0.173	UX	0	0.309	UX	0
		UY	0		UY	0		UY	0
		UZ	0		UZ	0		UZ	57.62
		RZ	81.81		RZ	87.67		RZ	0
6	0.154	UX	0	0.131	UX	0.11	0.289	UX	0.26
		UY	0		UY	0		UY	0
		UZ	0		UZ	0		UZ	0
		RX	3.27		RX	0		RX	N/A
7	0.135	UX	0	0.112	UX	0	0.266	UX	0
		UY	0		UY	13.60		UY	0
		UZ	0.97		UZ	0		UZ	0
8	0.121	UX	0	0.104	UX	0	0.264	UX	0
		UY	0		UY	0		UY	0
		UZ	0		UZ	1.51		UZ	4.97
		RX	2.49		RX	0		RX	N/A
9	0.110	UX	0	0.075	UX	0	0.155	UX	0
		UY	0		UY	0		UY	0
		UZ	0.60		UZ	0		UZ	0
		RZ	0.12		RZ	11.05		RZ	N/A
10	0.105	UX	0	0.064	UX	0	0.130	UX	0
		UY	12.27		UY	0		UY	0
		UZ	0		UZ	0		UZ	0
		RX	7.35		RX	1.98		RX	N/A

Table 2-18. Modal Analysis of Bridge Model with 30° skew

Mode	FE-SAP2000 Period, sec	Mass Participation, %		BS-SAP2000 Period, sec	Mass Participation, %		BS- DRAIN3DX Period, sec	Mass Participation, %	
1	0.617	UX	94.44	0.556	UX	94.44	0.541	UX	92.37
		UY	0		UY	0		UY	0.26
		UZ	0		UZ	0		UZ	0
		RY	16.03		RY	21.56		RY	N/A
2	0.448	UX	0.94	0.386	UX	3.30	0.383	UX	5.31
		UY	0.10		UY	14.73		UY	1.16
		UZ	0		UZ	0.10		UZ	0
		RY	42.29		RY	27.00		RY	N/A
3	0.367	UX	0	0.366	UX	0.73	0.357	UX	0
		UY	83.43		UY	69.17		UY	78.90
		UZ	0		UZ	0		UZ	0
		RX	53.23		RX	63.54		RX	N/A
4	0.331	UX	0	0.299	UX	0	0.317	UX	0
		UY	0		UY	0		UY	0
		UZ	66.83		UZ	45.69		UZ	0.49
		RX	20.30		RX	1.95		RX	N/A
5	0.164	UX	0	0.281	UX	0	0.293	UX	0
		UY	0		UY	0		UY	0
		UZ	0		UZ	18.68		UZ	57.4
		RY	6.15		RY	0.19		RY	0
6	0.164	UX	0	0.241	UX	0	0.289	UX	0.30
		UY	0		UY	0		UY	0
		UZ	0		UZ	0.65		UZ	0
		RZ	79.41		RZ	0		RZ	N/A
7	0.138	UX	0	0.207	UX	0	0.265	UX	0
		UY	0		UY	0		UY	0
		UZ	1.74		UZ	0		UZ	0
8	0.113	UX	0	0.193	UX	0	0.263	UX	0
		UY	0		UY	0		UY	0
		UZ	0		UZ	0		UZ	6.83
		RX	2.49		RX	0		RX	N/A
9	0.106	UX	0.12	0.169	UX	0	0.157	UX	0
		UY	12.45		UY	0		UY	0
		UZ	0		UZ	0		UZ	0
		RZ	0		RZ	89.00		RZ	N/A
10	0.105	UX	0	0.129	UX	0	0.129	UX	0
		UY	0		UY	0.1		UY	0
		UZ	0.28		UZ	0		UZ	0

Table 2-19. Modal Analysis of Bridge Model with 45° skew

Mode	FE-SAP2000			BS-SAP2000			BS- DRAIN3DX		
	Period, sec	Mass Participation, %		Period, sec	Mass Participation, %		Period, sec	Mass Participation, %	
1	0.654	UX	87.12	0.556	UX	97.00	0.554	UX	95.78
		UY	0		UY	0		UY	0
		UZ	0		UZ	0		UZ	0
		RY	11.88		RY	16.63		RY	N/A
2	0.378	UX	0.29	0.371	UX	0	0.364	UX	0.71
		UY	9.84		UY	81.36		UY	78.86
		UZ	0		UZ	0		UZ	0
		RY	35.74		RY	1.01		RY	N/A
3	0.361	UX	0	0.331	UX	1.32	0.337	UX	1.15
		UY	71.30		UY	2.92		UY	1.93
		UZ	0		UZ	0		UZ	0
		RX	46.71		RX	2.91		RX	N/A
4	0.306	UX	0	0.278	UX	0	0.318	UX	0
		UY	0		UY	0		UY	0
		UZ	63.3		UZ	1.26		UZ	0.63
		RX	18.96		RX	0		RX	N/A
5	0.183	UX	0	0.273	UX	0	0.293	UX	0.43
		UY	0.13		UY	0		UY	0.15
		UZ	0		UZ	63.67		UZ	0
6	0.165	UX	0	0.234	UX	0	0.279	UX	0
		UY	0		UY	0		UY	0
		UZ	0.22		UZ	0		UZ	48.49
		RZ	72.83		RZ	0		RZ	N/A
7	0.146	UX	0	0.213	UX	0	0.261	UX	0
		UY	0		UY	0		UY	0
		UZ	3.27		UZ	0		UZ	0
8	0.108	UX	0.34	0.197	UX	0	0.261	UX	0
		UY	12.77		UY	0		UY	0
		UZ	0		UZ	0		UZ	15.72
		RX	8.05		RX	0		RX	N/A
9	0.099	UX	0	0.172	UX	0	0.164	UX	0
		UY	0		UY	0		UY	0
		UZ	0		UZ	0		UZ	0
		RZ	0		RZ	88.60		RZ	N/A
10	0.095	UX	0	0.126	UX	0.11	0.129	UX	0
		UY	0		UY	0.48		UY	0.48
		UZ	0		UZ	0		UZ	0

Table 2-20. Modal Analysis of Benchmark Bridge (52° skew)

Mode	FE-SAP2000			BS-SAP2000			BS- DRAIN3DX		
	Period, sec	Mass Participation, %		Period, sec	Mass Participation, %		Period, sec	Mass Participation, %	
1	0.588	UX	93.65	0.592	UX	97.7	0.561	UX	96.3
		UY	0		UY	0		UY	0.03
		UZ	0		UZ	0		UZ	0
		RY	13.60		RY	14.85		RY	N/A
2	0.370	UX	0	0.376	UX	0	0.370	UX	0.52
		UY	70.80		UY	83.22		UY	81.08
		UZ	0		UZ	0		UZ	0
		RX	44.77		RX	75.74		RX	N/A
3	0.323	UX	0.78	0.317	UX	0.67	0.321	UX	0.73
		UY	8.92		UY	0.83		UY	0.54
		UZ	0		UZ	0		UZ	0
		RY	35.81		RY	37.71		RY	N/A
4	0.289	UX	0	0.280	UX	0	0.317	UX	0
		UY	0		UY	0		UY	0
		UZ	63.07		UZ	0.85		UZ	0.73
		RX	19.84		RX	0.05		RX	N/A
5	0.191	UX	0.18	0.267	UX	0	0.295	UX	0.56
		UY	0.67		UY	0		UY	0.21
		UZ	9.4E-6		UZ	64.06		UZ	0
6	0.170	UX	0	0.232	UX	0	0.277	UX	0
		UY	0		UY	0		UY	0
		UZ	0.31		UZ	0		UZ	40.43
		RZ	72.63		RZ	0		RZ	N/A
7	0.151	UX	0	0.215	UX	0	0.259	UX	0
		UY	0		UY	0		UY	0
		UZ	4.41		UZ	0		UZ	12.62
8	0.119	UX	0.34	0.198	UX	0	0.257	UX	0
		UY	12.58		UY	0		UY	0
		UZ	0		UZ	0.1		UZ	10.86
		RX	1.50		RX	0		RX	N/A
9	0.094	UX	0	0.178	UX	0	0.172	UX	0
		UY	0		UY	0		UY	0
		UZ	0		UZ	0		UZ	0
		RZ	0		RZ	89.17		RZ	N/A
10	0.089	UX	0	0.124	UX	0.09	0.131	UX	0
		UY	0		UY	2.09		UY	1.84
		UZ	0.13		UZ	0		UZ	0

Table 2-21. Modal Analysis of Bridge Model with 60° skew

Mode	FE-SAP2000 Period, sec	FE-SAP2000 Mass Participation, %		BS-SAP2000 Period, sec	BS-SAP2000 Mass Participation, %		BS- DRAIN3DX Period, sec	BS- DRAIN3DX Mass Participation, %	
1	0.638	UX	95.83	0.625	UX	98.07	0.574	UX	96.45
		UY	0.13		UY	0.01		UY	0
		UZ	0		UZ	0		UZ	0
		RY	10.51		RY	14.45		RY	N/A
2	0.371	UX	0	0.382	UX	0	0.385	UX	0
		UY	80.43		UY	84.33		UY	83.47
		UZ	0		UZ	0		UZ	0
		RX	50.55		RX	77.24		RX	N/A
3	0.277	UX	0	0.303	UX	0	0.319	UX	0
		UY	0.46		UY	0.20		UY	0
		UZ	0		UZ	0		UZ	1.73
		RY	37.07		RY	37.88		RY	N/A
4	0.251	UX	0	0.286	UX	0	0.309	UX	0.357
		UY	0		UY	0		UY	0.187
		UZ	55.04		UZ	1.59		UZ	0
		RX	17.28		RX	0		RX	N/A
5	0.203	UX	0	0.261	UX	0	0.300	UX	0.877
		UY	0.16		UY	0		UY	0.250
		UZ	0		UZ	63.68		UZ	0
6	0.174	UX	0	0.233	UX	0	0.276	UX	0
		UY	0		UY	0		UY	0
		UZ	0.34		UZ	0		UZ	31.21
		RZ	70.37		RZ	0		RZ	N/A
7	0.149	UX	0	0.226	UX	0	0.256	UX	0
		UY	0		UY	0		UY	0
		UZ	8.32		UZ	0		UZ	32.28
8	0.123	UX	0.33	0.204	UX	0	0.255	UX	0
		UY	13.53		UY	0		UY	0
		UZ	0		UZ	0.26		UZ	0
		RX	8.09		RX	0		RX	N/A
9	0.092	UX	0	0.198	UX	0	0.190	UX	0
		UY	0		UY	0.36		UY	0
		UZ	0		UZ	0		UZ	0
		RZ	0		RZ	90.22		RZ	N/A
10	0.087	UX	0	0.138	UX	0	0.137	UX	0
		UY	0		UY	12.80		UY	3.58
		UZ	3.02		UZ	0		UZ	0
		RX	0.95		RX	12.17		RX	N/A

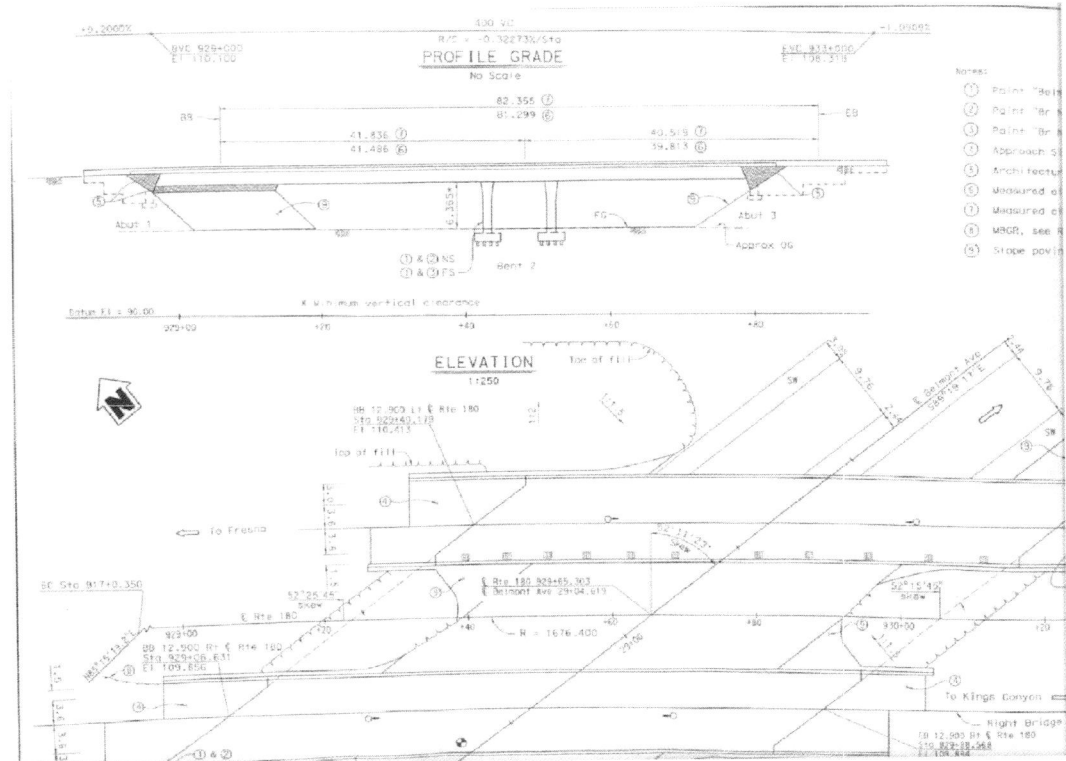

Figure 2-1. Bridge 42-04271 L/R Overview

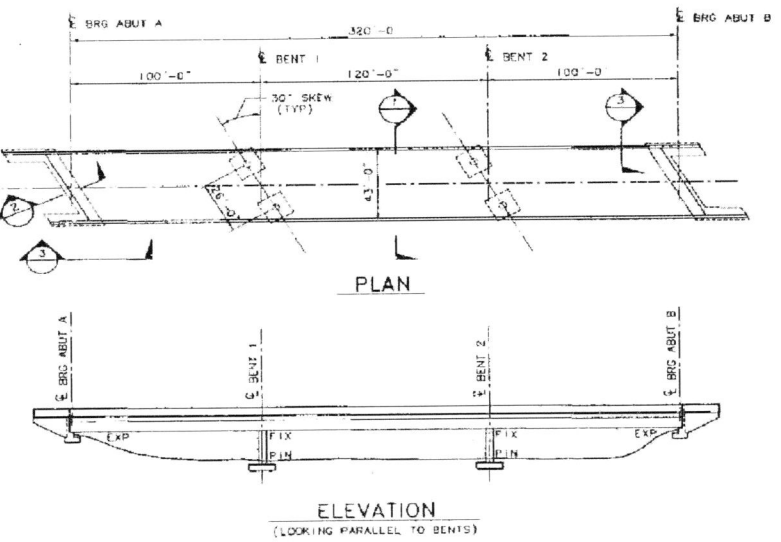

Figure 2-2. FHWA Bridge Overview

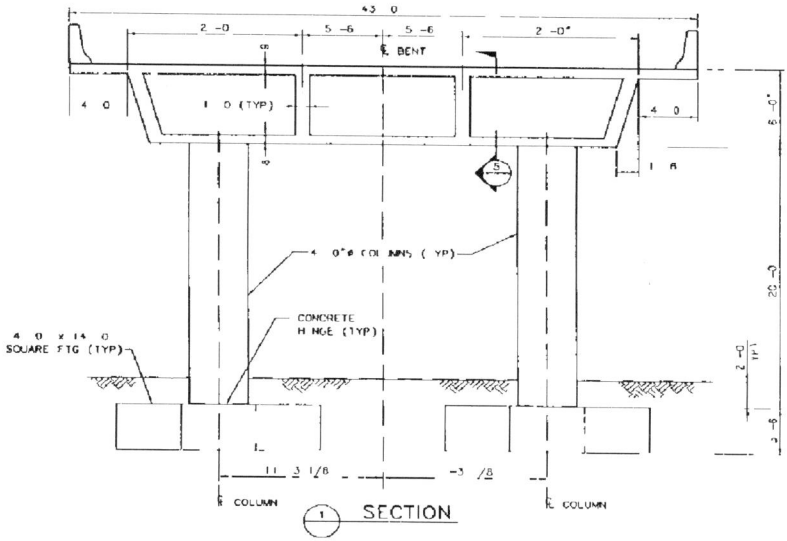

Figure 2-3. Cross-Section of FHWA Example 4 Bridge

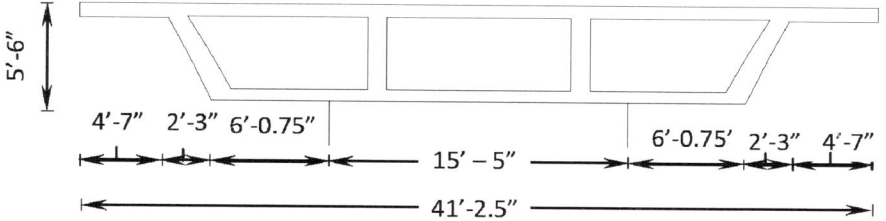

Figure 2-4. Cross-Section Dimensions of Bridge 42-04271 L/R (Aspect Ratio of 0.3)

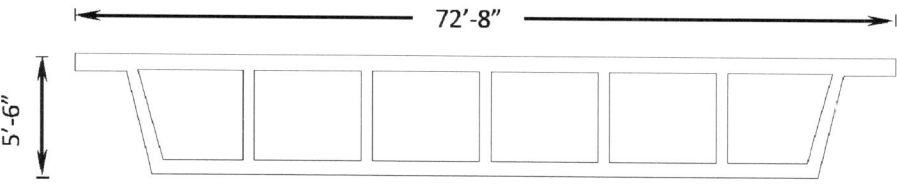

Figure 2-5. Cross-Section Dimensions of Bridge (Aspect Ratio of 0.54)

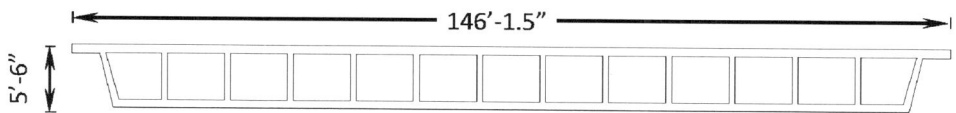

Figure 2-6. Cross-Section Dimensions of Bridge (Aspect Ratio of 1.1)

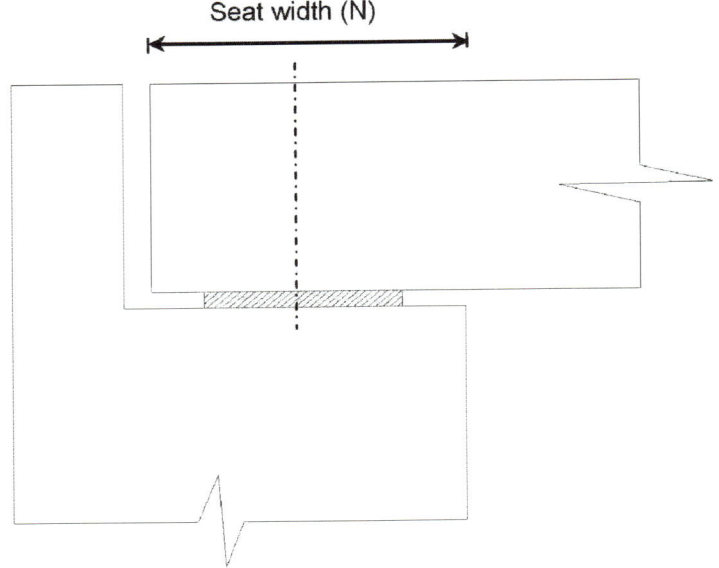

Figure 2-7. Abutment Seat width

Figure 2-8. FE Model of 42-04271 L/R Bridge

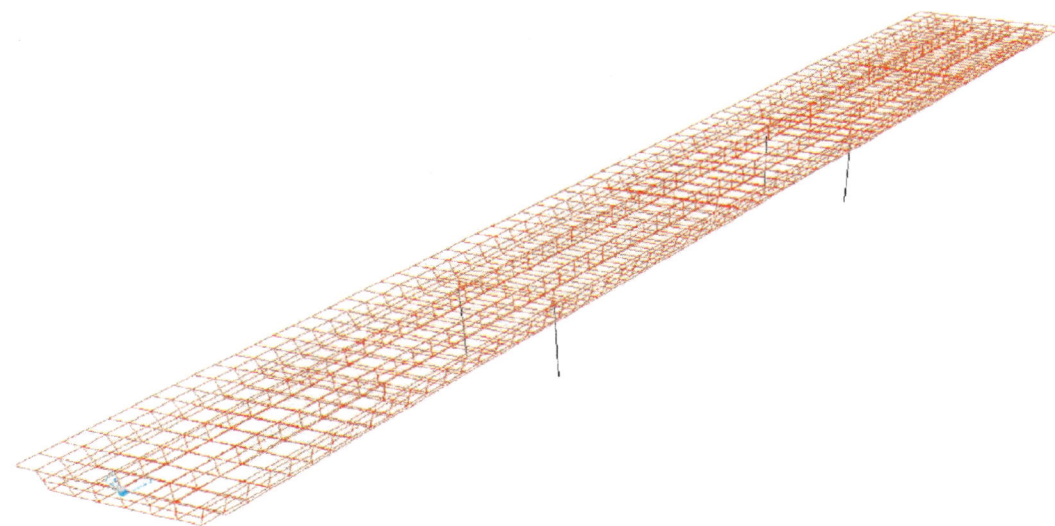

Figure 2-9. FE Model of FHWA Bridge

78

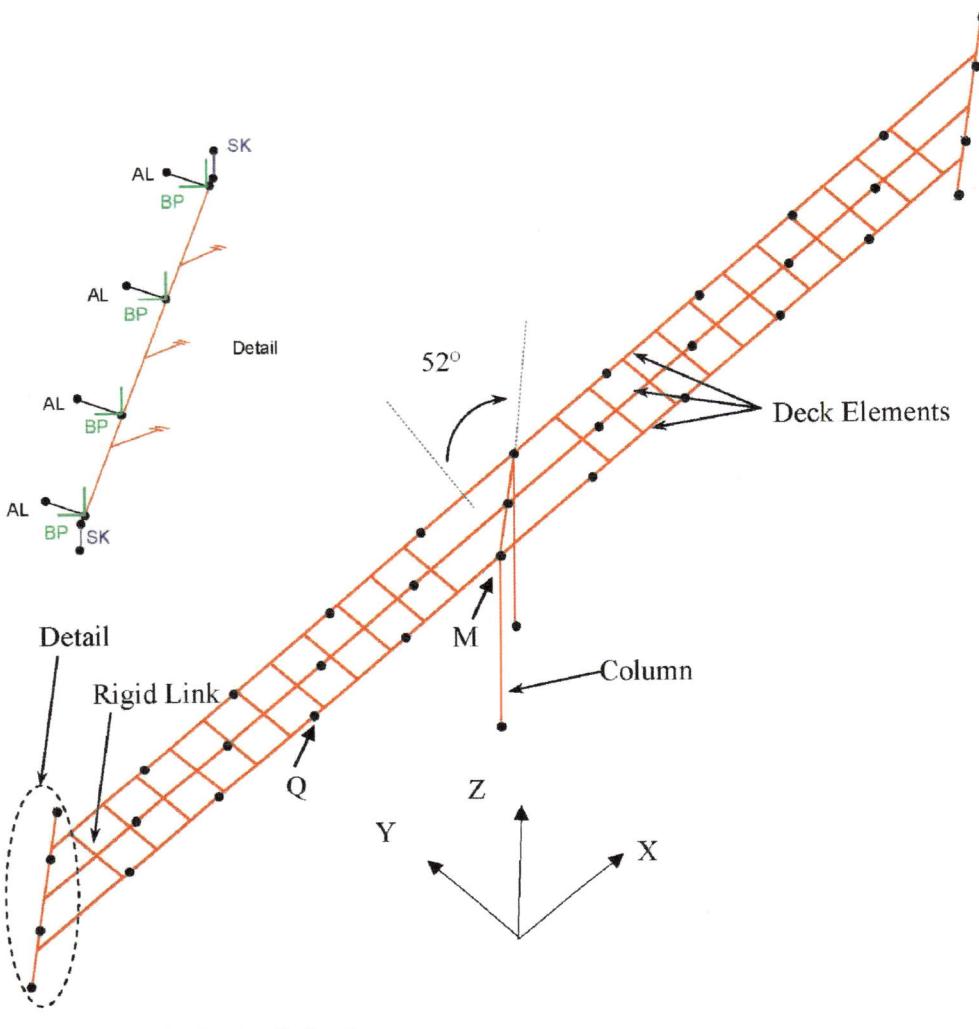

BP: Bearing Pad springs
SK: Shear key springs
AL: Abutment springs

Figure 2-10. Improved Beam-Stick Model of 42-04271 L/R Bridge

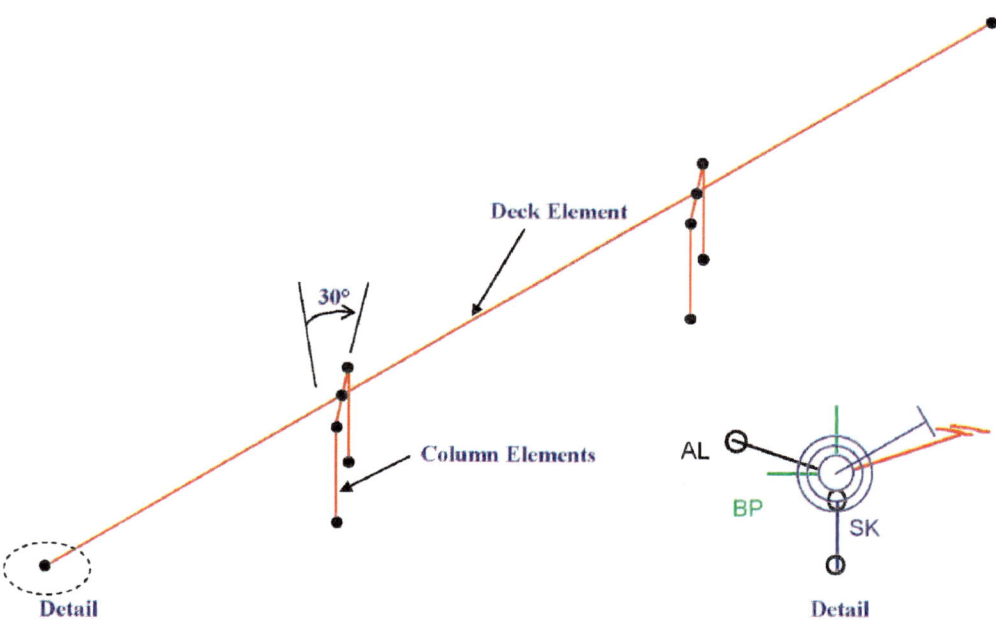

Figure 2-11. Single Spine Model of FHWA Example 4 Bridge

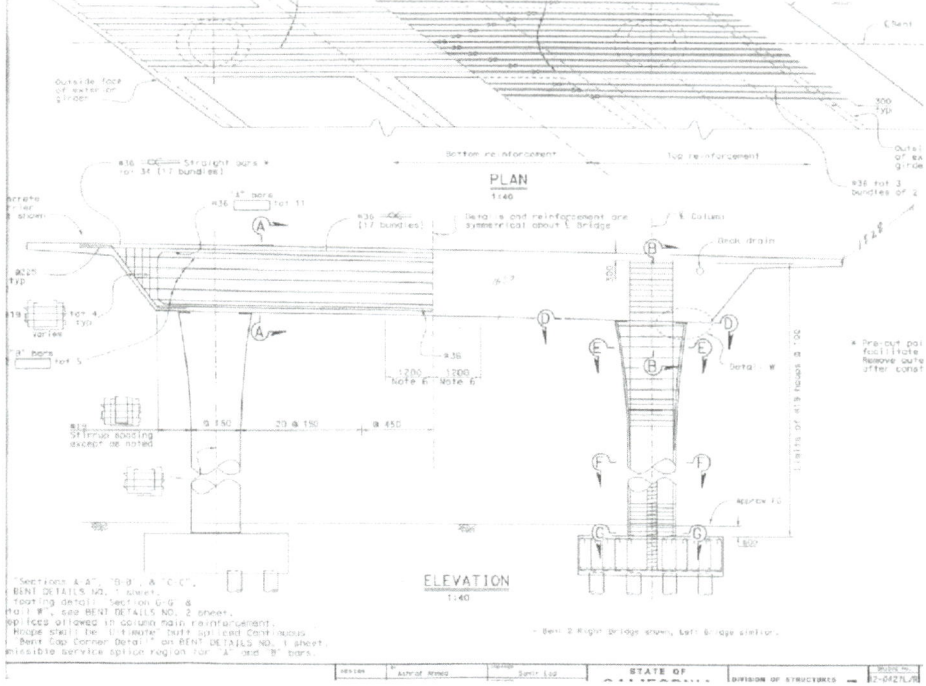

Figure 2-12. Details of Bent Cap Beam

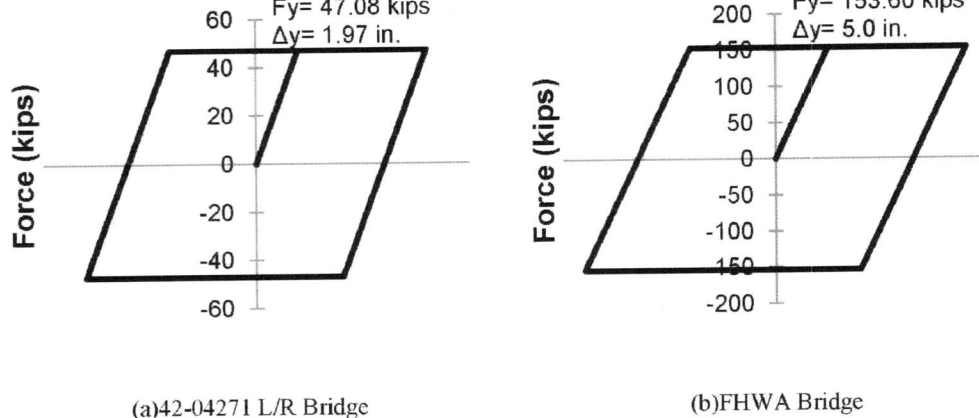

(a)42-04271 L/R Bridge   (b)FHWA Bridge

Figure 2-13. Hysteresis of Typical Bearing Pad

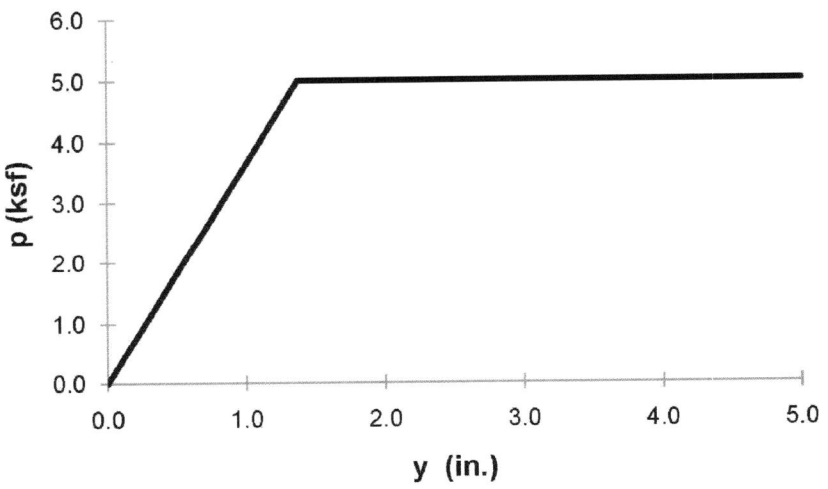

Figure 2-14. Soil Force-Displacement Relation

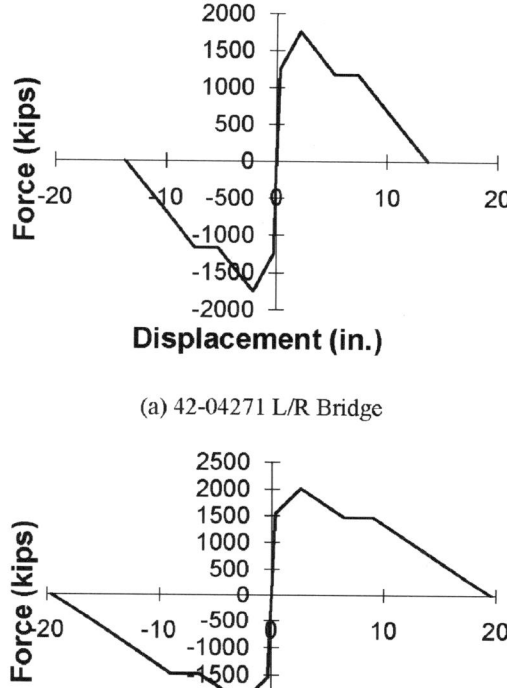

(a) 42-04271 L/R Bridge

(b)FHWA Bridge

Figure 2-15. Force-Displacement Relation of Shear Key

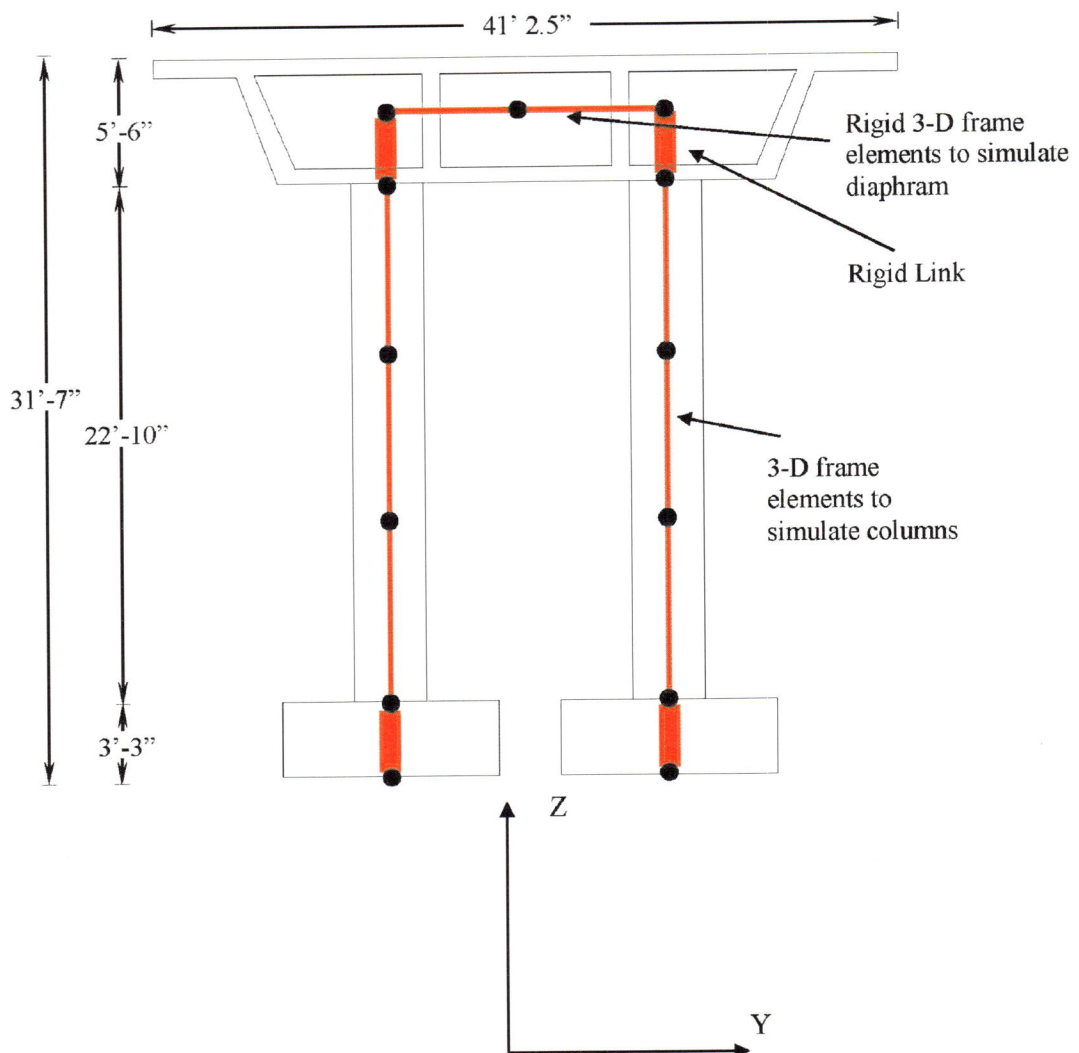

Figure 2-16. Improved Beam-Stick Model: Bent Elevation (Aspect Ratio of 0.3)

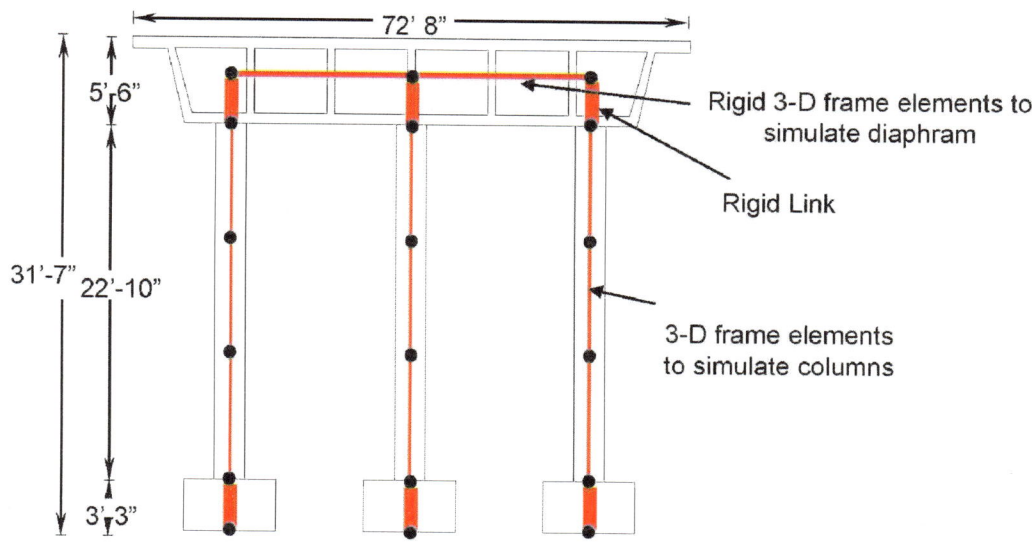

Figure 2-17. Improved Beam-Stick Model: Bent Elevation (Aspect Ratio of 0.54)

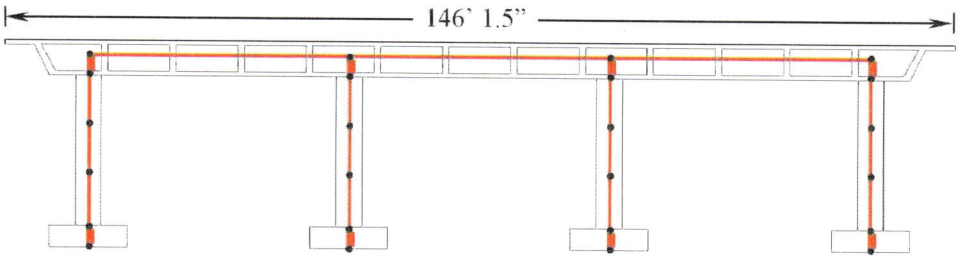

Figure 2-18. Improved Beam-Stick Model: Bent Elevation (Aspect Ratio of 1.1)

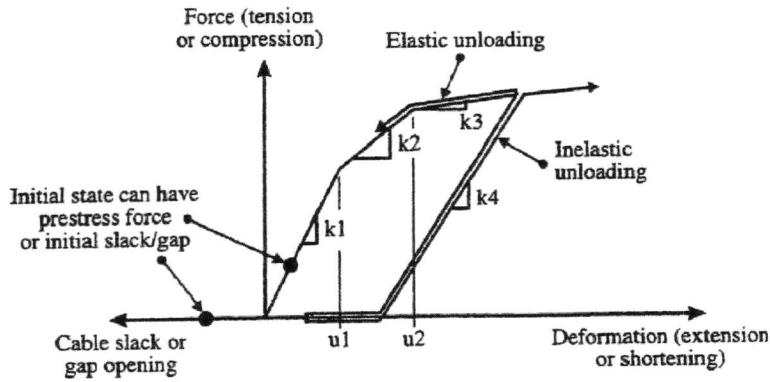

Note: k2 and/or k3 may be > k1.
k4 should be >= max(k1,k2,k3).

Figure 2-19. Hysteresis of Element 09 (Prakash et al., 1994)

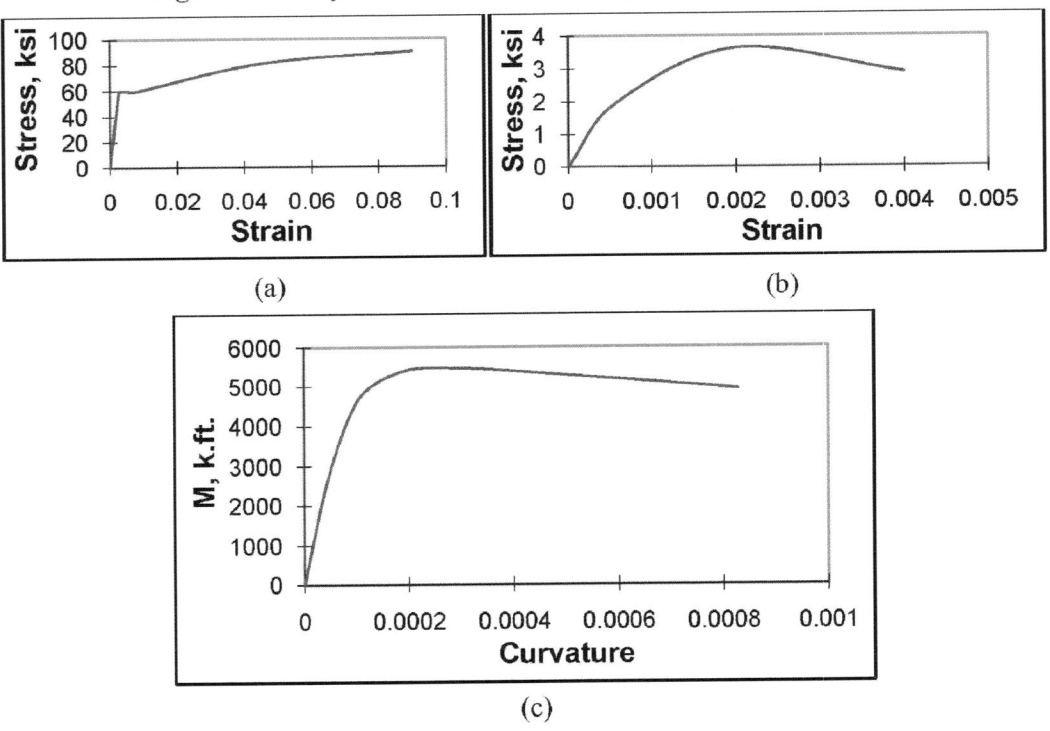

(a)

(b)

(c)

Figure 2-20. Characteristics of fibers (a) Stress-Strain relationship of steel, (b) Stress-Strain relationship of unconfined concrete, and (c) Moment Curvature relation of bent columns

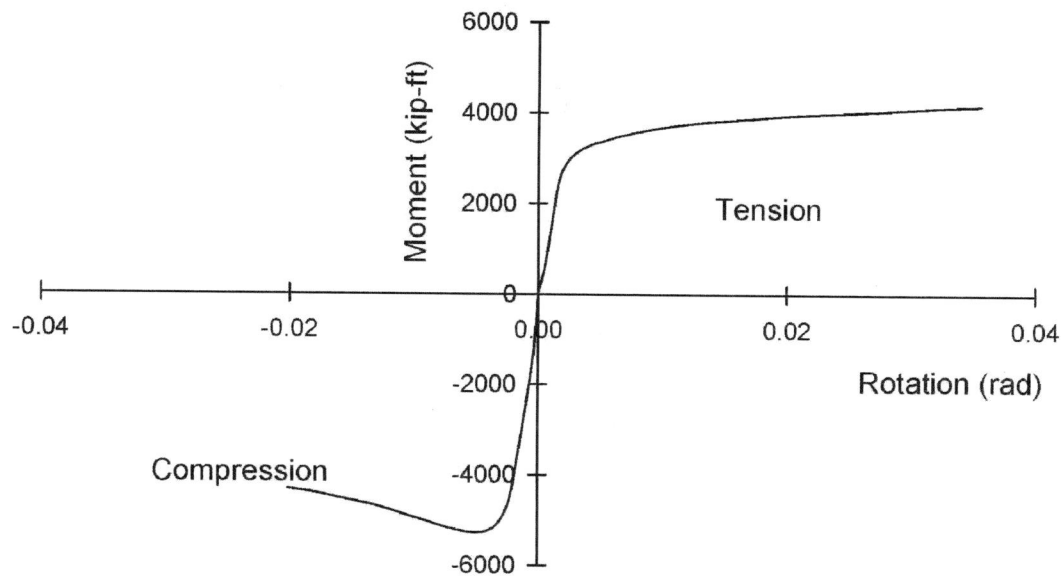

Figure 2-21. Lumped Moment Hinge Properties

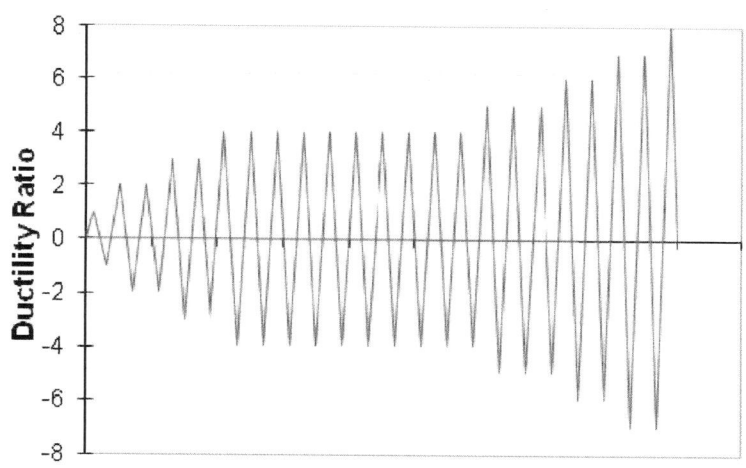

Figure 2-22. Lateral loading protocol (Cheok and Stone, 1999)

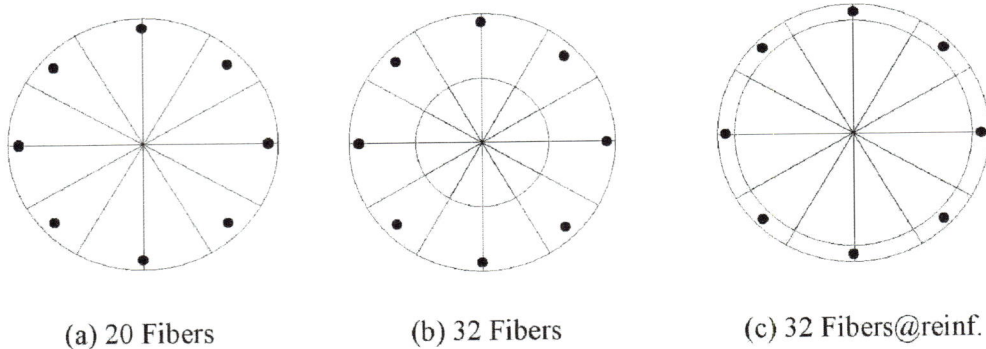

(a) 20 Fibers   (b) 32 Fibers   (c) 32 Fibers@reinf.

Figure 2-23. Proposed number and distribution of fibers

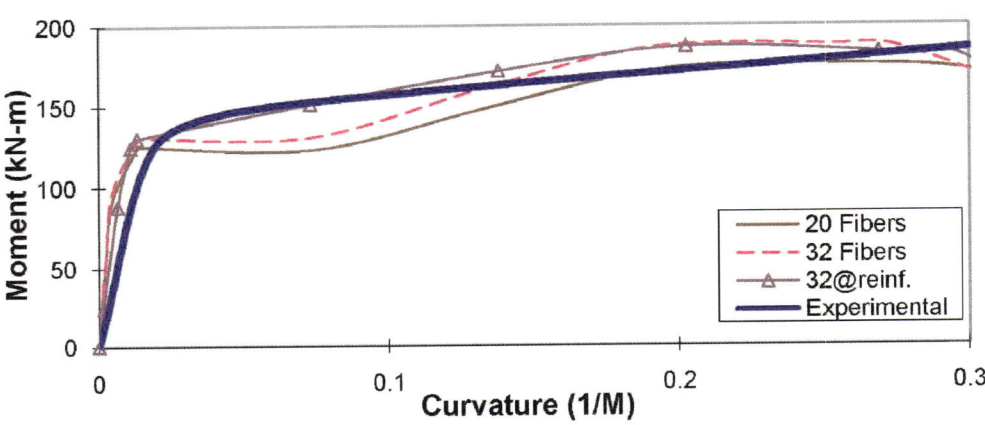

Figure 2-24. Comparison of analytical and experimental results: Specimen Four (Esmaeily and Xiao, 2002)

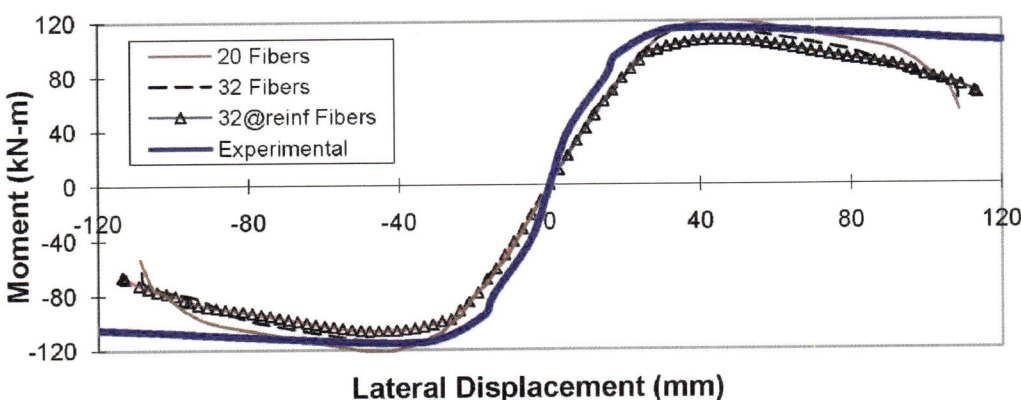

Figure 2-25. Comparison of analytical and experimental results: RC-2 (Sucuoglu and Baingo, 1999)

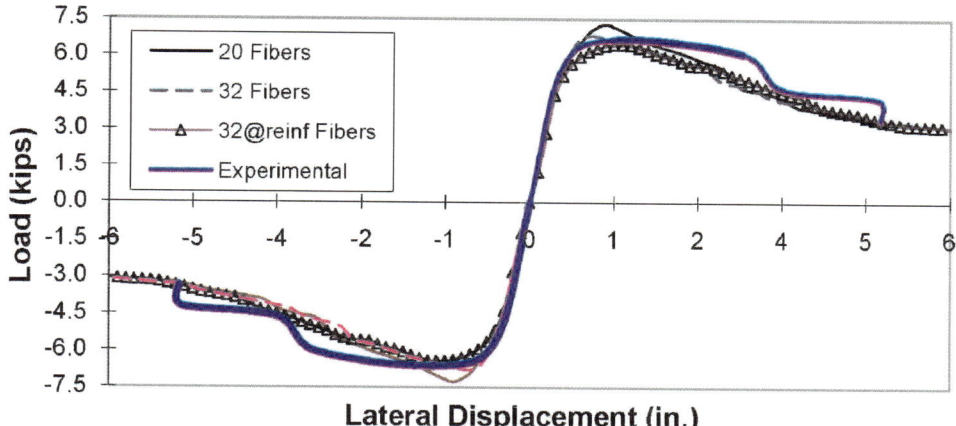

Figure 2-26. Comparison of analytical and experimental results: N6 (Cheok and Stone, 1990)

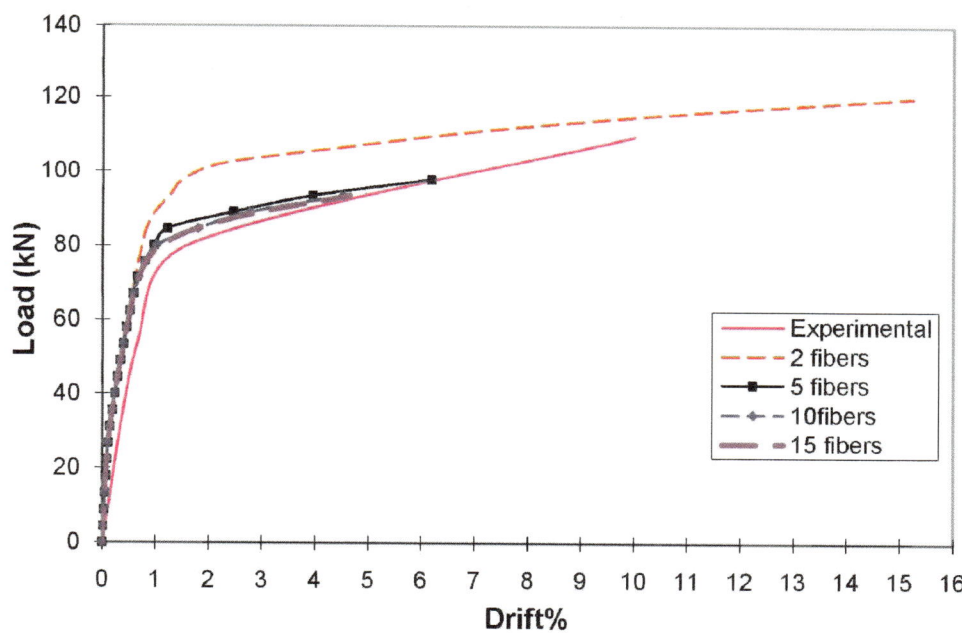

Figure 2-27. Load-Displacement relationship of bent of aspect ratio of 3.5 (Plastic hinge length= 0.05H)

88

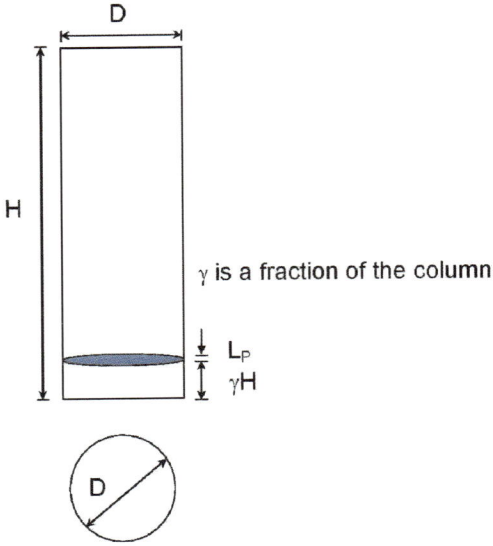

Figure 2-28. Plastic Hinge sketch

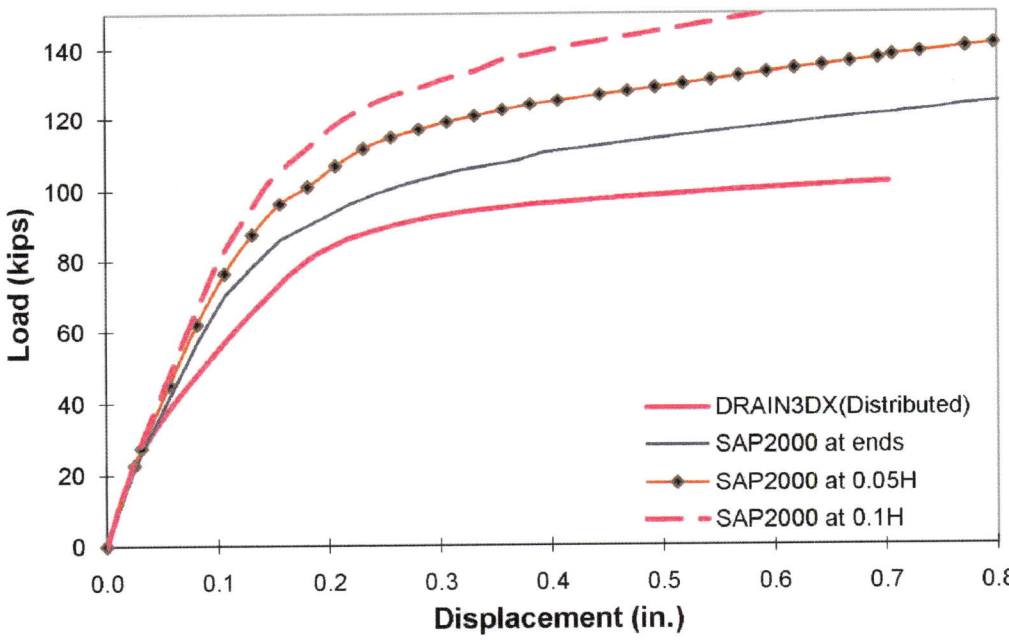

Figure 2-29. Load-Displacement relationship of bent of aspect ratio of 3.5 (Plastic hinge length= 0.05H)

89

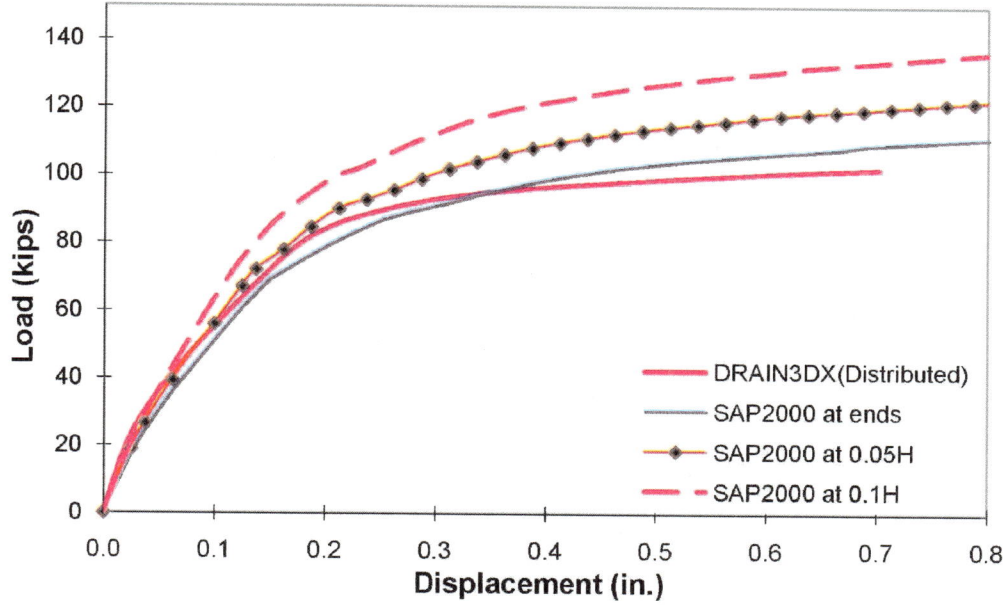

Figure 2-30. Load-Displacement relationship of bent of aspect ratio of 3.5 (Plastic hinge length= 0.10H)

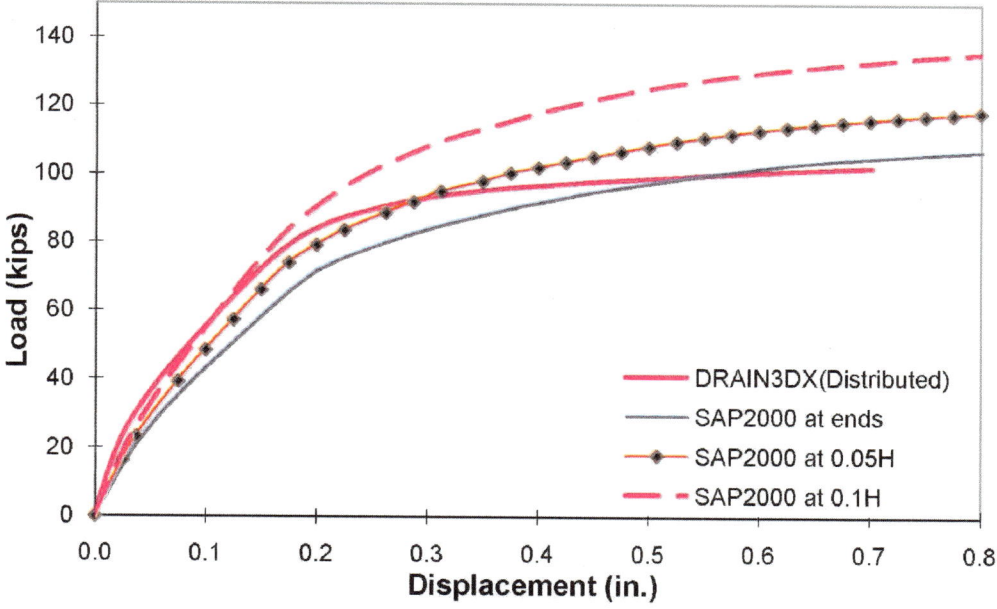

Figure 2-31. Load-Displacement relationship of bent of aspect ratio of 3.5 (Plastic hinge length= 0.15H)

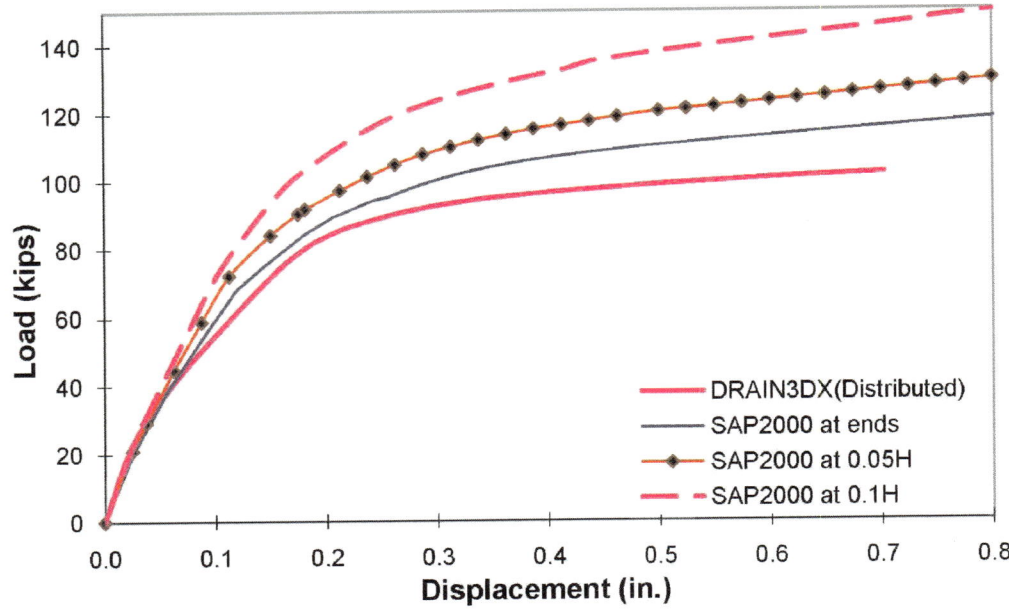

Figure 2-32. Load-Displacement relationship of bent of aspect ratio of 3.5 (Plastic hinge length= 0.25D)

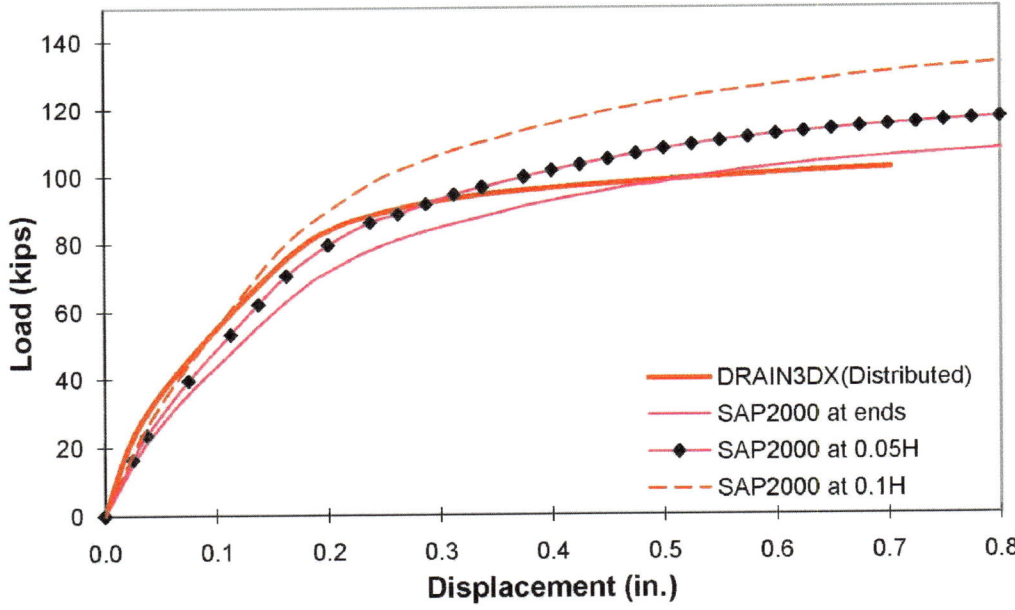

Figure 2-33. Load-Displacement relationship of bent of aspect ratio of 3.5 (Plastic hinge length= 0. 5D)

91

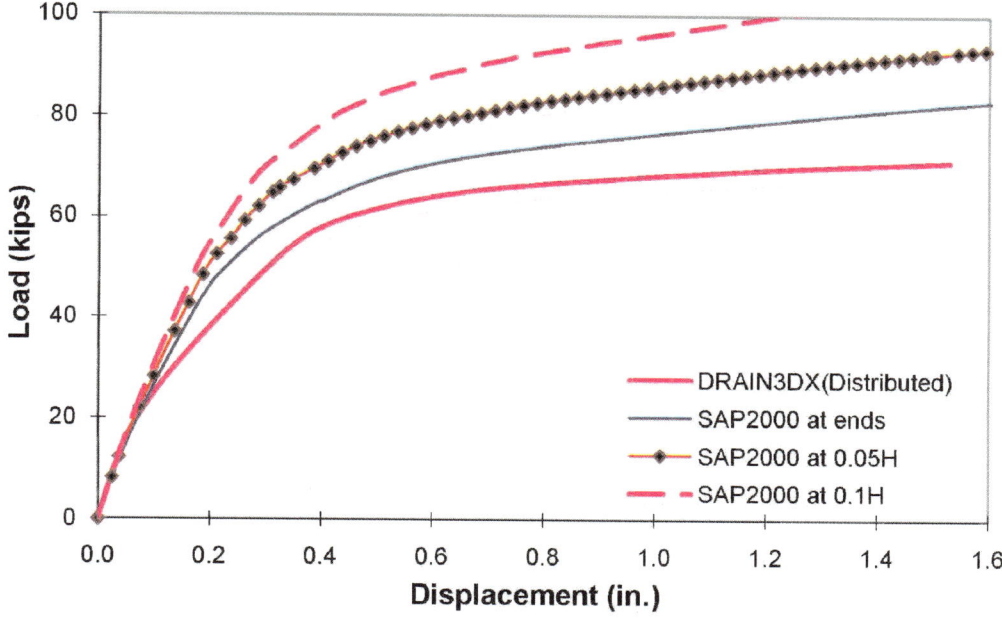

Figure 2-34. Load-Displacement relationship of bent of aspect ratio of 5.0 (Plastic hinge length= 0.05H)

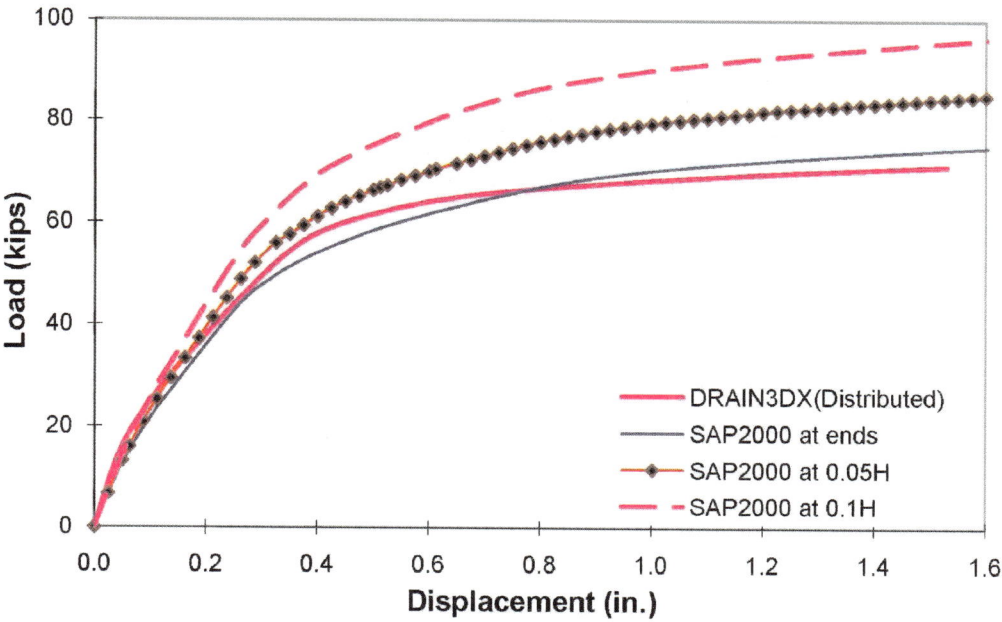

Figure 2-35. Load-Displacement relationship of bent of aspect ratio of 5.0 (Plastic hinge length= 0.10H)

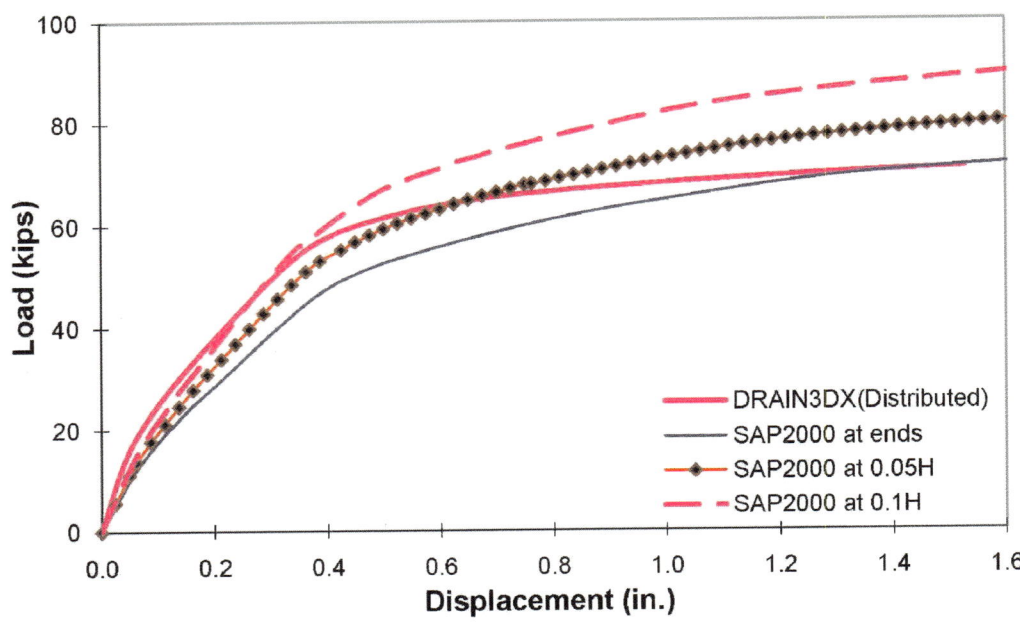

Figure 2-36. Load-Displacement relationship of bent of aspect ratio of 5.0 (Plastic hinge length= 0.15H)

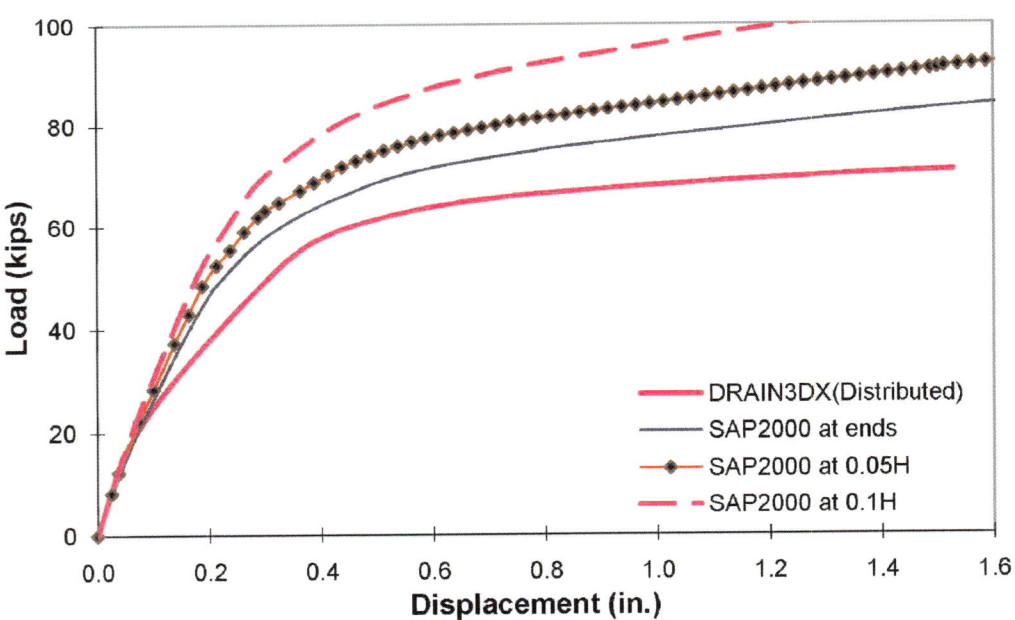

Figure 2-37. Load-Displacement relationship of bent of aspect ratio of 5.0 (Plastic hinge length= 0.25D)

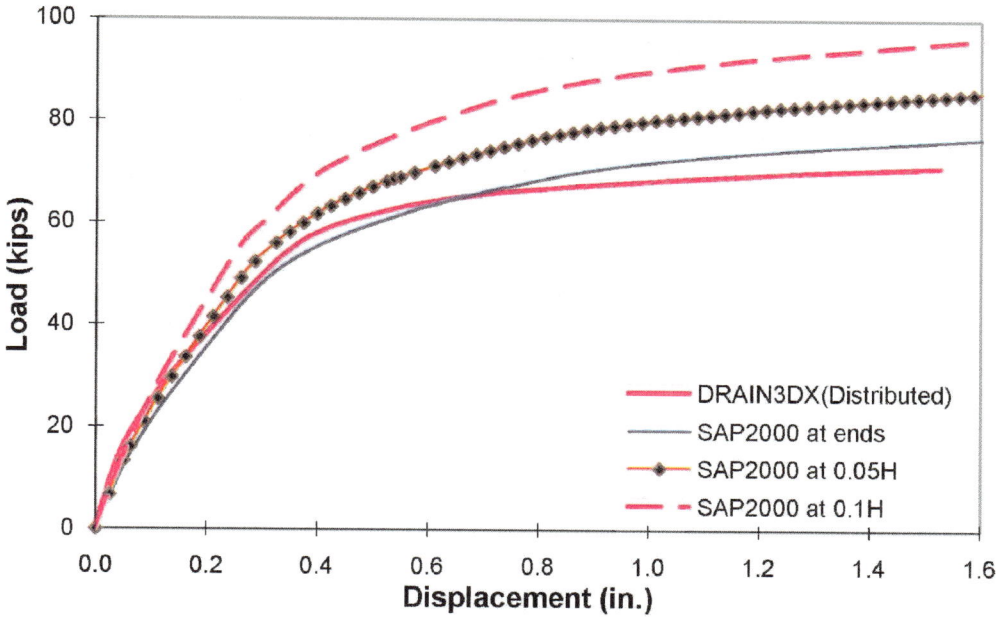

Figure 2-38. Load-Displacement relationship of bent of aspect ratio of 5.0 (Plastic hinge length= 0.5D)

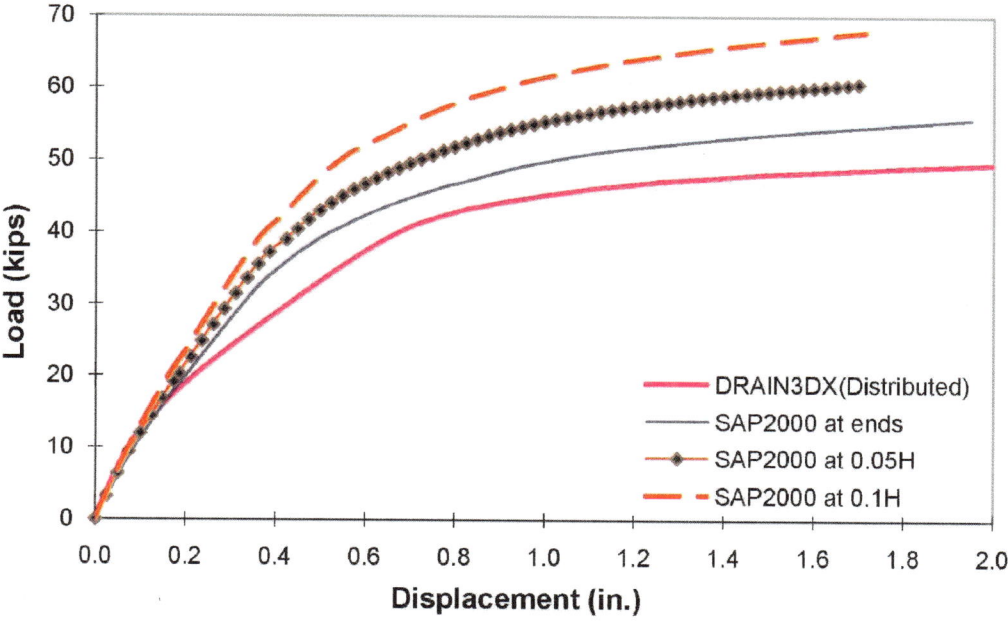

Figure 2-39. Load-Displacement relationship of bent of aspect ratio of 7.0 (Plastic hinge length= 0.05H)

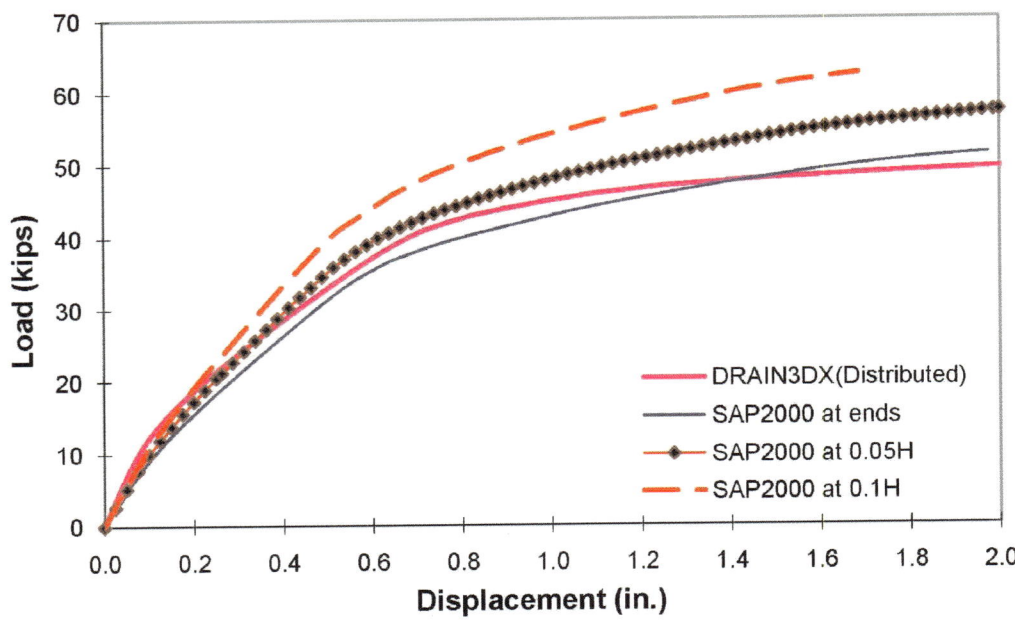

Figure 2-40. Load-Displacement relationship of bent of aspect ratio of 7.0 (Plastic hinge length= 0.10H)

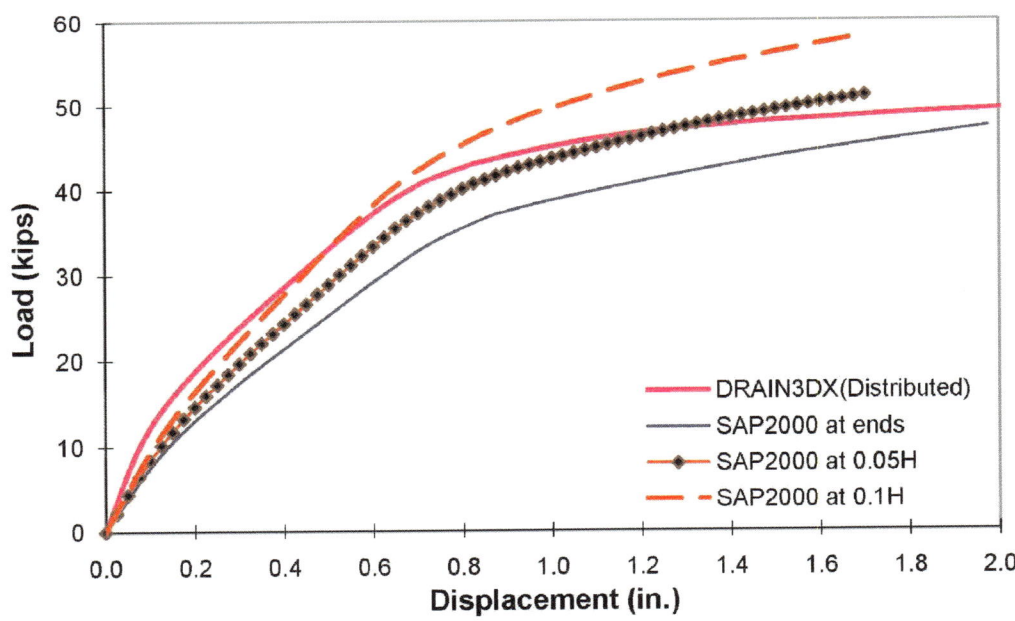

Figure 2-41. Load-Displacement relationship of bent of aspect ratio of 7.0 (Plastic hinge length= 0.15H)

95

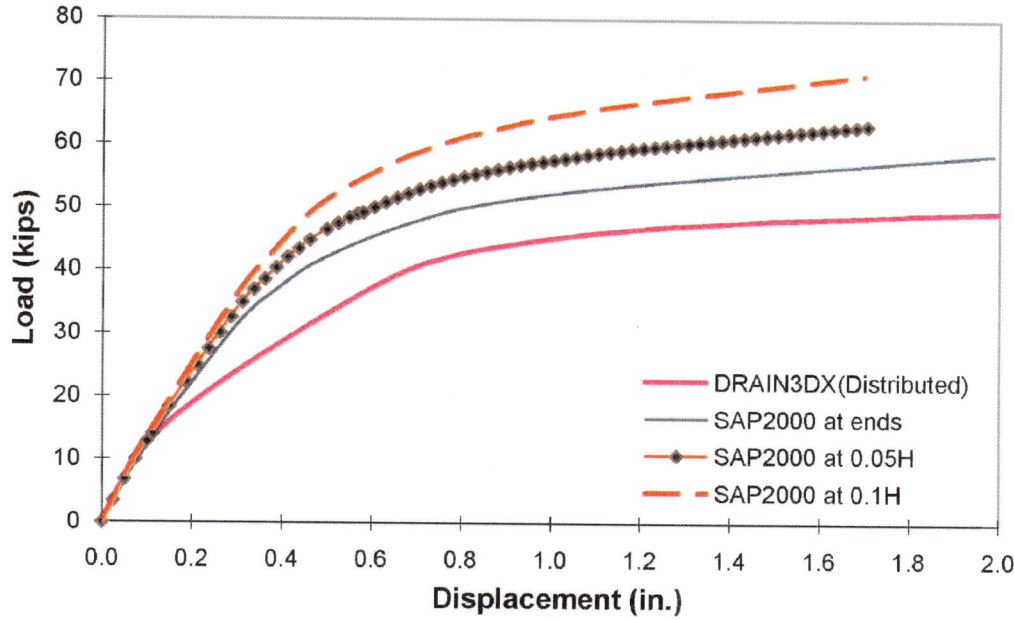

Figure 2-42. Load-Displacement relationship of bent of aspect ratio of 7.0 (Plastic hinge length= 0.25D)

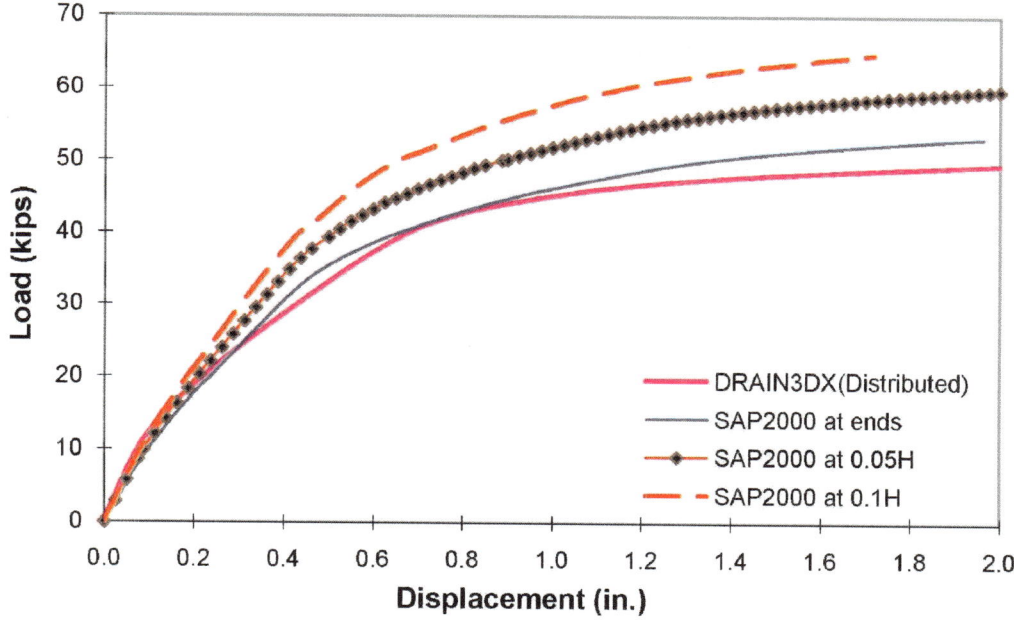

Figure 2-43. Load-Displacement relationship of bent of aspect ratio of 7.0 (Plastic hinge length= 0.50D)

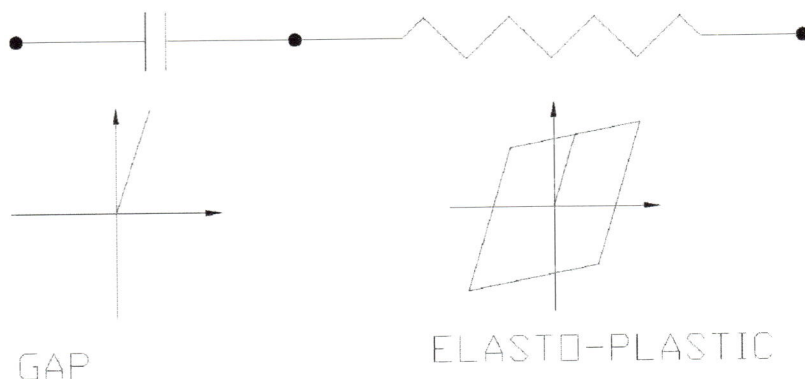

Figure 2-44. Modeling of Abutment-Soil Response for SAP2000

Unit	English	
p_{max}	5	(ksf)
y_h	h/48	(in)
For h = 5.5 ft		

y (in)	p (ksf)
0.000	0.00
0.250	0.91
0.500	1.82
0.750	2.73
1.000	3.64
1.250	4.55
1.375	5.00
1.500	5.00
2.000	5.00
3.000	5.00
5.000	5.00

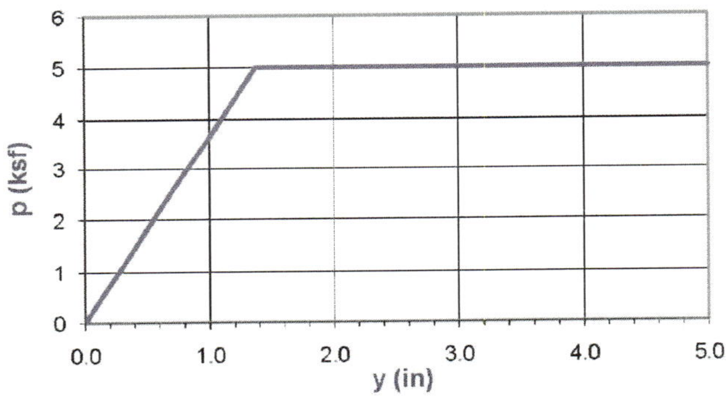

Figure 2-45. p-y Relation for Passive Pressure at Abutment Back Wall (provided by Caltrans)

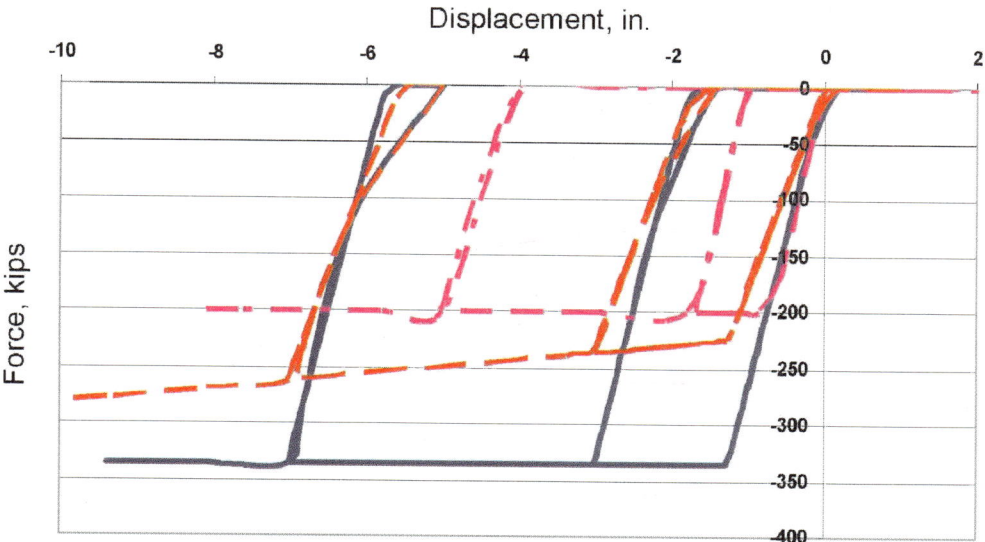

Figure 2-46. Modeling of nonlinear response of abutment soil: response to cyclic pushover

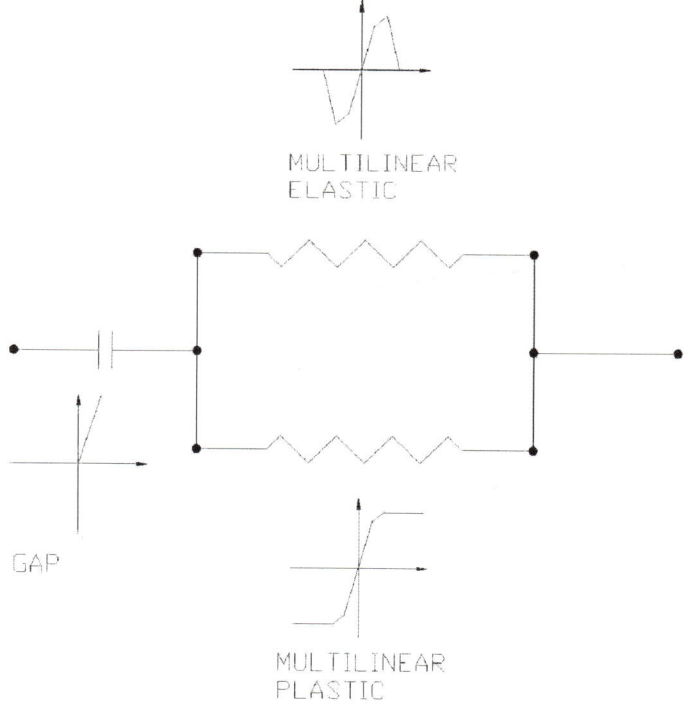

Figure 2-47. Modeling of Shear-key Response for SAP2000

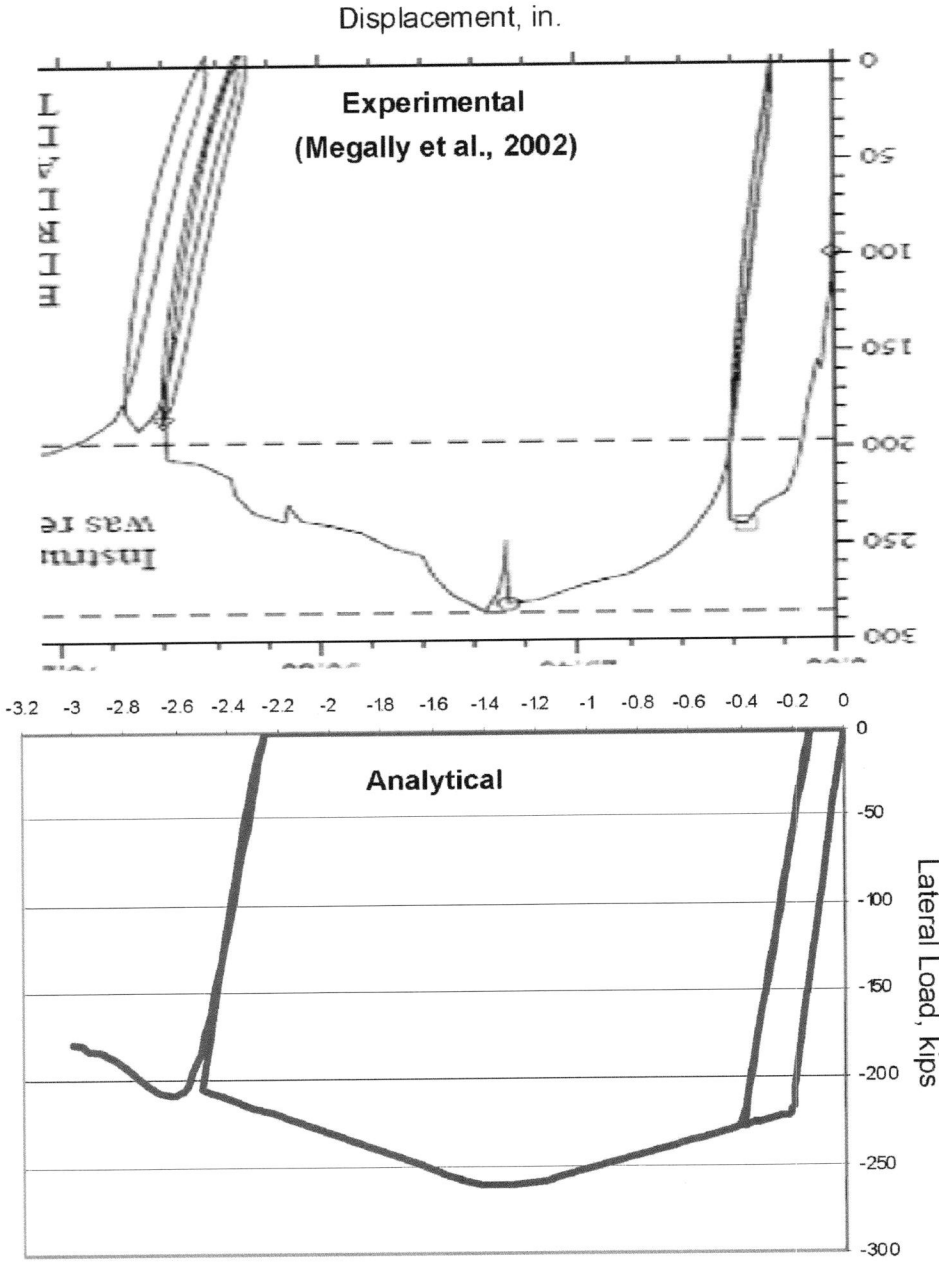

Figure 2-48. Comparison of analytical and experimental response of external shear-key

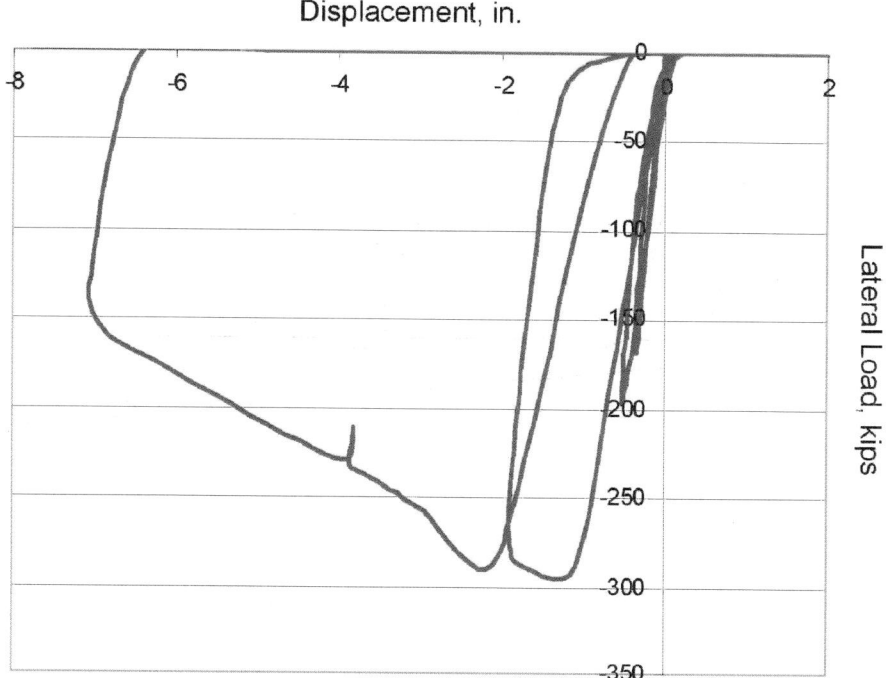

Figure 2-49. External shear key response: nonlinear time-history analysis

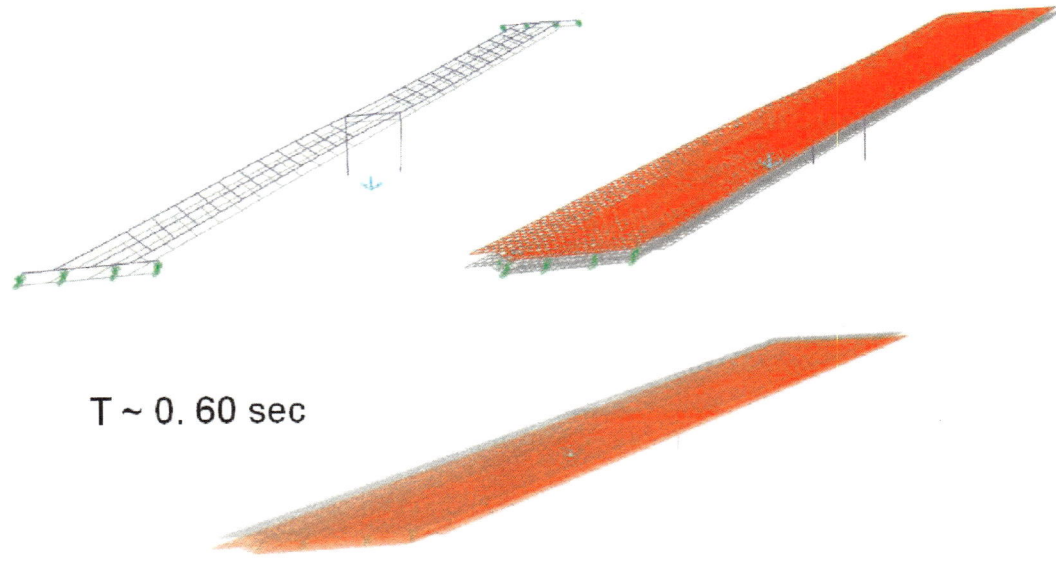

T ~ 0. 60 sec

Figure 2-50. 52 Degree skew mode 1

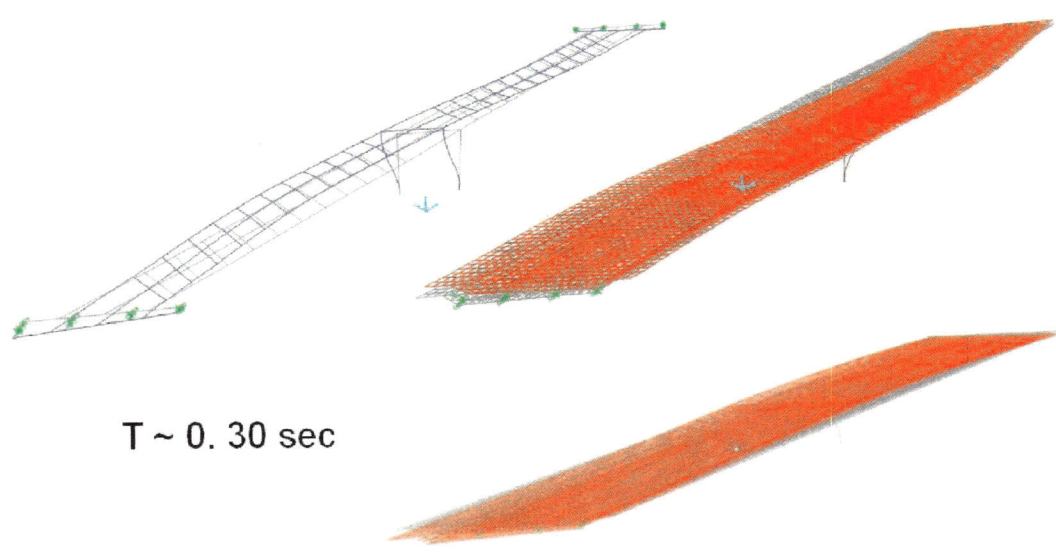

T ~ 0. 30 sec

Figure 2-51. 52 Degree skew mode 3

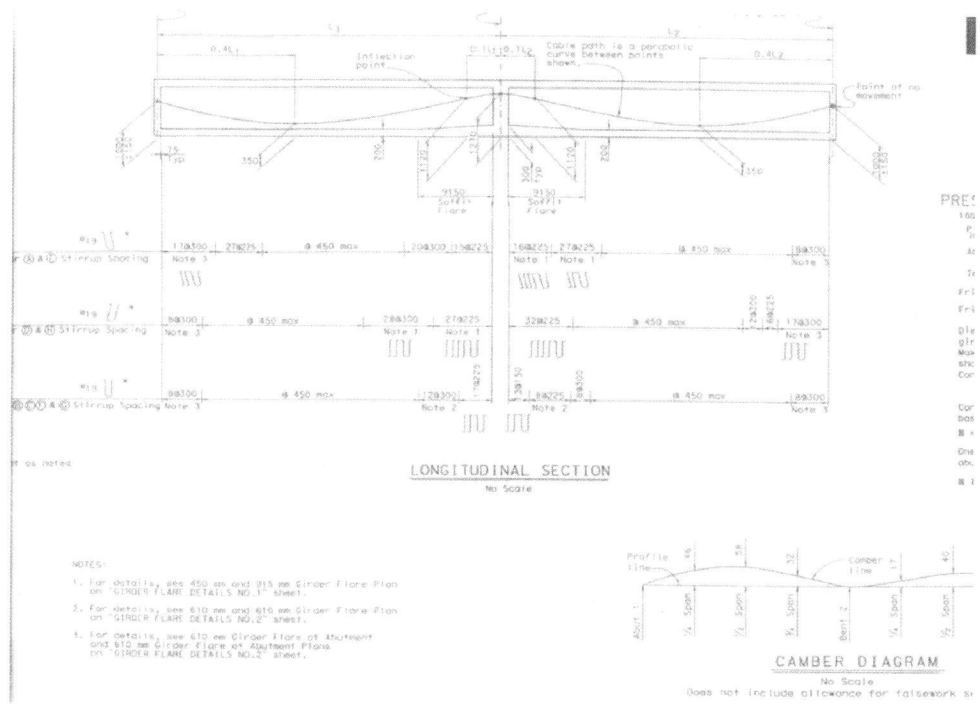

Figure 2-52. Longitudinal Section of the Benchmark Bridge

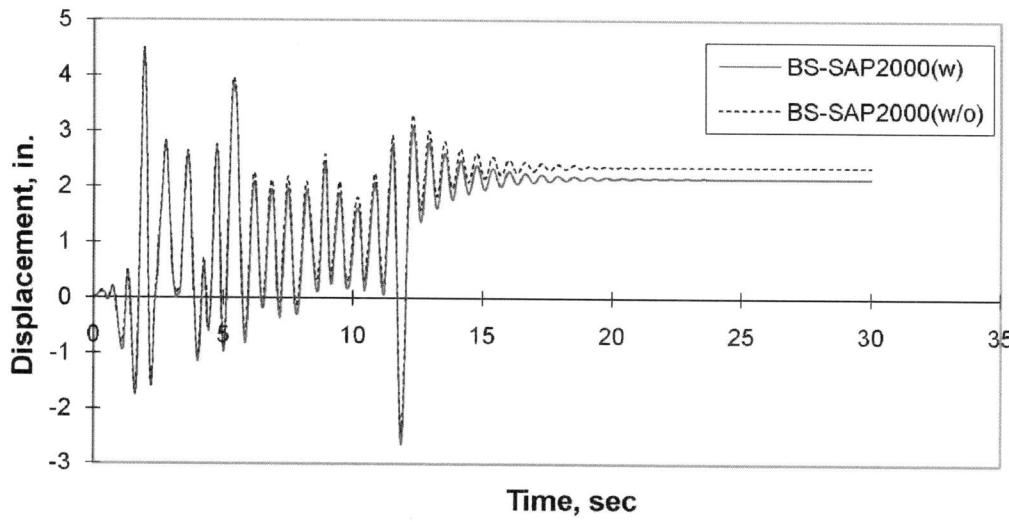

Figure 2-53. Longitudinal displacement of joint M with and without post-tensioning

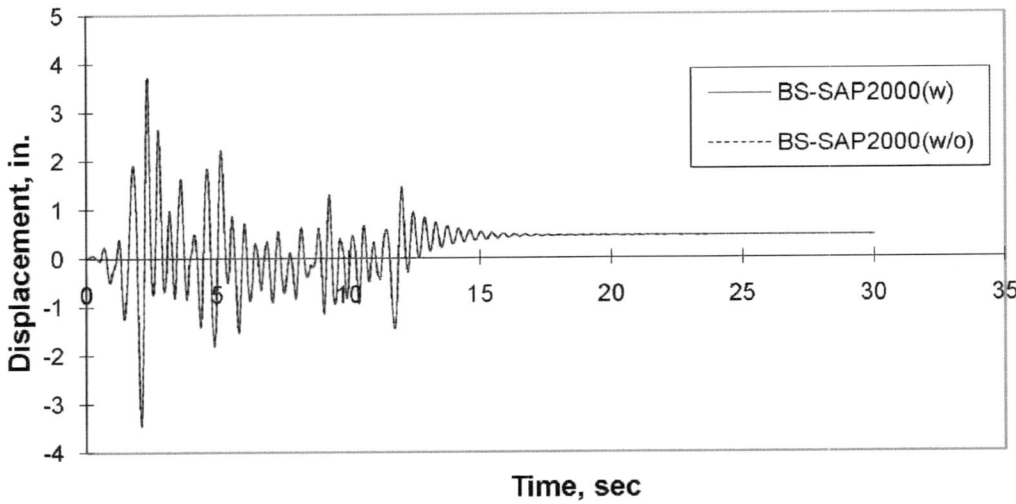

Figure 2-54. Transverse displacement of joint M with and without post-tensioning

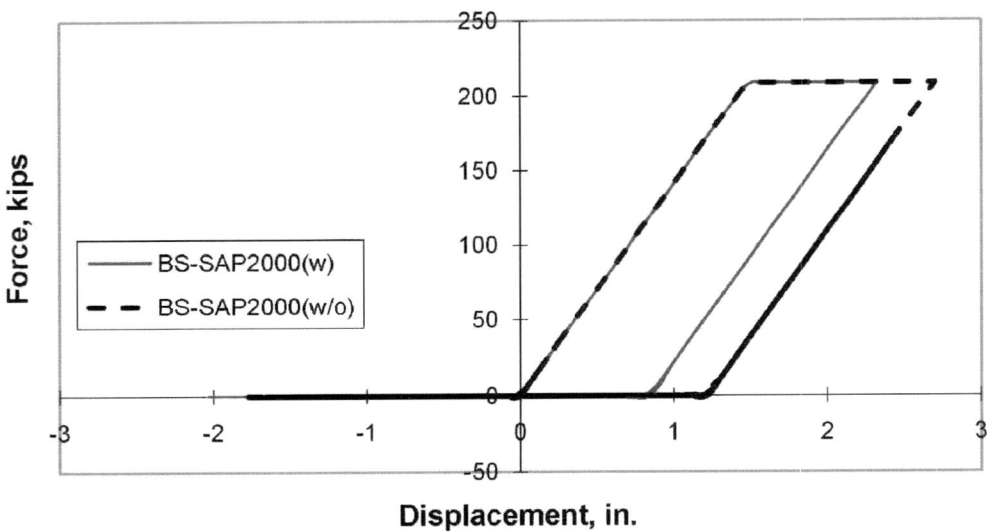

Figure 2-55. Force-Displacement relationship along spring AL of bridge model with and without post-tension

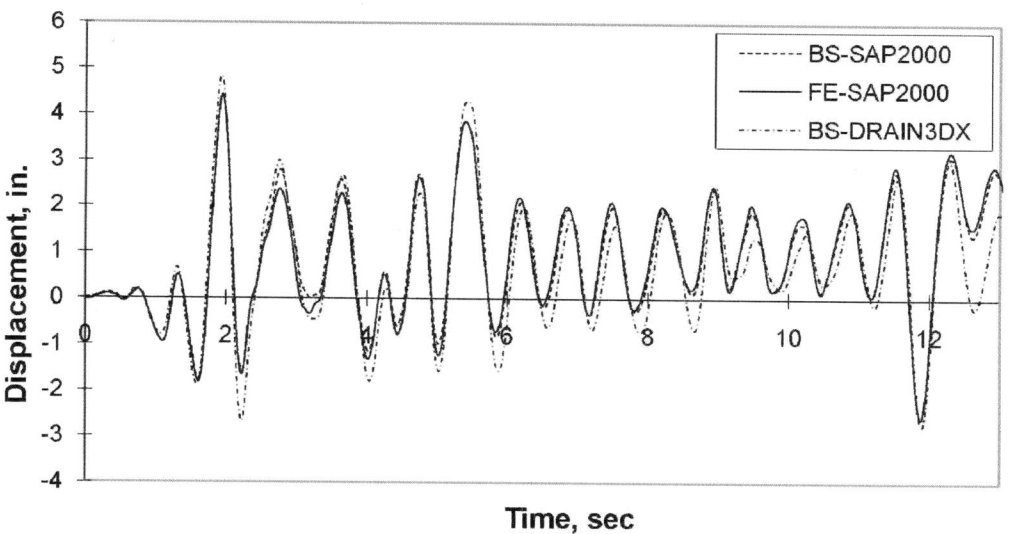

Figure 2-56. Displacement-Time history response at top of bent column (M) in longitudinal direction

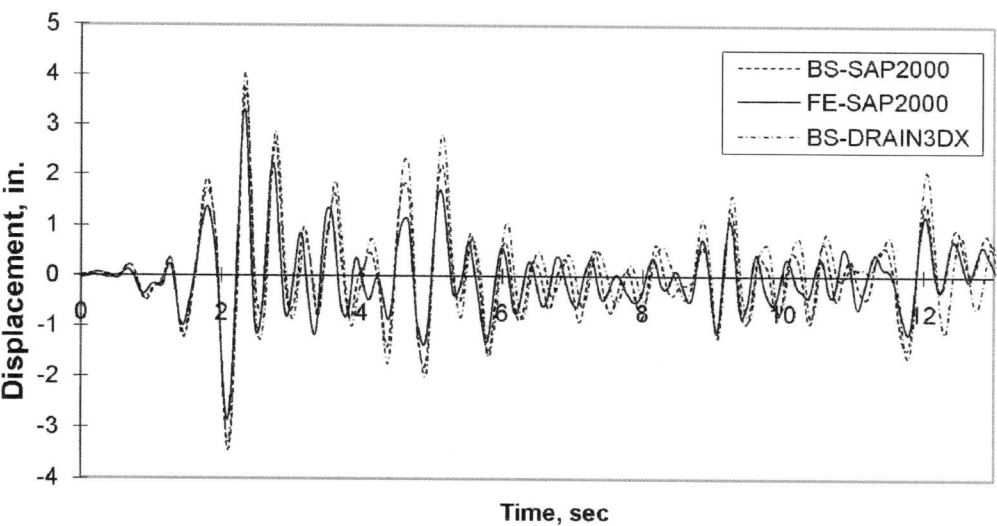

Figure 2-57. Displacement-Time history response at top of bent column (M) in transverse direction

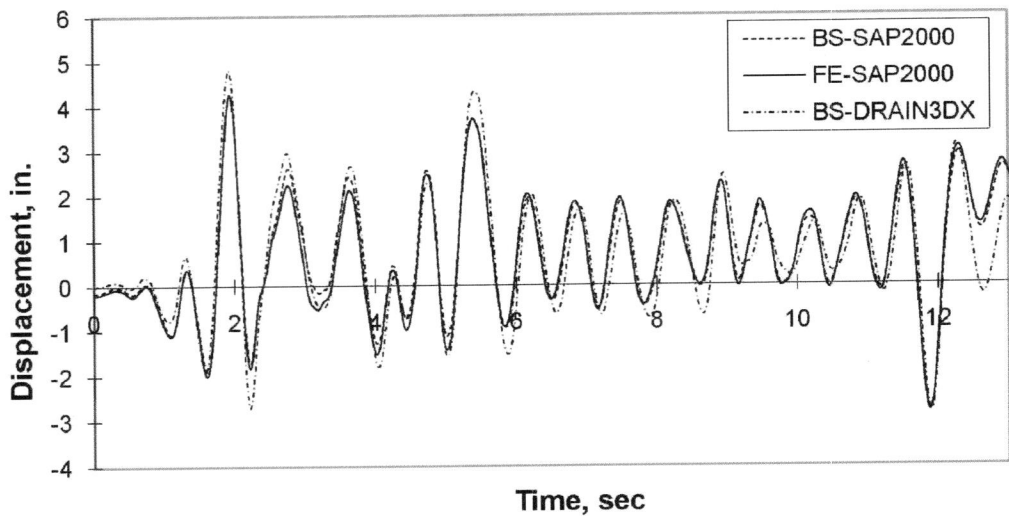

Figure 2-58. Displacement-Time history response at top of deck (Q) in longitudinal direction

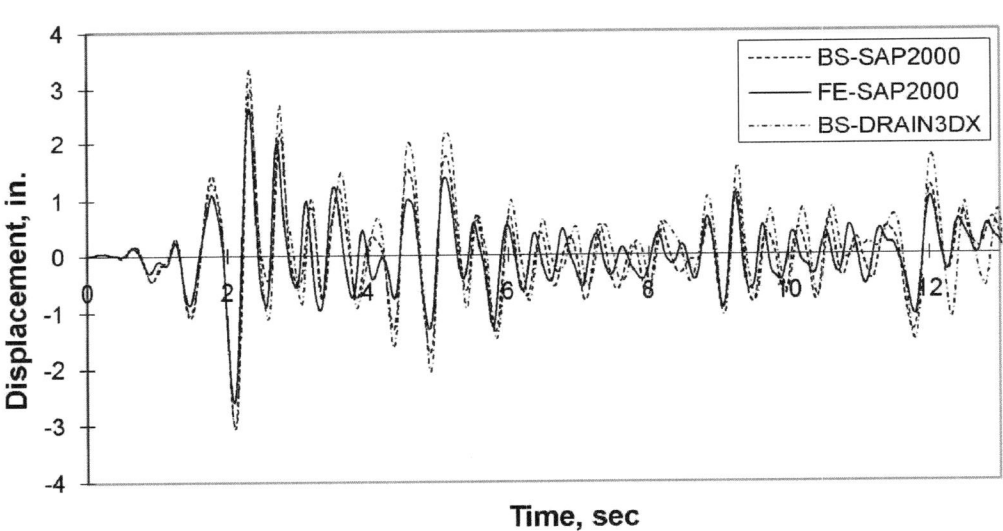

Figure 2-59. Displacement-Time history response at top of deck (Q) in transverse direction

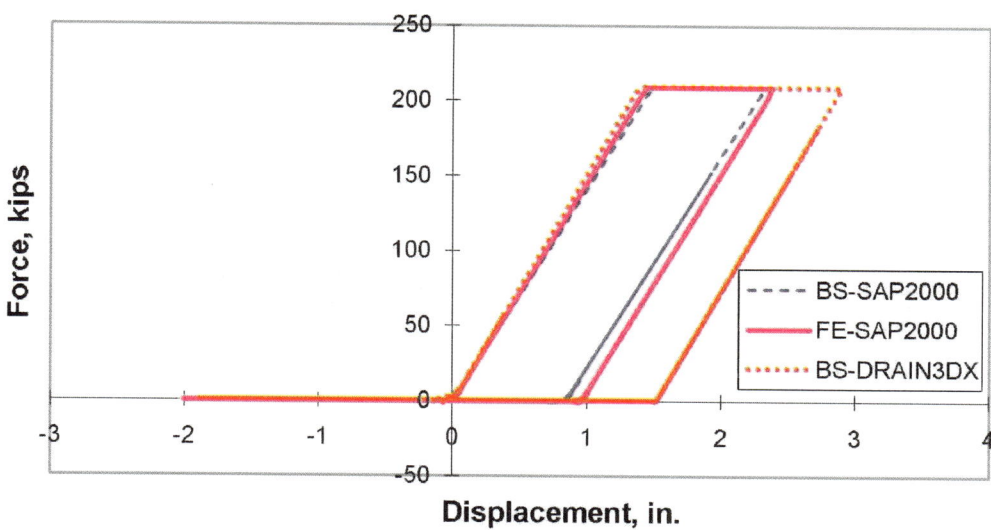

Figure 2-60. Force-Displacement Relationship along spring AL of Bridge Model

3. Comparative Seismic Response Assessment of Skewed Highway Bridges

3.1. Introduction

In this section, results of analysis performed on bridge models with skew angles 0, 20, 30, 45, 52, and 60 degrees is presented and discussed. The analysis follows the analytical matrix presented in chapter 2. A comprehensive set of nonlinear time history analyses was conducted on the improved beam-stick bridge models using DRAIN3DX. The objective of this part of the study is to investigate the effect of wide range of parameters on the seismic performance of skewed highway bridges. These parameters are the effect of intensity of ground motions, soil conditions, abutment support and bent support conditions, and aspect ratio interacting with the skew angle.

A comprehensive set of nonlinear time history analyses is presented herein to study the effect of these parameters on the seismic performance of skewed highway bridges. It should be mentioned that the improved beam stick models (Figure 3-1) were improved for these analyses in DRAIN3DX. The accuracy of the improved modified beam stick models were established earlier against more complex finite element models. Models of the bridges with small to large skew angles, namely 0, 20, 30, 45, 52, and 60 degrees were developed. Six pairs of ground motions were selected for each of the two soil types (Soil-D and –B) and applied in the transverse and longitudinal directions of the models. The objectives of the time history analyses were to investigate: 1) nonlinear response characteristics, 2) effect of ground motion intensity, 3) effect of skew angle, 4) effect of soil type, 5) effect of abutment support conditions (with and without shear keys springs), 6) effect of bent foundation condition, and 7) effect of aspect ratio on the seismic performance of the skewed highway bridges.

3.2. Selection of Ground Motions

The benchmark bridge (52° skew) was designed according to a site specific response spectra for soil type D with moment magnitude, M_W of 6.5 and peak ground acceleration (PGA) of 0.3g. For Soil type D and B, two PGA levels (0.3g and 0.6g) and six ground motions were selected for each of PGA level. The components of acceleration time histories were obtained from PEER Strong Motion Database (http://peer.berkeley.edu/smcat) for epicenteral distances of up to 30 kilometers. The average acceleration spectra of the weaker components and the stronger components were compared to the Caltrans design acceleration response spectra (ARS) in Figure 3-2 through Figure 3-5. The stronger component of each ground motion is applied in the transverse direction of the bridge models while the weaker component is applied in the longitudinal direction. It is noted that no significant effect of orientation of excitation with respect to the skew angle is expected based on the recent study by Schroeder (2006). The ground motions selected for soil type D are El Centro 1940, El Centro 1979, Loma Prieta 1989, Northridge 1994, Superstition Hills 1987, and Kocaeli Turkey 1999 and El Centro (Bond Corner) 1979, Duzce Turkey 1999, El Centro (Array #5) 1979, Loma Prieta 1989, Northridge (NewHall) 1994, and Northridge (Sylmar) 1994, respectively, and summarized in Table 3-1.

Similarly, the ground motions selected for soil type B are Castaic 1971, Duzce Turkey 1999, Lake Hughes 1971, Loma Prieta 1989, Morgan Hill 1984, and Tabas Iran 1978, and Coalinga 1983, Duzce Turkey 1999, Kobe 1995, Loma Prieta 1989, Northridge (Castaic) 1994, and Northridge (Katherine) 1994 respectively, and summarized in Table 3-2.

3.3. Results and Discussion of the Nonlinear Time History Analysis

Six bridge models with skew angles 0, 20, 30, 45, 52, and 60 degree were subjected to 12 pairs of ground motions; 6 pairs of ground motions have PGA of 0.3g and the other 6 pairs ground motions have PGA of 0.6g. It should be mentioned that the modified beam stick models were used for these analyses using DRAIN3DX.

3.3.1. Effect of Ground Motion Intensity

For this phase of the study, all of the skewed bridge models have pinned bent foundations with abutment-soil interaction and shear keys modeled explicitly. Several parameters are monitored and reported with respect to the skew angle. These parameters are:

1- Displacement in the longitudinal direction (Ux) and the transverse direction (Uy) of three nodes along the middle girder of the deck. As shown in Figure 3-6, these nodes are designated as 90 (at the bent), 74 (at 40% of the bridge span), and 64 (at the abutment),

2- For the bent columns (C1 and C2), bending moment about the axis that is along-to-skew (Myy), bending moment about the axis that is normal-to-skew (Mzz), maximum axial force developed in the columns, maximum shear force in columns along the transverse direction (qy), and shear force in columns along the longitudinal direction

(qz). Also, the curvature ductility ($\mu = \dfrac{\psi_T}{\psi_y}$) in both directions with respect to the

skew angle is reported.

3- Abutment-soil interaction

4- Shear key response

A comparison of the response of bridge models under the two levels of intensity of ground motions was conducted to examine the effect of increasing the level of the peak ground acceleration (PGA) for the same type of soil condition (D).

Figure 3-7 presents sample results of nonlinear time history analyses due to 0.6g ground motions. The figure shows the transverse deck displacement responses at the bent. The maximum response values for each ground motion are reported and the average of these maxima is determined for each skew. Figure 3-8 shows sample abutment-soil interaction and shear key force-deformation responses. Time history analysis results are presented in terms of response indices which are defined as the ratio of the absolute average maximum response (displacement, bending moment, axial force, etc) of the bridges with various skew angles to the absolute average maximum of the same response quantity of the straight bridge.

$$Response\ Index = \frac{R\max_{\alpha}}{R\max_{0}} \tag{3-1}$$

Figure 3-9 through Figure 3-33 present the response indices of the parameters mentioned earlier as a function of the skew angle for the two levels of ground motions. The average response is represented by a dashed line while the response due to each ground motion was represented by scattered points. The average of maximum response quantities was used to present the effect of different parameters. However, the scatter implies significant uncertainty and variability. Figure 3-34 through Figure 3-44 show a comparison between the actual values of the response quantities as a function of the skew angle for the two levels of ground motion intensities, namely, PGA of 0.3g and 0.6g. As was mentioned earlier, all the foundations are pinned while both the abutment-soil interaction and the shear keys are modeled explicitly. In the following sections, the effect of 0.3g ground motions and the effect of 0.6g ground motions are presented using the response indices. However, comparison between the effect of the two levels on ground motion intensities is presented using the actual response quantities.

Applying the weaker level of ground motions (0.3g) is not expected to cause yielding in any of the components of the bridge models since the bridge is designed originally to this level of seismic demand. However, it is instructional to study the linear elastic response of the bridges and the effect of skew angle on the general response characteristics.

3.3.1.1. Response to Ground Motions with PGA=0.3g

Figure 3-9 through Figure 3-20 show the results due to 0.3g intensity level. Figure 3-9 and Figure 3-10 present the displacement response index for the deck in the longitudinal (x) and transverse (y) directions due to the application of six ground motions with PGAs of 0.3g. In the longitudinal direction (Figure 3-9), it can be observed that the three nodes on the deck show similar trends regardless of the skew angle; as the skew angle increases, the displacement response index increases. The average displacement of the deck at 60 degree skew is approximately 30% larger than that with no skew. This observation suggests that the model becomes more flexible as the skew increases which is directly related to the diminishing contribution of both abutment-soil interaction and shear keys with the skew to the longitudinal response. The shear keys are aligned in the transverse direction while the abutment-soil springs are perpendicular to the deck. Therefore, as the skew angle increases, abutment-soil interaction contribution in the longitudinal direction becomes lower.

In the transverse direction (Figure 3-10), a slight variation, as the skew increases, was observed for bent and deck nodes, 90 and 74 respectively in an average sense. The average transverse displacement of node 64 (at the abutment) increases as the skew angle increases; however, it is noted that the actual displacement values are not significantly large (Figure 3-34 and Figure 3-35). The difference in trend between the deck nodes and the node at abutments is attributed to the location of the node at abutment therefore it may be affected by gap opening and closing. In addition, this behavior suggests that the effectiveness of shear keys decreases as skew angle increases, however, response quantities are still small.

Figure 3-11 and Figure 3-12 present the bending moment response index at the top of columns (C1 and C2) about the y-y axis (M_{yy}) and the z-z axis (M_{zz}). As was observed earlier, bridges with larger skew angles experienced larger deformations, which in turn, resulted in larger ductility demands, therefore M_{yy} and M_{zz} are expected to be affected.

For Myy, it can be observed that the response index generally increases with increasing skew angle. The maximum response index is calculated for 52 and 60 degree skew which is almost 30% larger than that with no skew. For Mzz, it follows the same trend as Myy while the response index at 52 degree is up to 40% larger than that with no skew case. There is not a significant variation in the bending moments in an average sense, but large amount of scatter is evident.

Figure 3-13 shows the response index for the axial force in columns with respect to skew angle. No significant variation was recorded. As can be seen in the figure, axial forces in the columns varied by +/- 5% between 0 to 60 degree skew angles.

Figure 3-14 and Figure 3-15 show the response index for shear in the bent columns (q_y and q_z) as a function of the skew angle. For both q_y, and q_z, the maximum response index was for the 60 degree skew case which indicates approximately up to 40% larger shear force in the direction along the skew, and 25% larger shear force in the direction perpendicular to the skew.

In summary, bridges with larger skew angles experienced larger deformations, especially in the transverse direction, however the bent forces were affected. These observations were made considering only the average response quantities, however results of individual ground motions suggest significant variability. In addition, forces in both bent columns were almost the same for a given ground motion with no significant difference.

Figure 3-16 and Figure 3-17 show the response index of deformation at the abutment due to abutment-soil interaction in terms of deformations "into" the soil (U+) and "away" from the soil; gap opening- (U-) with respect to skew angle. Only the results of the four abutments springs on one side of the bridge are presented; identical observations are made for the abutment-soil springs at the other abutment. For U+, it can be observed that all four abutment springs show similar trend. No skew and 20 degree skew cases show similar levels of deformation. This can be due to the fact that "yielding" of abutment-soil springs starts to occur at small skew angles (0 to 20 degree skew), however, yielding may take place for larger skew angles as well. As the skew angle increases, the index decreases while the index for 60 degree skew is approximately 24% lower than the no skew case. The presence of shear keys reduces the demand on abutment-soil springs, especially, for larger skew angles. For U-, (i.e. gap opening) an overall assessment of Figure 3-17 suggests that expected gap opening near the obtuse angle is comparable to the no skew case. An overall assessment of Figure 3-17a through Figure 3-17d also suggests that the gap opening is more pronounced near the obtuse angle whereas gap opening gradually decreases with skew angle at the acute angle. This may also imply larger in-plane rotations at the abutments as the skew angle increases. Figure 3-21 shows sample deck rotations at the abutment. For a given skew angle, it is evident that the gap opening increases from acute to obtuse corner. For example, in case of 60 degree skew, the gap opening index is 30% larger at the obtuse corner. The trend of gap opening is not consistent with that of transverse displacement at the abutment which increases with the skew. Using the well known transformation matrix to determine the deformation along the abutment-soil interaction springs from the abutment deformation in the longitudinal and transverse directions, it can be confirmed that the gap opening at the abutment decreases with the skew since the transverse displacement at the abutment is small. shear keys played an important role in reducing the demand on abutment-soil springs for larger skew angles. Therefore, failure of the shear keys can lead to a different trend as the transverse displacement at the abutment is expected to increase followed by a larger gap opening as the skew increases.

Figure 3-18 presents the response index of shear key deformations as a function of the skew angle. It can be observed that shear keys on the diagonals show similar trend but with no significant variation as a function of the skew angle which can be attributed to elastic behavior of shear keys under the smaller level of ground motions.

Figure 3-19 and Figure 3-20 present the curvature ductility demand with respect to the skew angle in both directions (μ_{yy} and μ_{zz}). From both figures, the ductility factors are less than unity which confirms that both columns remain elastic when subjected to ground motions with the PGAs of 0.3g. In addition, yielding of abutment-soil springs took place at small skew angles (0 and 20 degrees) while no yielding in shear keys took place.

3.3.1.2. Response to Ground Motions with PGA=0.6g

Figure 3-22 through Figure 3-33 show the results due to 0.6g intensity level. Figure 3-22 and Figure 3-23 present the displacement of the deck in the longitudinal (x) and transverse (y) directions, respectively due to the application of six ground motions with the PGAs of 0.6g. In the longitudinal direction, it can be observed that the same trend was followed by the three nodes on the deck. As the skew angle increases, the response index increases. This increase can be attributed to both abutment-soil interaction and shear keys. Shear keys are aligned in the transverse direction while the abutment-soil springs are perpendicular to the deck. Therefore as skew increases, abutment-soil interaction becomes less pronounced in the longitudinal direction. In the transverse direction, the same trend was observed as in 0.3g case.

Figure 3-24 and Figure 3-25 present the response index for bending moments at the top of columns (C1 and C2) about the y-y axis (Myy) and the z-z axis (Mzz). For Myy, it can be observed that there is not a significant variation as skew increases which suggests that most of the columns have yielded or close to yield. Mzz increases with the skew; the maximum response index was at 60 degree skew which is approximately 30% larger than no skew case.

Figure 3-26 shows the response index for axial forces in columns with respect to skew angle. It follows the same trend as in 0.3g case.

Figure 3-27 and Figure 3-28 show the response index for shear forces in columns in both directions (q_y and q_z) as a function of the skew angle; skew increases, q_y increases, and it can be observed that there is not a significant variation as skew increases.

Figure 3-29 and Figure 3-30 show the response index for the deformation of the abutment-soil springs, as deformations "into" the soil (U+) and "away" from the soil (U-) with respect to skew angle. Only the results of the four abutments springs on one side of the bridge are presented. For U+, it can be observed that all four abutment springs show similar trend. As the skew angle increases, the index decreases. The lowest value of the index occurred at 60 degree skew which is lower than that with no skew by about 44%. It should be noted that most of the abutment-soil springs have "yielded" (Figure 3-41). For U-, it follows same trend as in 0.3g case.

Figure 3-31 presents the response index for the shear key deformations as a function of the skew angle. It can be observed that shear keys on the diagonals show similar trend. Also, all of the shear keys yielded but they did not fail. It is evident that the demand on shear keys increases as skew angle increases.

Figure 3-32 and Figure 3-33 present the curvature ductility demand with respect to the skew angle in both directions (μ_{yy} and μ_{zz}). For μ_{yy}, the ductility factors are larger than unity which confirms that both columns yielded under 0.6g intensity level of ground motions. For μ_{zz}, the ductility factors are smaller than unity for small skew angles in an average sense. This suggests that the presence of shear keys reduces the demand on the columns in the transverse direction especially for small skew angles. Also, the demand on columns increases as the skew angle increases.

3.3.1.3. Comparison of Response to PGA=0.3g and 0.6g

Figure 3-34 through Figure 3-44 present comparisons between the response of bridges with different skew angles under the two levels of ground motions (0.3g and 0.6g) for all the monitored parameters. These comparisons essentially highlight the difference between linear-elastic and post-elastic response characteristics for different skew angles.

Figure 3-34 and Figure 3-35 present the displacement of the deck in the longitudinal and transverse directions (x and y) respectively, under the application of twelve pairs of ground motions with PGAs of 0.3g and 0.6g (six motions per each PGA). In the longitudinal direction, the same trend was observed for the two levels of ground motions. As the skew angle increases, the displacement increases as was discussed earlier. The three nodes on the deck translated together with almost the same displacement longitudinally for the same skew angle.

In the transverse direction, insignificant variation was observed for nodes 90 and 74 while displacement of node 64 (at abutment) increased with skew angles, however, it is small. Transverse deck displacement due to the application of ground motions with PGA of 0.6g was approximately twice those due to the application of ground motions with PGA of 0.3g.

Figure 3-36 and Figure 3-37 present the bending moments at the top of columns (C1 and C2) about the y-y axis (Myy) and the z-z axis (Mzz). For Myy, similar trend was observed. Moment developed due to the application of larger level of intensity was about two times that developed due to the application of the lower level. Mzz increases as skew angle increases. The presence of uncertainties in the response quantities is evident in the scatter shown in the plots presented.

Figure 3-38 shows the axial forces in columns with respect to the skew angle. It follows the same trend between the two levels of intensity with a factor of 1.13. No significant variation was recorded. As can be seen in the figure, axial forces in the columns varied by +/- 13% between 0 to 60 degree skew angles.

.

Figure 3-39 and Figure 3-40 show the shear forces in columns in both directions (q_y and q_z) as a function in skew angle. For q_y, the same trend between the two levels of intensity was noticed with a factor of 1.8. For q_z, the same trend between the two levels of intensity was noticed with a factor of 2.

Figure 3-41 and Figure 3-42 show the deformation of abutment-soil springs, as deformations "into" the soil (U+) and "away" from the soil (U-) with respect to skew angle. Only the results of the four abutment springs on one side of the bridge are presented. As skew angle increases, the displacement decreases. There is a factor of about 3 due to application of larger level of intensity. For U-, the behavior due to application of two levels of intensity showed similar trend with a factor of about 2. However, a more pronounced in-plane rotation at the abutments as the skew increases, is evident from these figures.

Figure 3-43 presents the shear key deformations as a function of the skew angle. It can be observed that trend due to 0.6g intensity level was different from that of 0.3g intensity due to yielding of the shear keys during the larger intensity motions.

Figure 3-44 presents the envelope of the average curvature ductility demand with respect to the skew angle for both directions (μ_{yy} and μ_{zz}). For μ_{max}, the ductility factors are larger than unity which confirms that both columns yielded under 0.6g level of ground motions while they remained elastic under 0.3g. Also, there is a factor of more than 4 between the two levels of motions. However, it is interesting to note that the maximum ductility demand remains relatively constant for all skew angles.

3.3.2. Effect of Soil Type

The effect of two soil types (Soil-D and Soil-B) on the seismic performance of skewed bridges is investigated and discussed in what follows:

As presented earlier, for each soil type 12 pairs of ground motions were selected (six per each intensity). The ground motions selected were presented earlier for the PGAs of 0.3g and 0.6g. Figure 3-45 through Figure 3-55 present the nonlinear time history average response of bridges with various skew angles for the two levels of ground motions for both soil types (D and B). Models included both abutment-soil interaction and shear keys. Figure 3-56 through Figure 3-65 show a comparison between the average response quantities for the two levels of intensities, namely, PGAs of 0.3g and 0.6g for the two soil types, but this time assuming no shear keys present. Here, the purpose is to establish boundaries for response quantities by considering true (full hysteresis) shear key response and no shear keys present.

Figure 3-45 and Figure 3-46 present the average displacement of the deck in the longitudinal (x) and transverse (y) directions due to the application of twelve pairs of ground motions with the PGAs of 0.3g and 0.6g (six per each). In the longitudinal direction, x, it can be observed that the displacement of the three nodes on the deck show similar trends and response quantities as a function of the skew angle. However, slightly larger displacements – as the skew angle increases – are evident for bridges on Soil-D. On the other hand, for Soil-B, variation of displacement response was not significant. It can be concluded that the larger response was experienced by bridges on Soil-D than those on Soil-B by a factor about 1.5 for 0.6g. Also, it should be noted that increasing the intensity of the ground motions did not affect the response trend while it increased the response quantities as concluded earlier. For Soil-D, the average displacement of the deck at 60 degree skew is approximately 17% larger than that with no skew. This increase can be attributed to both abutment-soil interaction and shear keys.

In the transverse direction, a slight variation in displacements as a function of the skew angle was observed for the bent and deck nodes, 90 and 74 respectively. The average transverse displacement of node 64 (at the abutment) increases as the skew angle increases; however, the actual displacements are not significant. For a given skew angle, the transverse displacement decreases towards the abutments. This observation was valid for both soil types while bridges on Soil-B showed smaller response values. Also, the response trend seemed to remain similar for higher intensity motions as well.

Figure 3-47 and Figure 3-48 present the bending moment response at the top of columns (C1 and C2) about the y-y axis (Myy) and the z-z axis (Mzz). For Myy, it can be observed that the response increases with increasing skew angles. The response at 30 degree and 45 degree skew was almost the same. The maximum response took place at 30 and 45 degree skew which is almost 13% larger than that with no skew for both soil types. On the other hand, slightly lower values were observed for 52 and 60 degree skew. Bridges on Soil-B experienced similar trend with lower values. For Mzz, the response increases as the skew angle increases. Columns of bridges on Soil-B experienced smaller column moments. Also, the increase in moment with skew angle suggests that the effectiveness of shear keys reduces as skew angle increases. However, for large skew angles (> 30 degree) shear keys may not have any significant effect. This observation is confirmed later in the next section when the effect of abutment boundary conditions is presented. The two columns experienced very similar moments in both directions and for the same skew angle and soil type.

Figure 3-49 shows the axial force response in columns with respect to skew angle. No significant variation was recorded. As can be seen in the figure, axial forces in the columns varied by +/- 5-7% between 0 to 60 degree skew angles. It is obvious that soil type did not affect axial force in columns significantly. Also, both columns experienced very similar axial forces for a given skew angle and soil type.

Figure 3-50 and Figure 3-51 show shear response in the bent columns (q_y and q_z) as a function of the skew angle. q_y showed similar response trend to Mzz response while q_z showed similar response trend to Myy response. Smaller response values were experienced by bridges on Soil-B. Increasing the ground motion intensities did not affect the response trend.

Figure 3-52 and Figure 3-53 show the deformation response of abutment-soil interaction springs, as deformations "into" the soil (U+) and "away" from the soil (U-) with respect to the skew angle. Only the results of the four abutments springs on one side of the bridge are presented; identical observations are made for the abutment-soil springs at the other abutment. For U+, it can be observed that all four abutment springs show similar trend for both soil types. No skew and 20 degree skew cases show similar levels of deformation for 0.3g level of ground motions. This can be due to the fact that "yielding" of abutment-soil springs started to occur at small skew angles (0 to 20 degree skew), however, yielding at larger skew angles can take place. As the skew angle increases, the displacement decreases. At 60 degree skew, abutment displacement is lower than that of no skew case by about 51% for the PGA of 0.6g and Soil-D while it is about 45% for Soil-B. It should be mentioned that larger values were recorded as the abutment-soil interaction in bridges on Soil-D. This indicates that the larger deformation demand at abutments is due to Soil-D. For U-, (i.e. gap opening), Figure 3-53 suggests that expected gap opening near the obtuse angle is lower than the no skew case by about 37% whereas gap opening gradually decreases with skew angle near the acute angle. This also implies larger in-plane rotations at the abutments as the skew angle increases. Also, smaller gap opening was implied by smaller abutment-soil spring deformations in bridges on Soil-B, following a similar trend to that of bridges on Soil-D.

Figure 3-54 presents the response of shear key deformations as a function of the skew angle. It can be observed that shear keys on the diagonals show similar trend. But there is not a significant variation in the shear key deformation as a function of the skew angle for lower level of ground motions. At lower level of ground motions shear keys did not experience any yielding in an average sense. For both types of soil, shear keys experienced yielding under the PGA of 0.6g but they did not fail. From the results of 0.3g, the demand on shear keys increase as skew increases therefore yielding of shear keys are likely to occur at large skew angles (> 30 degree) which may explain the possible inefficiency of these elements as skew angle increases.

Figure 3-55 presents the envelope of average curvature ductility demand with respect to the skew angle of both directions (μ_{yy} and μ_{zz}) for both soil types. From the figure, for the PGA of 0.3g, the ductility factors are less than unity which confirms that both columns remain elastic when subjected to this level of ground motions. For the PGA of 0.6g, μ_{max} slightly increases as skew angle increases while μ_{zz} increases as skew angle increases which shows the effect of shear keys on column demand. The demand on columns increases as the skew increases. No yielding was observed in columns of bridges on Soil-B therefore there is lower demand on columns due to the effect of soil type.

In summary, soil type does not have any significant effect on the response trends. Bridges on Soil-B showed smaller response values for all parameters compared to those on Soil-D. Abutment-soil interaction springs and shear keys yielded for both soil types when subjected to ground motions with PGAs of 0.6g. The effectiveness of shear keys decreases as the skew angle increases. Consequently, the demand on shear keys increases with the skew. No yielding took place of columns of bridge on Soil-B.

Figure 3-56 through Figure 3-65 present the summary results for "without shear key" case on the two different soil types and two different levels of intensity. It is noted that further discussions and the response comparisons of the "with and without shear key" cases are provided in section 3.3.3. Figure 3-56 and Figure 3-57 present the displacement of the deck in the longitudinal (x) and transverse (y) directions, respectively due to the application of ground motions with the PGAs of 0.3g and 0.6g. In the longitudinal direction, it can be observed that same trend was followed by the three nodes on the deck. Regardless of location of the node, nodes experienced the same displacement in the longitudinal direction for the same skew angle and soil type in an average sense. For Soil-D, as the skew angle increases, the response increases. For Soil-B, a slight variation in the response was observed as skew increases. In the transverse direction, nodes away from the abutment experienced smaller displacements. In addition, changing soil type did not affect the trend. However, larger displacement response is evident on Soil-D. The deck displacements of the bridges on Soil-B do not seem to be affected by larger skew angles as much as those of bridges on Soil-D. For node 64, at abutment location, a significant increase in displacement was observed as skew angle increases with a maximum value of about 8 in. (Figure 3-57c). It can be observed that complete removal of the shear key led to significant increase in the transverse displacement.

Figure 3-58 and Figure 3-59 present the bending moment response at the top of columns (C1 and C2) about the y-y axis (Myy) and the z-z axis (Mzz). For Myy, it can be observed that there is a slight variation as skew increases for Soil-D with the PGA of 0.6g which suggests that most of the columns have yielded or close to yield. For Soil-B, the response increases with skew. For Mzz, yielding is expected to take place for bridges on Soil-D at the larger level of ground motions as a slight variation was observed in moment values with skew. Nonetheless, response values decreases as skew angles increases. Columns of bridge on Soil-B showed lower response values following a similar trend to those on Soil-D.

Figure 3-60 shows the response of axial force in columns with respect to skew angle. It follows the same trend as "with shear key" case. No significant variation was recorded for the axial force in columns with the skew.

Figure 3-61 and Figure 3-62 show the response of shear force in columns in both directions (q_y and q_z) as a function of skew angle. q_y followed same trend as Mzz while q_z followed same trend as Myy.

Figure 3-63 and Figure 3-64 show the deformation of abutment-soil springs, as deformations "into" the soil (U+) and "away" from the soil (U-) with respect to skew angle. Only the results of the four abutment springs on one side of the bridge are presented. For U+, it can be observed that all four abutment springs show similar trend. As the skew angle increases, the response value decreases. However, the gap opening is expected to increase significantly with the skew due to the absence of shear keys. The lowest value of the response occurred at 60 degree skew. It should be noted that most of the abutment-soil springs have "yielded" for Soil-D with PGA of 0.6g. For Soil-B, abutment-soil springs attached to bridges away from the obtuse angle with larger skew angles did not yield. For U-, bridges on Soil-D showed larger response values regardless of the level of intensity. Largest average gap opening was about 8 in.

Figure 3-65 present the envelope of average curvature ductility demand with respect to the skew angle in both directions (μ_{yy} and μ_{zz}). For μ_{max}, under 0.6g intensity level of ground motions the ductility factors for soil type D are larger than unity which confirms that both columns yielded. Lower response values were observed for Soil-B for same level of ground motions. Columns of bridges on Soil-B experienced no yielding.

3.3.3. Effect of Shear Key

In order to account for the two extreme cases with respect to the shear keys, two sets of analyses were conducted of the bridges with and without shear keys. When the shear keys are modeled, they are modeled explicitly with full hysteresis definition as presented in chapter 2. The results for two different soil types (Soil-D and Soil-B) are presented for the bridge models under study (with aspect ratio of 0.3). The aspect ratio is defined as the ratio of the width of the bridge including the overhang over the span length. In addition, results for bridge models with larger aspect ratio (0.54) and on Soil-D are presented and discussed herein.

Figure 3-66 through Figure 3-75 present the results for bridges with 0.3 aspect ratio, on Soil-D, and pinned bent foundations for the case "with shear keys" and "without shear keys" under the two levels of ground motions. Figure 3-66 and Figure 3-67 present the displacement of the deck in the longitudinal and transverse directions (x and y), respectively under the application of twelve pairs of ground motions with the PGAs of 0.3g and 0.6g (six motions per each). In the longitudinal direction, the same trend was observed for the two levels of ground motions as explained earlier in section 3.3.1.3; as the skew angle increases, the displacement increases. The deck of the bridges without shear keys experienced slightly larger displacements than those with shear keys. However, for straight bridges displacements in the longitudinal direction do not seem to be affected by the absence of shear keys. It is anticipated that deformations in the longitudinal direction will not be affect significantly by the absence of shear keys while it is expected that transverse deformations will be affected significantly.

In the transverse direction, insignificant variation of transverse displacements with the skew angle was observed at nodes 90 and 74 while displacement at the node 64 (at abutment) increased significantly with the skew angles due to the effect of shear keys. Transverse deck displacements at the abutments for "without shear key" case were about four times than those "with shear keys". Also, the average transverse deck displacements for bridges "without shear keys", were larger than those "with shear keys" for a given skew angle. At 60 degree skew, transverse displacements for "without shear key" case were larger than no skew case by about 60%. This clearly shows the effect of shear keys on the response and furthermore it shows that the demand on shear keys increase as skew increases. It can be concluded that shear keys help reduce the transverse deck displacements, especially at the abutments, by a factor of more than 4. Shear keys do not affect the longitudinal deck displacement significantly. Increasing level of intensity of ground motions did not affect the trend.

Figure 3-68 and Figure 3-69 present the bending moment at the top of columns (C1 and C2) about the y-y axis (Myy) and the z-z axis (Mzz). For Myy, similar trend between the two cases was observed. Similar response values were obtained suggesting that shear keys are not effective in reducing the moment about y-y axis. Response values generally increase as skew angles increases. For Mzz, some discrepancy was observed in terms of trend for bridges with and without shear keys. Response of bridges with shear keys increases with skew while response of those without shear keys decreases with skew showing larger response. This observation confirms that the effectiveness of shear keys reduces as skew angles increases. It is also noted that shear keys affected the response of Mzz while it did not affect Myy significantly. This is owing to the presence of shear keys which restraint deformations in the transverse direction. It is clear that shear keys can reduce the demand on columns of skewed bridges with small and moderate skew angles, marginally. The same conclusion was drawn for the two levels of ground motions.

Figure 3-70 shows the axial force in columns with respect to the skew angle. It follows the same trend while no significant difference was observed in both cases.

Figure 3-71 and Figure 3-72 show the shear forces in columns in both directions (q_y and q_z) as a function of skew angle. q_y follows the same trend as Mzz while q_z follows the same trend as Myy. Response of bridges with shear keys increases with skew while response of those without shear keys decreases with skew showing larger response. This observation confirms that the effectiveness of shear keys reduces as skew angles increases. It is clear that shear keys can reduce the demand on columns of skewed bridges with small and moderate skew angles, marginally.

Figure 3-73 and Figure 3-74 show the deformation of abutment-soil springs, as deformations "into" the soil (U+) and "away" from the soil (U-) with respect to skew angle. Only the results of the four abutment springs on one side of the bridge are presented. For U+, as the skew angle increases, the displacement decreases. The absence of shear keys is expected to increase abutment-soil interaction. Abutment springs for bridges with shear keys showed slightly larger deformation throughout. For U-, a different trend was observed for without shear key case as gap opening increases with skew. Removing the transverse restraint (i.e. shear key) led to larger gap opening as the skew increases. This can be attributed to the fact that abutment-soil springs are aligned to be perpendicular to skew therefore, at large skew angles, abutment deformation is more pronounced toward the transverse direction. It was established earlier that transverse displacement increases with the skew therefore the demand on abutments increase.

However, a more pronounced in-plane rotation at the abutments is evident from these figures. In summary, removing the shear key may lead to larger gap opening and more pronounced in-plane rotation.

Figure 3-75 present the envelope of average curvature ductility demand with respect to the skew angle in both directions (μ_{yy} and μ_{zz}). Under 0.6g level of ground motions, the ductility factors are larger than unity which confirms that both columns yielded while columns remained elastic under 0.3g level. The shear keys have a considerable effect in reducing the ductility demand on columns of skewed bridges.

In summary, the absence of shear keys did not affect the displacements in the longitudinal direction significantly while it increased those in the transverse direction dramatically by a factor of 4 at the abutments. Both columns have yielded in both cases under 0.6g. The effectiveness of shear keys to reduce the demand on columns reduces as the skew angle becomes larger, but shear key is effective in reducing the gap opening at abutment for large skew angles. In both cases, abutment-soil springs yielded while in "without shear key" case gap openings were larger and increased with skew. However, if the current seismic criteria for abutment seat width, which is measured normal to the centerline of the bearing (Figure 3-76), were followed, no unseating would take place in these bridges as will be discussed in chapter 4.

Figure 3-77 through Figure 3-86 present the results of bridges with 0.3 aspect ratio, on Soil-B, and pinned bent foundations for the cases "with shear keys" and "without shear keys" under the two levels of ground motions. The same observations as Soil-D case can be drawn accompanied by lower values of response. The largest transverse displacement at abutment location was observed at 60 degree skew with a value about 6 in. for without shear key case. Deformation into abutments was lower compare to those due to Soil-D. Lower abutment deformation values were recorded for "without shear keys" case. At large skew angles, for without shear keys case abutment-soil springs may not yield. Nonetheless, larger gap openings were observed for without shear key case with a maximum value about 6 in. Both columns remained elastic in both cases. It suggests that abutment-soil springs or columns of skewed bridges without shear keys on Soil-B may not yield even under large ground motions.

The effect of the aspect ratio on the seismic performance of skewed highway bridge is studied in more detail later in this document. The effect of the presence of shear keys is also investigated for larger aspect ratio (0.54). This was done to approach a unified conclusion which can be applicable to a wide range of skewed bridges. In this section, however, only the results for one soil type (Soil-D) are presented for the bridge models under study (with aspect ratio of 0.54) for all skew angles. It is noted that the results of the 52 degree skew were not included because the analyses could not numerically convergence for the entire range of the ground motions. Also, only the larger levels of excitation with PGA of 0.6g are discussed in the following.

Figure 3-87 through Figure 3-96 present comparisons between the response of bridge models on Soil-D and pinned bent foundation with different skew angles under the larger level of ground motions (0.6g) with and without shear keys for the larger aspect ratio (0.54). Figure 3-87 and Figure 3-88 present the displacement of the deck in the longitudinal and transverse directions (x and y), respectively. In the longitudinal direction, as the skew angle increases, the displacement increases. Deck of the bridges without shear keys experienced larger displacements than those with the shear keys. The percentage of increase in the longitudinal displacement due to the absence of the shear keys are larger than that experienced by the bridges with the lowest aspect ratio (0.3). It is anticipated that deformations in the longitudinal direction will not be affected significantly by the absence of shear keys while it is expected that transverse deformations will be affected.

For the transverse direction, insignificant variation with skew angle was observed for nodes 90 and 74 while displacement of node 64 (at abutment) increased significantly with skew angle due to the flexibility of the deck and effect of shear keys. Transverse deck displacement at abutment for "without shear key" case was about five times that "with shear key". Also, the rate of increase in displacement was larger for without shear key case. This clearly confirms the conclusion drawn earlier that the demand on shear keys increases as the skew angle increases. Also, it can be concluded that shear keys have noticeable effect in reducing the transverse deck displacement, especially at the abutments. The presence of shear keys tends to affect the longitudinal deck displacement for larger aspect ratio.

Figure 3-89 and Figure 3-90 present the bending moment at the top of columns (C1 and C2) about the y-y axis (Myy) and the z-z axis (Mzz). It is anticipated here that the absence of shear keys may affect both of Myy and Mzz since it affects both longitudinal and transverse displacements. For Myy, moment increases as skew angle increases, but larger response values were recorded for columns without the shear keys. For Mzz, bridges with and without shear keys achieved different trends. Figure 3-89 suggests that the two columns behave similarly and the effectiveness of shear keys are more pronounced for Mzz.

Figure 3-91 shows the axial force in columns with respect to the skew angle. No significant variation was recorded. As can be seen in the figure, axial force in the columns varied by +/- 5% between 0 to 60 degree skew angles.

Figure 3-92 and Figure 3-93 show the shear forces in columns in both directions (q_y and q_z) as a function of skew angle. For q_y and qz, a slight change of response values were observed by changing the skew angle while the absence of shear keys seem to not affect either the response values or the trend.

Figure 3-94 and Figure 3-95 show the deformation of abutment-soil springs, as deformations into the soil (U+) and away from the soil (U-) with respect to skew angle. Only the results of the four abutment springs on one side of the bridge are presented. For U+, as skew angle increases, the displacement decreases. The absence of shear keys is expected to increase abutment-soil interaction. Abutment springs for the bridges with shear keys showed larger deformation throughout especially for those with large skew angles. For U-, a different trend was observed for without shear key case as gap opening increased with skew. It was established earlier that transverse displacement increases with skew therefore demand on abutments increase. A large gap opening takes place close to the acute angle while it reduces toward the obtuse angle. This produced a more pronounced in-plane rotation at the abutments.

Figure 3-96 presents the envelope of average curvature ductility demand with respect to the skew angle in both directions (μ_{yy} and μ_{zz}). A larger ductility demand was observed for those without shear keys. From the figure, the effectiveness of shear keys in reducing the demand on columns increases as skew angle increases. This contradicts with the observation considering the effectiveness of shear keys for bridge with aspect ratio of 0.3 due to the extensive yielding of shear keys.

In summary, the absence of shear keys affects displacements in the longitudinal direction slightly while it increased those in the transverse direction dramatically by a factor of 5 at the abutment locations. Both columns have yielded in both cases. The demand on shear keys increases as the skew angle becomes larger. Abutment-soil springs were yielded for the case with shear keys for all skews while in without shear key case they did not yield at large skews in some cases, however, gap openings were larger and increases with skew. As was mentioned before, although these values are large, if the current seismic design criteria for abutment seat width (Figure 3-76) were followed, no unseating would take place in these bridges.

Figure 3-97 through Figure 3-107 present comparisons between the response of bridge models on Soil-D with different skew angles under the two levels of ground motions (0.3g and 0.6g) with and without shear keys. However, the bent foundations are assumed to be fixed for the columns. Figure 3-97 and Figure 3-98 presents the displacement of the deck in the longitudinal and transverse directions (x and y), respectively. In the longitudinal direction, as the skew angle increases, the displacement increases. Deck of the bridges without the shear keys experienced slightly larger displacements than those with shear keys. It is anticipated that absence of shear keys may not affect the deformations in the longitudinal direction while it may significantly affect the transverse deformations.

In the transverse direction, insignificant variation with skew angle was observed for nodes 90 and 74 while displacement of node 64 (at abutment) increased significantly with skew angles. Transverse deck displacements at the abutments for without shear key case were about four times than those with shear keys. Also, average transverse deck displacements for bridges without shear keys, were larger than those with shear keys for a given skew angle. Figure 3-98 shows that demand on shear keys increase as skew increases. It can be concluded that shear keys have noticeable effect in reducing the transverse deck displacement, especially at the abutments.

Figure 3-99 and Figure 3-100 present the bending moment for columns (C1 and C2) about the y-y axis (Myy) and the z-z axis (Mzz). For Myy, similar trend was observed between the two cases. For the lower level of ground motions, larger response values were experienced by bridges without shear keys. A slight difference between the response values for the two cases was observed for small skew angles while it increases as the skew angle increases. It suggests that the effectiveness of shear keys in reducing Myy increases as skew angle increases. For Mzz, the two cases achieved different trends for the lower level of ground motions. Response of bridges with shear keys increases constantly with skew while response of those without shear keys is almost the same for all skew angles. Hence, closer values of moment were achieved for larger skew angles between the two cases. This observation confirms that the effectiveness of shear keys to reduce moments on columns reduces as skew angles increases. It can be concluded that shear keys can reduce the demand on columns of skewed bridges with small and moderate skew angles as long as they remain intact.

Figure 3-101 shows the axial force in columns with respect to the skew angle. No significant variation was recorded for without shear key case while a considerable variation was recorded for with shear key case.

Figure 3-102 and Figure 3-103 show the shear forces in columns in both directions (q_y and q_z) as a function in skew angle. q_y follows the same trend as Mzz while q_z follows the same trend as Myy.

Figure 3-104 presents the torsion in columns (C1 and C2). The trend of torsion with skew of bridges with shear keys tends to increase as skew increases. For the lower level of ground motions, larger response values were experienced by bridges without shear keys.

It increases from no skew case to 20 degree skew case afterwards it begins to level out. For larger level of ground motions, significantly larger response values were observed and the response values increases with the skew. It can observed that the presence of the shear keys play an important role in reducing the torsion on columns especially for larger skew angles. Torsion values are not large.

Figure 3-105 and Figure 3-106 show the deformation of the abutment-soil springs, as gap closure and opening (U+ and U-) with respect to skew angle. Only the results of the four abutment springs on one side of the bridge are presented. For U+, as skew angle increases, the displacement decreases. The absence of shear keys is expected to increase the demand on abutment-soil springs. Abutment springs for bridges with shear keys showed slightly larger deformation throughout. For U-, a different trend was observed for without shear key case as gap opening increases with skew compared to that with shear keys. It was established earlier that transverse displacement increases with skew therefore demand on abutments increase. Larger gap opening takes place closer to the acute angle and it reduces as moving towards the obtuse angle. Therefore, a more pronounced in-plane rotation at the abutments took place.

Figure 3-107 presents the envelope of average curvature ductility demand with respect to the skew angle in both directions (μ_{yy} and μ_{zz}). Under 0.3g ground motions level, for the without shear key case the ductility factors are almost constant as skew increases and is approximately unity while with shear key case the response values increase with the skew. Therefore, a smaller difference in ductility demand can be observed for larger skew angles. This suggests that the effectiveness of shear keys to reduce the ductility demand reduces for larger skew angles. For larger level of ground motions, both columns reached yielding with large ductility demand on those without shear keys.

In summary, the absence of shear keys did not affect displacement in the longitudinal direction significantly while it increased that in the transverse direction dramatically by a factor of about 4 at abutment location. Both columns have yielded in both cases. The effectiveness of shear keys to reduce the bent forces on columns reduces as the skew angle becomes larger. For abutment-soil springs, for "without shear key" case, gap openings were larger and increases with skew and in-plane rotation took place. Shear keys were found to be effective to reduce torsion on columns especially for large skew angles by a factor about 3.

3.3.4. Effect of Foundation Boundary Condition

The effect of foundation boundary conditions was studied considering the two extreme cases, as is the usual practice; namely pinned and fixed bent foundations. The nonlinear time history analyses were performed for only Soil-D, with two levels of ground motions and with nonlinear shear key elements and abutment-soil springs.

Introducing fixed base is anticipated to reduce deformation of bridge deck, increase demand on columns and develop torsion, and reduce the demand on both abutment-soil springs and shear key springs. Figure 3-108 and Figure 3-109 present the displacement of the deck in the longitudinal and transverse directions (x and y), respectively under the application of twelve pairs of ground motions with PGA of 0.3g and 0.6g (six motions per each PGA). In the longitudinal direction, as the skew angle increases, the displacements increase. Deck of the bridges with pinned bent foundation experienced larger displacements than that with fixed bent foundation by about 22% for PGA of 0.6g regardless of skew angle. The effect of introducing fixity was larger for the larger level of ground motions; leading to larger difference between response values for any skew angle for 0.6g case.

In the transverse direction, insignificant variation was observed for nodes 90 and 74 while displacement of node 64 (at abutment) increased significantly with skew angles. Transverse deck displacements for pinned bent foundation case were about 26% larger than those with fixed bent foundation for nodes 90 and 74. Also, the increase in displacement was larger for without shear key case. At abutment location (node 64), displacement of fixed bent foundation case was smaller than that of pinned bent foundation case. This clearly shows the effect of fixity on the response. In addition, increasing level of intensity of ground motions did not affect the trend.

Figure 3-110 and Figure 3-111 present the bending moments of columns (C1 and C2) about the y-y axis (Myy) and the z-z axis (Mzz). For Myy, similar trend was observed between the two cases. For fixed bent foundation case, noticeable larger response was experienced especially for lower level of ground motions. For Mzz, introducing fixity led to larger response with the skew. Response of bridges with fixed bent foundation was significantly larger than that with pinned bent foundation. This shows the effect of the fixity on increasing the bending moment on columns leading to increase in demand on columns. It seems that introducing the fixity have the same effect on bending moments on columns regardless of the skew angle.

Figure 3-112 shows the axial force in columns with respect to the skew angle. It follows the same trend while larger values were observed for fixed bent foundation case. No significant variation was recorded for pinned bent foundations while a variation was recorded for fixed bent foundations. The average axial force for bridges with fixed base was larger than those with pinned case for a given skew angle. It should be mentioned that the two columns experienced similar axial forces.

Figure 3-113 and Figure 3-114 show the shear forces in columns in both directions (q_y and q_z) as a function of skew angle. q_y follows the same trend as Mzz while q_z follows the same trend as Myy. Significantly larger response was observed in columns of the bridges with fixed bent foundation; by a factor greater than 2.5.

Figure 3-115 and Figure 3-116 show the deformation of abutment-soil springs, as deformations "into" the soil (U+) and "away" from the soil (U-) with respect to the skew angle. Only the results of the four abutment springs on one side of the bridge are presented. Abutment-soil springs attached to bridges on pinned bent foundation showed larger displacement either into or away from abutments by largest percentage of about 36%. As the skew angle increases, both the gap opening and closing reduces. At the PGA of 0.6g, abutment springs have yielded for all skew angles for both bent foundation boundary conditions. Also, Figure 3-116 suggest larger in-plane deck rotations since larger gap opening can be observed toward the obtuse angle at the abutment with increased skew angles greater than 45 degree. However, this observation is more pronounced for pinned bent foundation case.

Figure 3-117 presents the response of shear key deformations as a function of the skew angle. It can be observed that shear keys on the diagonals show similar trend. A slight difference was noticed among the four shear keys showing tendency for having uneven distribution of forces which could lead to progressive collapse. Larger deformations of shear keys were noticed at larger skew angles for lower ground intensities. For fixed bent foundation and pinned bent foundation cases, yielding of shear keys were reported in an average sense under PGA of 0.6g with lower shear key deformations for the fixed case.

Figure 3-118 presents the curvature ductility demand with respect to the skew angle in both directions (μ_{yy} and μ_{zz}). The larger forces developed in bent columns of bridges with fixed bent foundation led to larger ductility ratios in both directions. The trend followed that of pinned base case. Therefore, introducing fixity increased the ductility demand in columns.

In summary, introducing fixed bent foundation condition reduced displacements in the longitudinal and transverse directions of the bridge deck significantly. The bent forces on columns increased which led to the demand on columns to increase significantly. On the other hand, the deformation and force demands at the abutments and on the shear keys reduced noticeably. Also, pinned bent foundation assumption led to larger in-plane deck rotations as the skew angle increases compared to the bridges with fixed bent foundation.

3.3.5. Effect of Bridge Aspect Ratio

For this part of the study, the benchmark bridge, which had an aspect ratio of 0.3, was altered to produce aspect ratios of 0.54 and 1.1. Six bridge models were developed for each aspect ratio and with skew angles of 0, 20, 30, 45, 52, and 60 degree. It should be also noted that the bridge models with aspect ratio of 0.54 were modeled using three stick beams for the deck and three columns. Whereas, the bridge models with aspect ratio of 1.1 were modeled using four stick beams for the deck to model the stiffness and mass distribution and with four columns. Shear key and abutment-soil elements were modeled explicitly as explained earlier. DRAIN3DX was used to perform the nonlinear time history analyses of the twelve models. The analyses were conducted for only one soil type (Soil-D) and for the larger level of ground motions (0.6g) and for pinned bent foundations case.

Increasing the aspect ratio is anticipated to increase the in-plane rotation of the bridge decks with skew. This may lead to a more complex response for the skewed bridges. The demand on columns may increase due to the increase of bent forces. Also, it is crucial to monitor the response with respect to the skew angle in order to examine the effect of aspect ratio on the seismic performance. The time history analyses were performed for the two aspect ratios and the results are presented in the form of comparison among these two aspect ratios and the original aspect ratio (0.3).

Figure 3-119 and Figure 3-120 present the average displacement of the deck in the longitudinal and transverse directions (x and y), respectively subjected to the six ground motions with the PGAs of 0.6g. In the longitudinal direction, for all aspect ratios, the displacement increases as the skew angle increases. Deck of the bridges with aspect ratios of 0.54 and 1.1 experienced larger displacements than that with aspect ratio of 0.3 by about 13% and 58% respectively, at 60 degree skew. However, increasing the aspect ratio did not affect the trend with the skew angle as mentioned earlier for other response quantities.

In the transverse direction, significant increase was observed for the deck displacements at nodes 90 and 74 while as the aspect ratio increases– although small – the displacement of node 64 (at abutment) increased with skew angle. As the aspect ratio increases, the transverse deck displacement increases. Also, the transverse deck displacement becomes smaller from the bent location to the abutments for all aspect ratios. For all aspect ratios, there is an increase in the transverse displacement with the skew, especially, for bridges with skew larger than 30 degree. Nonetheless, at abutment location (node 64), deck displacement with aspect ratio of 0.54 was slightly larger than that with aspect ratio of 0.3 whereas that with aspect ratio of 1.1 was significantly larger by about 300% regardless of the skew angle. This can be attributed to the failure of shear keys for bridges with aspect ratio of 1.1.

Figure 3-121 and Figure 3-122 present the bending moment of the external columns (C1 and C2) about the y-y axis (Myy) and the z-z axis (Mzz). For Myy, similar trend was observed among all the cases. Similar values of Myy with the skew was achieved by bridges with the aspect ratios of 0.3 and 0.54 while larger response was achieved by bridges with the aspect ratio of 1.1. For Mzz, the response increases with the skew. The response achieved by the bridges with the aspect ratio of 1.1 was consistently larger compared to the lower aspect ratios.

Figure 3-123 shows the axial force in two columns with respect to the skew angle. No significant variation was recorded regardless of the aspect ratio.

Figure 3-124 and Figure 3-125 show the shear forces in columns in both directions (q_y and q_z) as a function in skew angle. There is not any significant variation in the shear force response as the skew increases or as the aspect ratio increases.

Figure 3-126 and Figure 3-127 show the deformation of abutment-soil springs, as deformations into the soil (U+) and away from the soil (U-) with respect to the skew angle. Only the results of the four abutment springs on one side of the bridge are presented. For U+, all models followed similar trend as the response decreases when the skew angle increases except for the aspect ratio of 1.1. Similar response was achieved by bridges with aspect ratios of 0.54 and 0.3 while abutment-soil springs of bridges with aspect ratio of 1.1 experienced larger response values for most of the cases. Also, yielding of abutment-soil springs was observed for all of the aspect ratios. In-plance rotations became more pronounced as the aspect ratio increases. For U-, the gap opening tends to decrease with the skew while the bridges with aspect ratio of 0.54 show larger gap opening than those with aspect ratio of 0.3 especially for large skew angles. Gap opening response for bridge with aspect ratio of 0.54 tends to level out with the skew for some of the cases and constantly increases with the skew for bridge with aspect ratio of 1.1. The constant increase can be attributed to the loss of shear keys. But more importantly it is evident that the in-plane rotation of the deck becomes more pronounced for larger skew angles and larger aspect ratios.

However, an interesting observation was related to the general trend for the gap opening in bridges with and without shear keys. Bridges modeled with explicit shear key elements experienced larger gap openings at the obtuse corner that became smaller toward the acute corner. This behavior was observed regardless of the aspect ratio, however, only in bridges with a skew angle 45 degree or larger. For smaller skew angles, a rather uniform gap opening can be seen in the associated figures. It should be noted that although the shear key elements yielded and provided only marginal resistant in the transverse direction in most of the cases, the observation regarding the gap opening was still valid. On the other hand, skewed bridges that were modeled without the shear keys experienced larger gap openings at the acute corner that became smaller toward the obtuse corner. And the latter observation was valid for all skew angles and for all aspect ratios as well. And finally, the gap openings were significantly smaller in bridges modeled with the shear key elements.

Figure 3-128 presents shear key deformations as a function of the skew angle. It can be observed that shear keys on the diagonals show similar trend. The deformation response as a function of the skew angle decreases for larger aspect ratios or remain nearly constant for smaller aspect ratios. However, larger deformation demand on shear keys for larger aspect ratios is evident in the figure. For aspect ratio of 0.54, the response starts larger than response for aspect ratio of 0.3 at no skew case afterwards it converges at large skew angles. This suggests that the effectiveness of shear keys diminish with the skew for bridges with larger aspect ratios. For all aspect ratio, the yielding of shear keys took place throughout which reveals the larger demand on shear keys due to the developed in-plane rotations. Bridges with aspect ratio of 1.1 experienced failure of shear keys in most of the cases. A slight difference was noticed among the four shear keys showing tendency for having uneven distribution of forces.

Figure 3-129 presents the envelope of average curvature ductility demand with respect to the skew angle in both directions (μ_{yy} and μ_{zz}). For all aspect ratios, yielding took place in columns. Introducing the larger aspect ratio led to the increased ductility demand on columns especially for larger skew angles.

In summary, introducing larger aspect ratios slightly increased displacement in the longitudinal and significantly increased displacement in the transverse direction of the bridge deck. The bent forces on columns increased which led to the increased demand. However, complex response behavior was observed in bridges with larger aspect ratios. On the other hand, gap opening and closing at the abutments become more pronounced and the demand on shear keys increased, and increased in-plane deck rotations with skew was evident. Uneven distribution of abutment forces may take place which can lead to progressive failure, and complex global system behavior which is evident from the unequal deformations among the abutment-soil interaction springs and among the shear key springs. In addition, Figure 3-130 through Figure 3-139 present a comparison between all aspect ratios with and without the shear keys and for all the response parameters and the figures confirm the observation which is presented above.

Table 3-1. Considered Ground Motions (Mw = 6.5 ~ 7.4; soil type = D)*

Record Name	Station Name	Distance (mi)	PGA (g)	PGV (in./s)	PGD (in.)	Duration (sec)
El Centro 1940	El Centro Array#9	5 15	0 313, 0 215	11 7, 11 9	5 25, 9 41	40 0
El Centro 1979	El Centro Array#3	5 78	0 266, 0 221	18 4, 15 7	7 45, 9 18	39 5
Loma Prieta 1989	Anderson Dam	13 3	0 244, 0 240	8 0, 7 2	3 05, 2 65	39 6
Northridge 1994	Century City	15 97	0 256, 0 222	8 3, 9 9	2 63, 2 25	40 0
Superstition Hills 1987	El Centro Imp	8 64	0 358, 0 258	18 3, 16 1	6 89, 7 95	40 0
Turkey 1999	Duzce	7 89	0 312, 0 358	23 1, 18 3	17 37, 6 93	27 1
El Centro 1979	Bonds Corner	1 55	0 588, 0 775	17 8, 18 1	6 60, 5 86	37 6
Duzce Turkey 1999	Duzce	5 1	0 348, 0 535	23 6, 32 9	16 55, 20 31	25 8
El Centro#5 1979	El Centro Array#5	0 62	0 519, 0 379	18 5, 35 6	13 92, 24 81	39 2
Loma Prieta 1989	Gilroy Array#3	8 95	0 555, 0 367	14 1, 17 6	3 23, 7 58	39 9
Northridge 1994	NewHall	4 41	0 583, 0 590	29 7, 38 3	6 92, 14 98	40 0
Northridge 1994	Sylmar	3 98	0 604, 0 843	30 8, 51 0	6 32, 12 87	40 0

*PGA, PGV, and PGD are given for the two components of the ground motions

Table 3-2. Considered Ground Motions (Mw = 6.5 ~ 7.4; soil type = B)*

Record Name	Station Name	Distance (mi)	PGA (g)	PGV (in./s)	PGD (in.)	Duration (sec)
San Fernando 1971	Castaic	15 47	0 324, 0 268	6 2, 10 2	0 91, 1 84	30 0
Duzce Turkey 1999	Lamont	8 27	0 114, 0 257	4 4, 6 4	3 9, 2 9	42 33
San Fernando 1971	Lake Hughes	12 62	0 366, 0 283	6 7, 5 0	0 65, 1 2	36 6
Loma Prieta 1989	Gilroy Array#7	15 05	0 226, 0 323	6 5, 6 5	1 0, 1 3	39 95
Morgan Hill 1984	Gilroy Array#6	1 93	0 434, 0 316	19 37, 9 65	3 06, 1 52	29 98
Tabas Iran 1978	Dayhook	10 5	0 328, 0 406	8 11, 10 4	4 94, 3 37	23 84
Coalinga 1983	Oil City	5 1	0 866, 0 447	16 6, 9 8	2 4, 0 9	21 24
Duzce Turkey 1999	Lamont	5 1	0 97, 0 514	14 37, 7 95	2 16, 2 94	41 5
Kobe 1995	KJM	0 37	0 821, 0 644	32 0, 29 3	7 0, 7 9	48 0
Loma Prieta 1989	Corralitos	3 17	0 644, 0 479	21 7, 17 8	4 28, 4 5	39 95
Northridge 1994	Castaic	14 05	0 568, 0 514	20 5, 20 6	1 66, 0 95	40 0
Northridge 1994	Katherine	9 1	0 877, 0 64	16 1, 14 9	2 1, 2 0	24 99

*PGA, PGV, and PGD are given for the two components of the ground motions

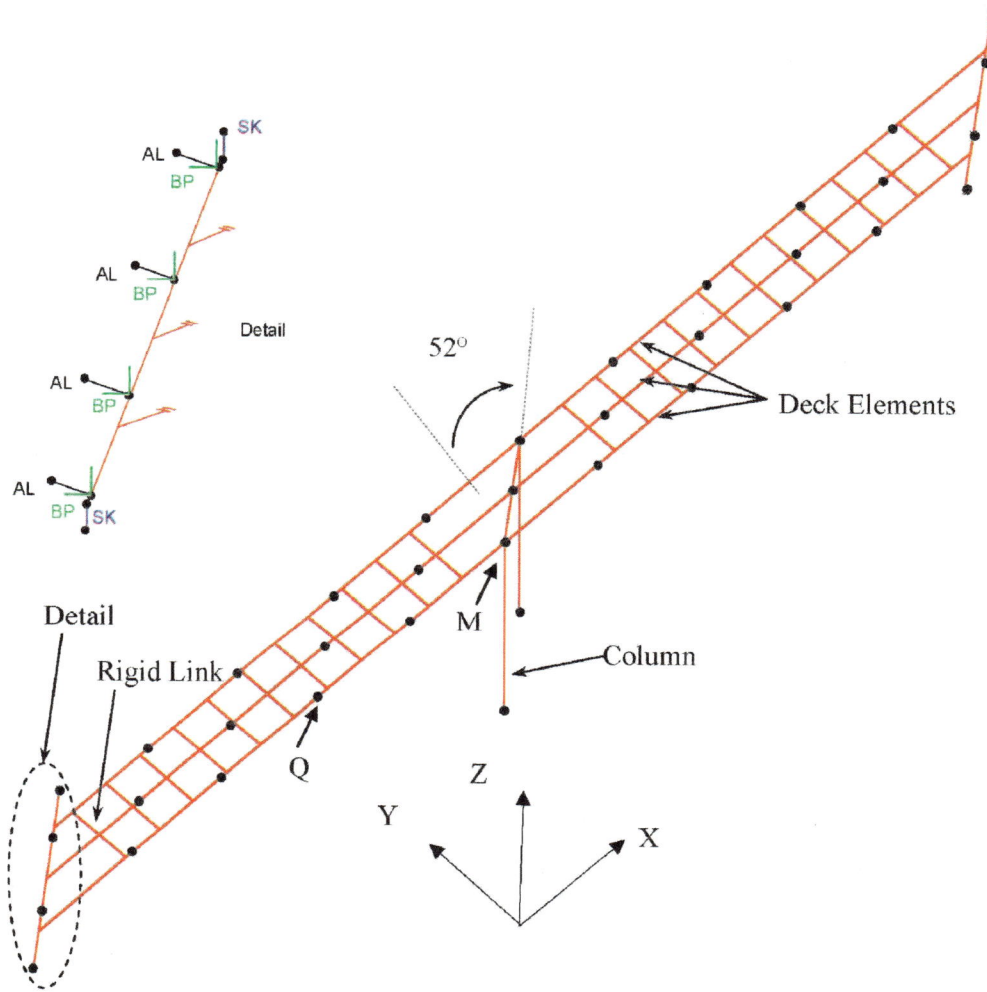

BP: Bearing Pad springs
SK: Shear key springs
AL: Abutment springs

Figure 3-1. 3-D view of Benchmark Bridge

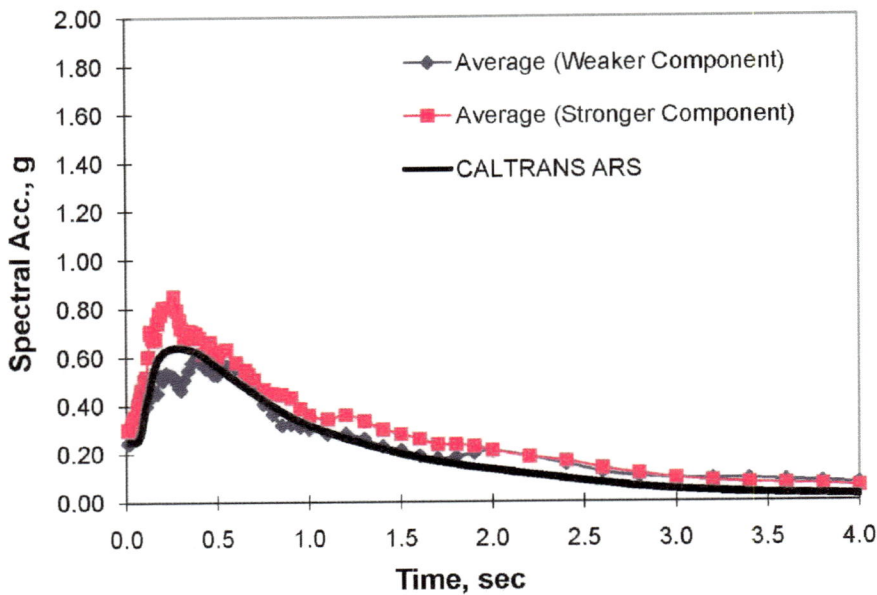

Figure 3-2. ARS of Ground Motions (PGA= 0.3g, Soil-D)

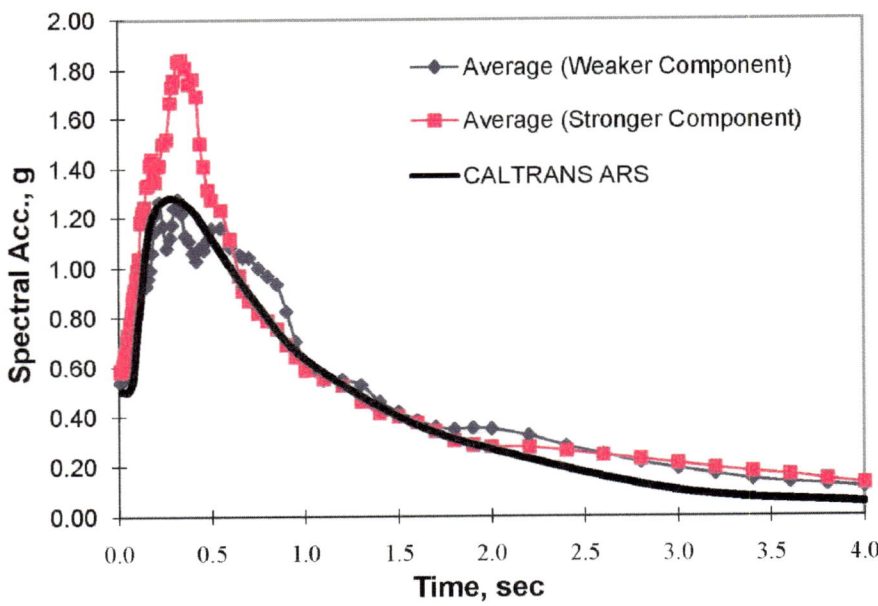

Figure 3-3. ARS of Ground Motions (PGA= 0.6g, Soil-D)

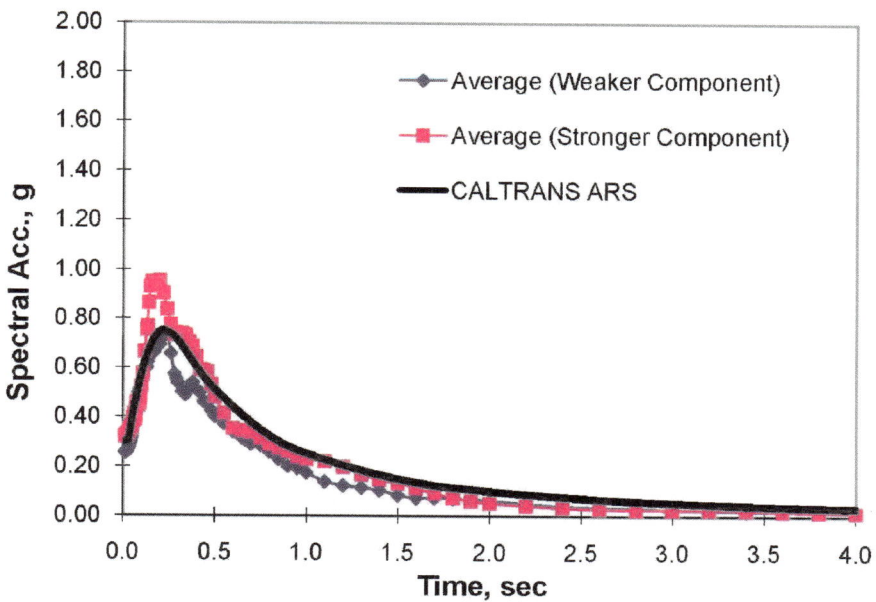

Figure 3-4. ARS of Ground Motions (PGA= 0.3g, Soil-B)

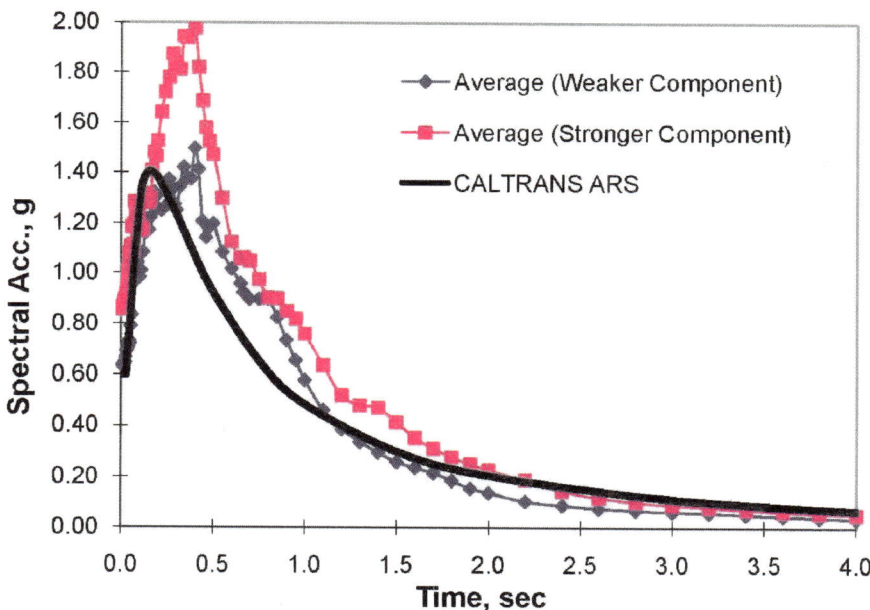

Figure 3-5. ARS of Ground Motions (PGA= 0.6g, Soil-B)

152

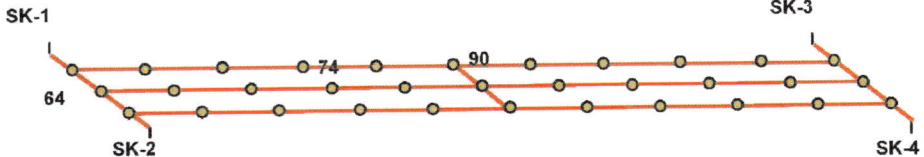

Figure 3-6. Plan of Benchmark Skewed Bridge

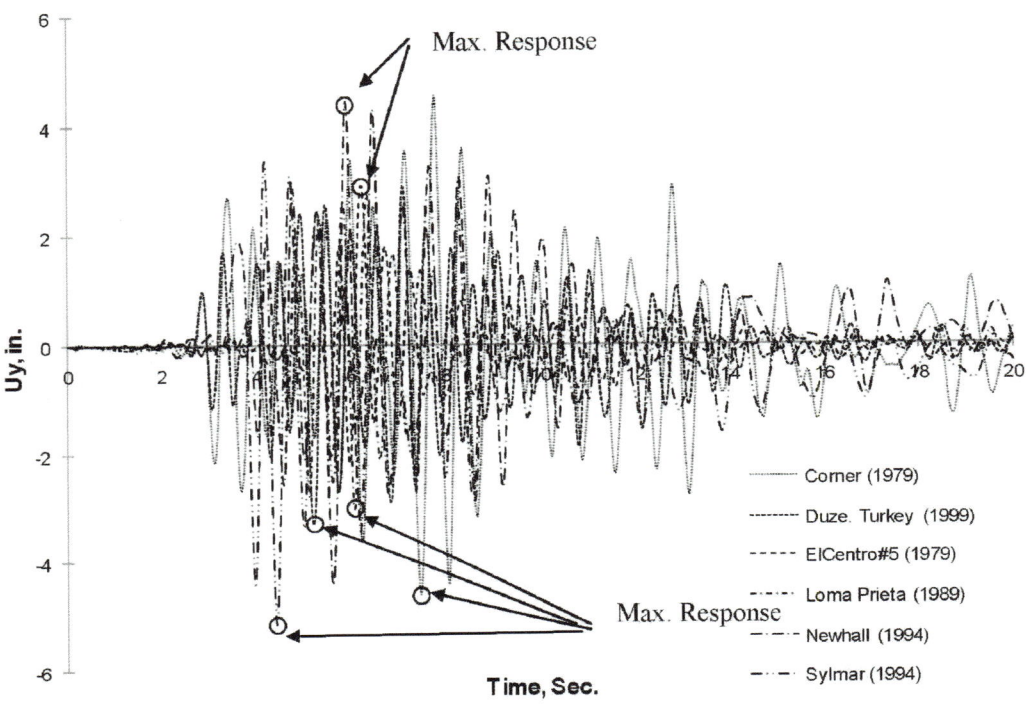

Figure 3-7. Transverse Displacement time history at Bent Location for Bridges with Pinned Bent Foundation and Shear-keys under 0.6g

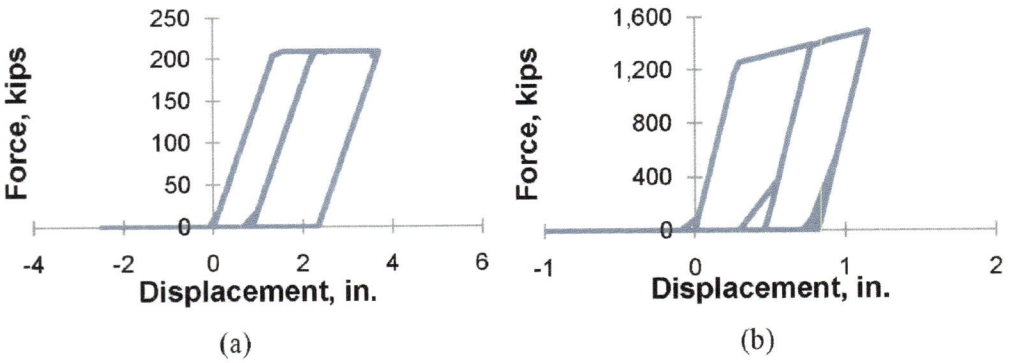

(a)   (b)

Figure 3-8. Example of Spring Hysteresis under THA for Benchmark Bridge with Pinned Bent Foundation under Sylmar(1994) (a) Abutment-Soil and (b) Shear Key

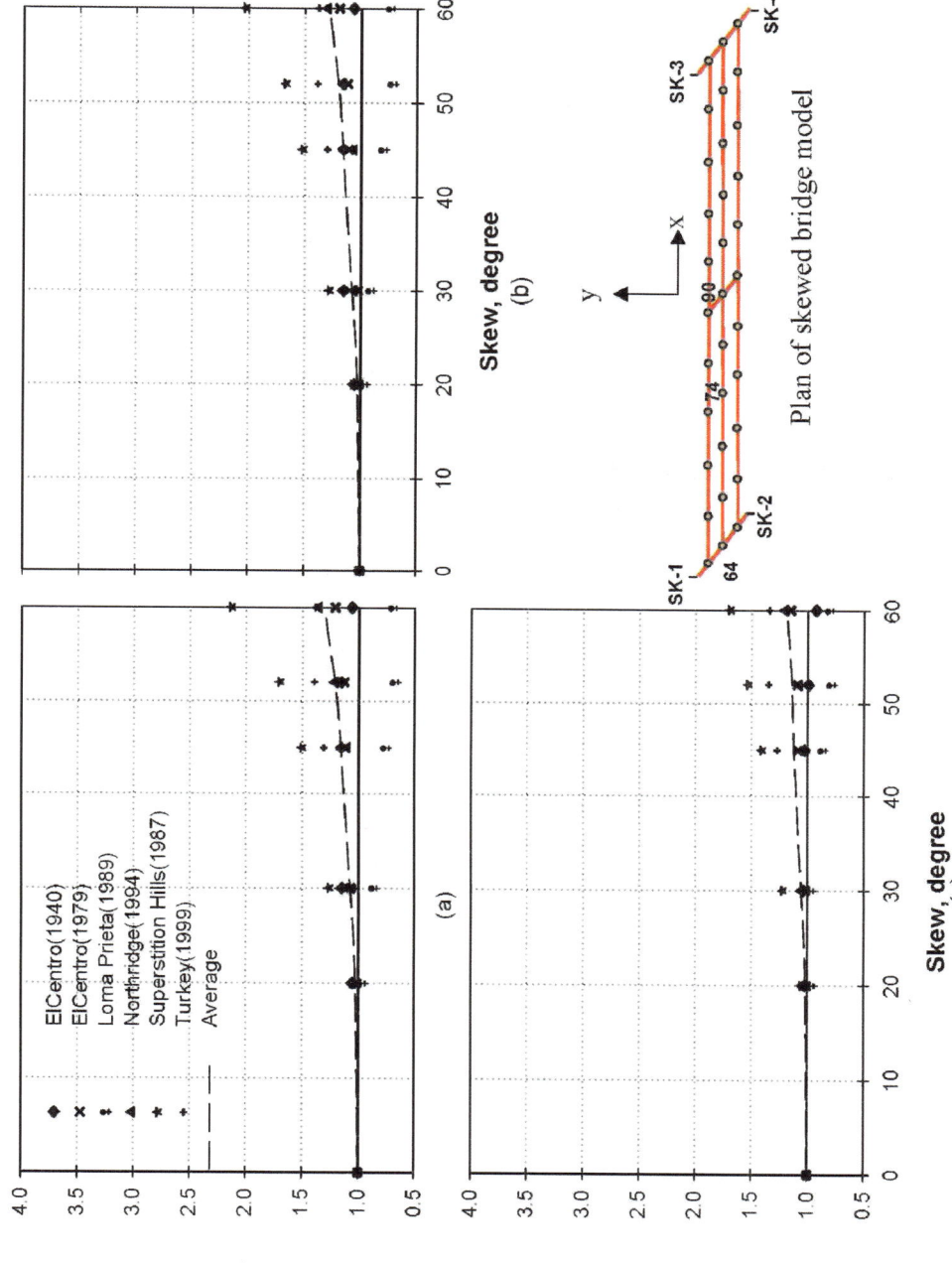

Figure 3-9. Displacement Response Index in X-Direction for nodes (a) 90, (b) 74, and (c) 64 (0.3g)

153

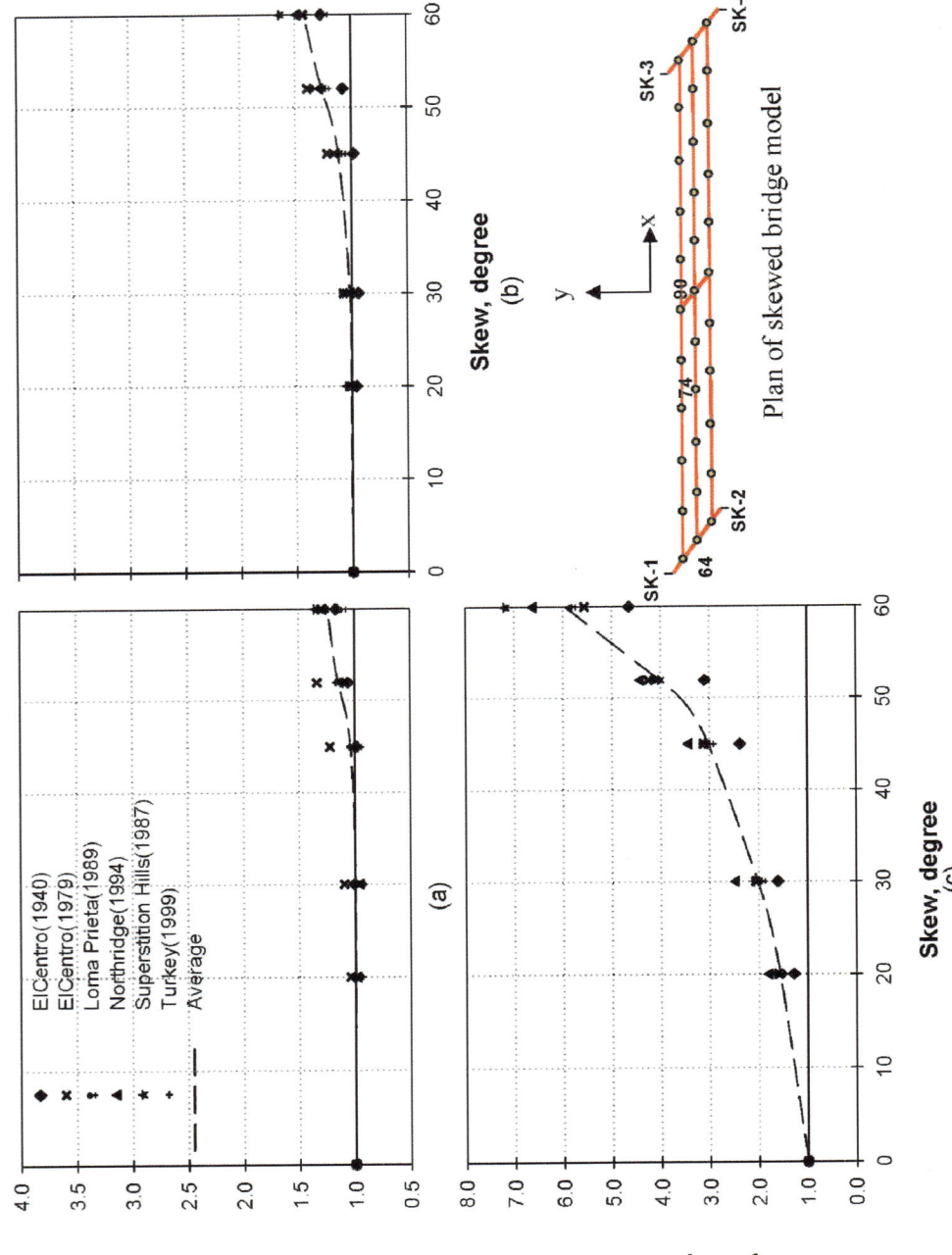

Figure 3-10. Displacement Response Index in Y-Direction for nodes (a) 90, (b) 74, and (c) 64 (0.3g)

154

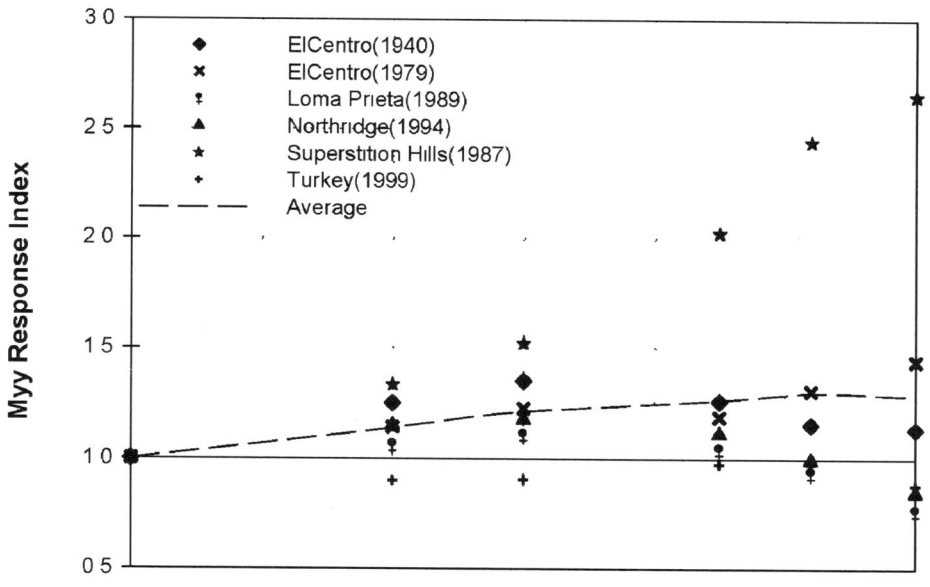

(a)

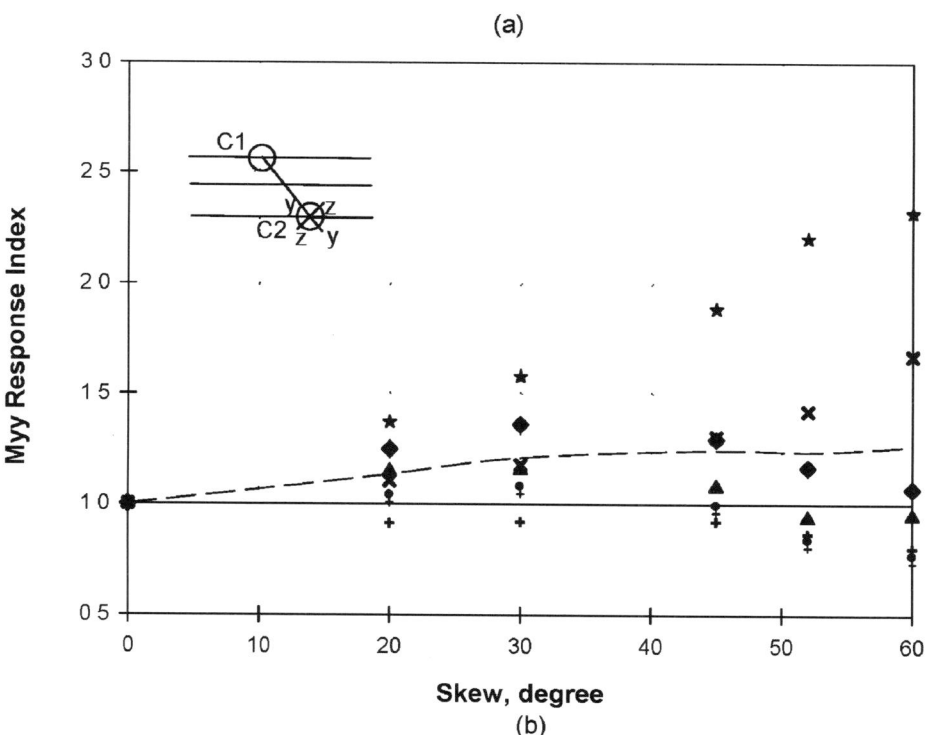

Skew, degree

(b)

Figure 3-11. Moment Response Index in Y-Direction of (a) C1, and (b) C2 (0.3g)

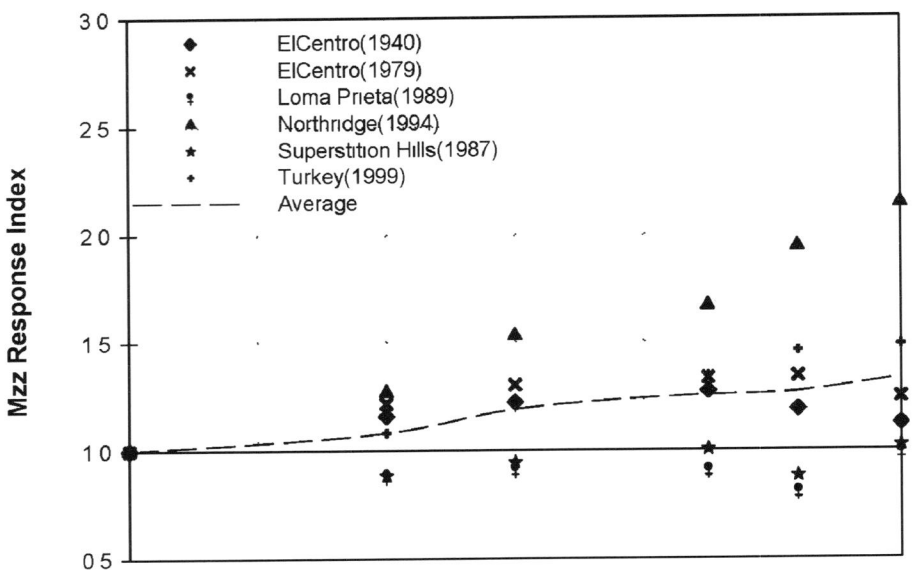

(a)

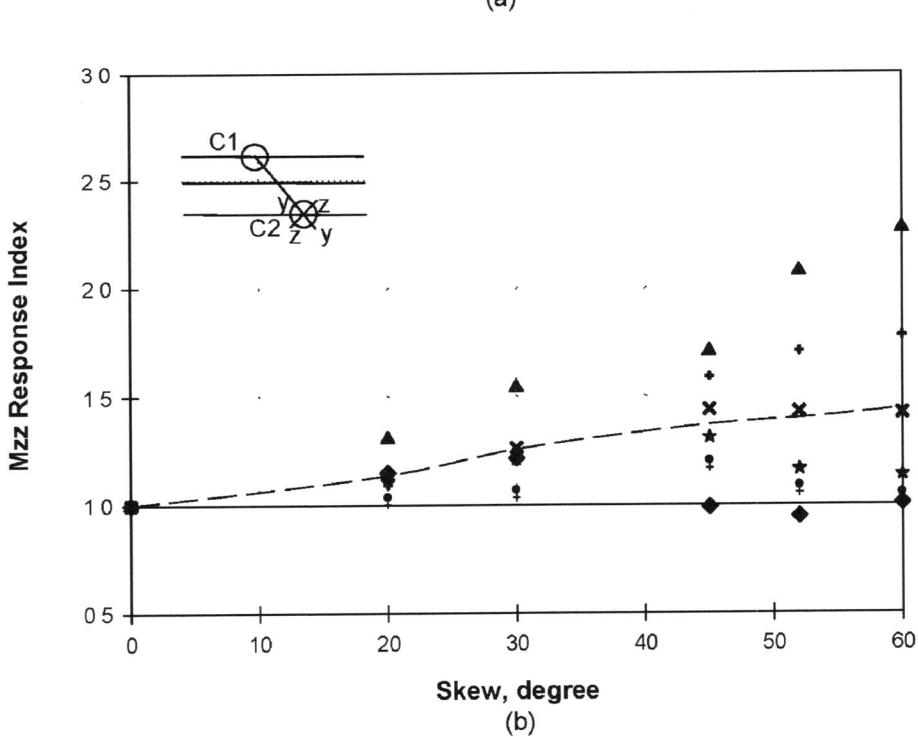

Skew, degree

(b)

Figure 3-12. Moment Response Index in Z-Direction of (a) C1, and (b) C2 (0.3g)

157

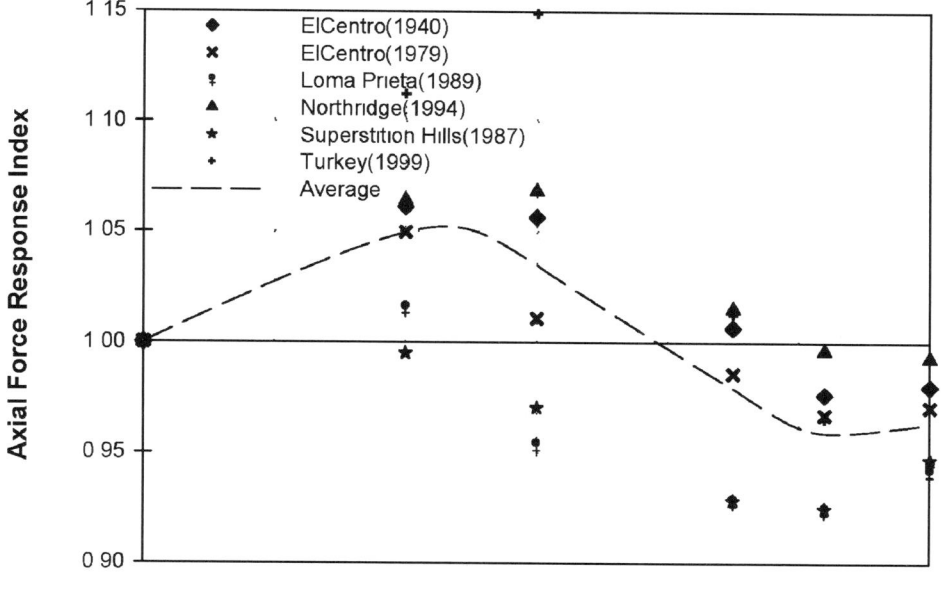

(a)

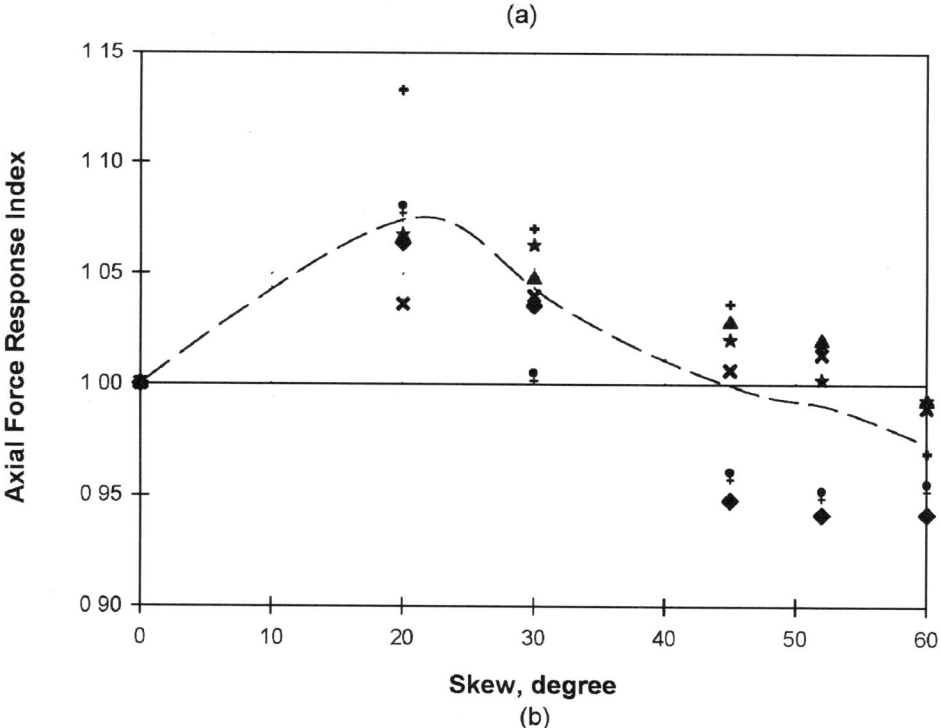

Skew, degree

(b)

Figure 3-13. Axial Force Response Index in (a) C1, and (b) C2 (0.3g)

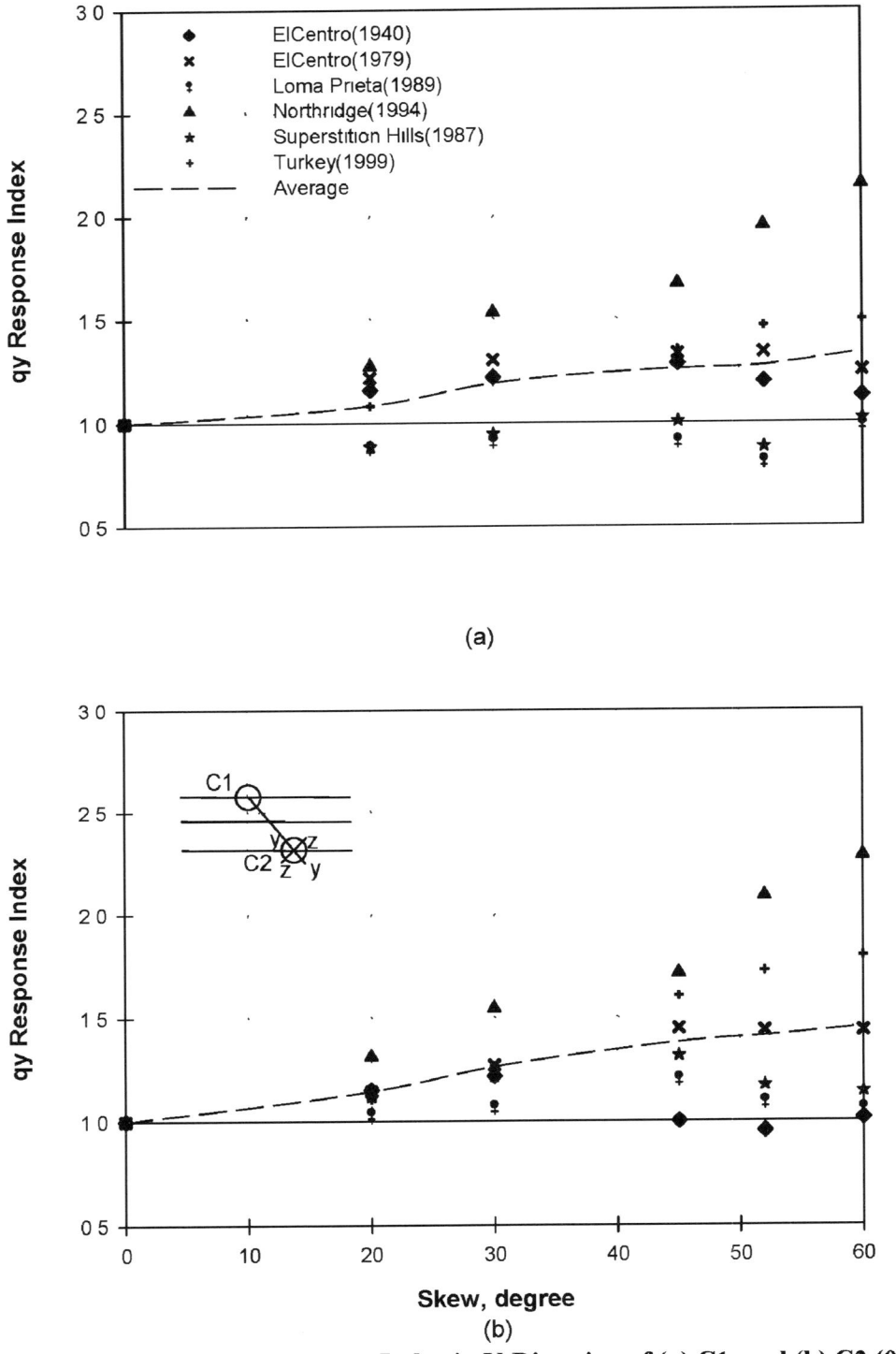

(a)

(b)

Figure 3-14. Shear Force Response Index in Y-Direction of (a) C1, and (b) C2 (0.3g)

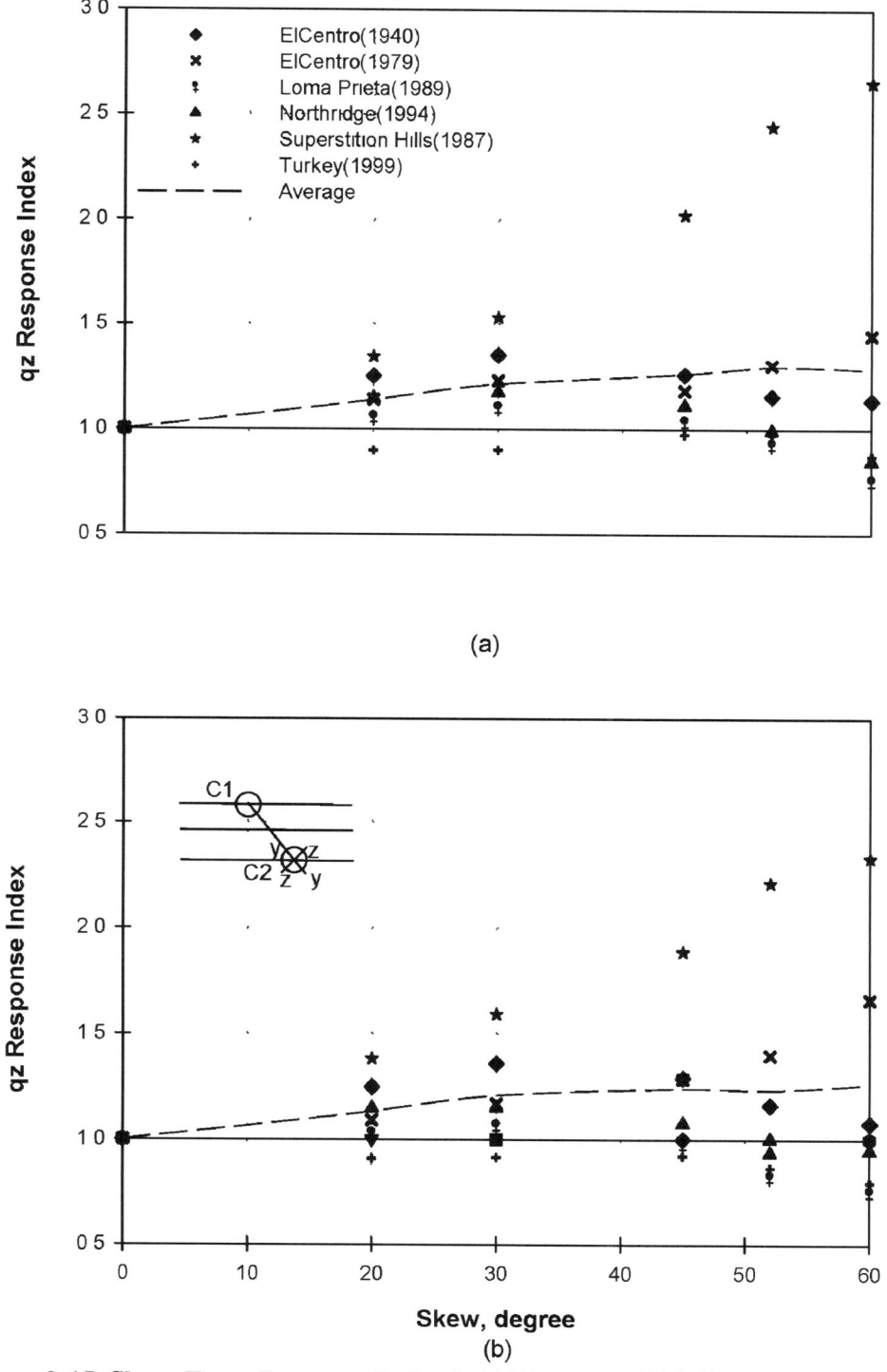

(a)

(b)

Figure 3-15. Shear Force Response Index in Z-Direction of (a) C1, and (b) C2 (0.3g)

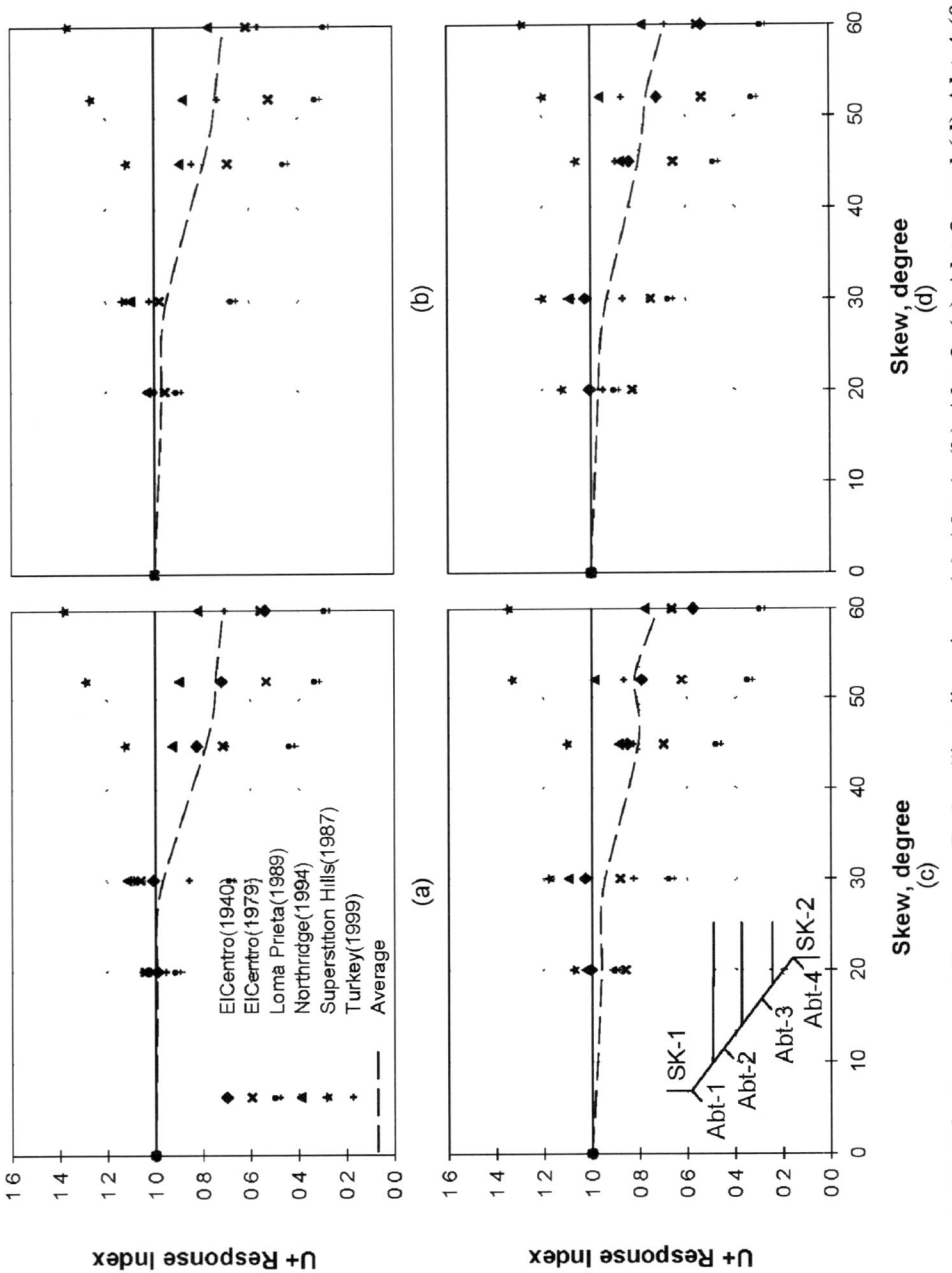

Figure 3-16. Displacement Response Index "into" springs (a) Abt-1, (b) Abt-2, (c) Abt-3, and (d) Abt-4 (0.3g)

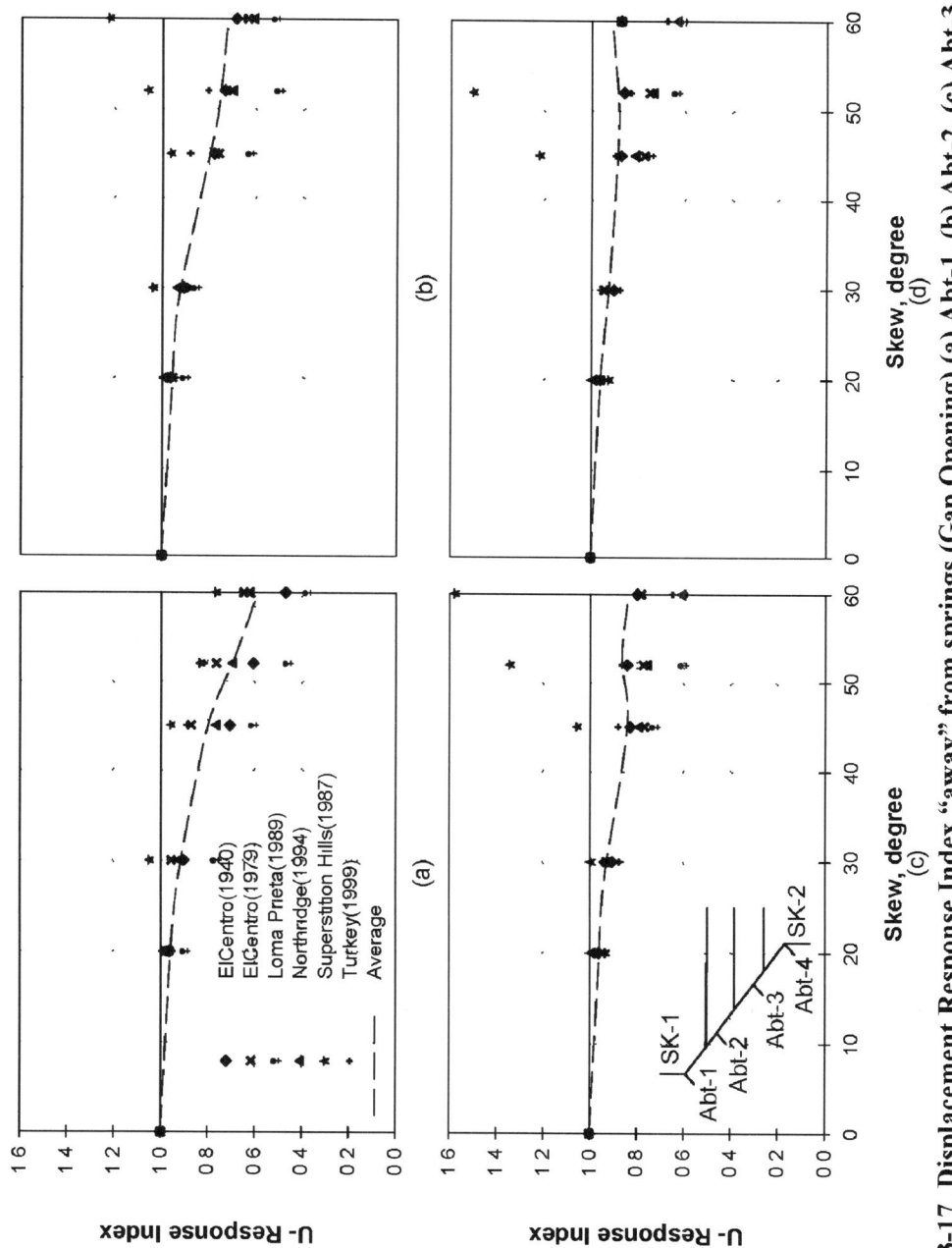

Figure 3-17. Displacement Response Index "away" from springs (Gap Opening) (a) Abt-1, (b) Abt-2, (c) Abt-3, and (d) Abt-4 (0.3g)

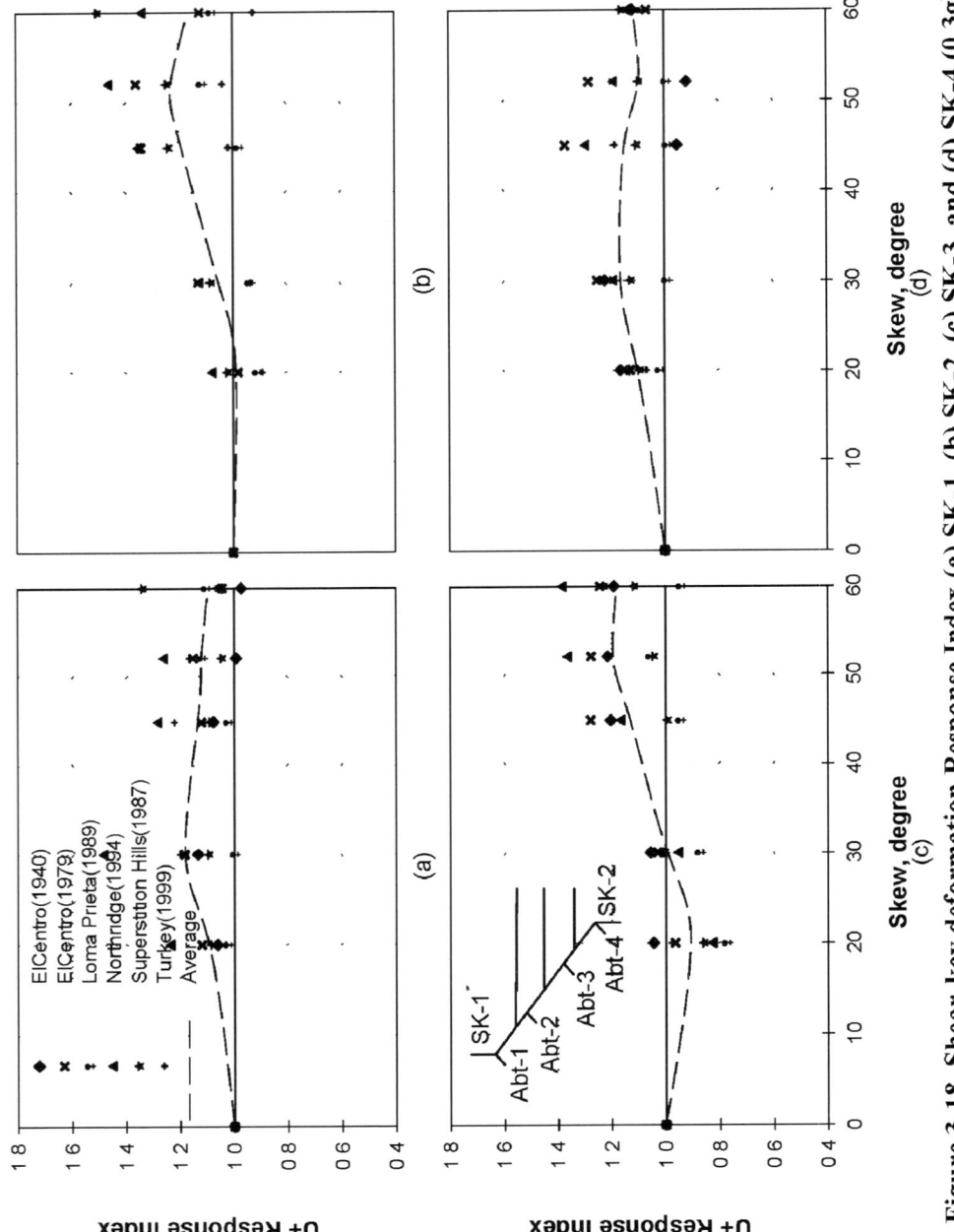

Figure 3-18. Shear-key deformation Response Index (a) SK-1, (b) SK-2, (c) SK-3, and (d) SK-4 (0.3g)

*SK-3 and SK-4 are at the other abutment

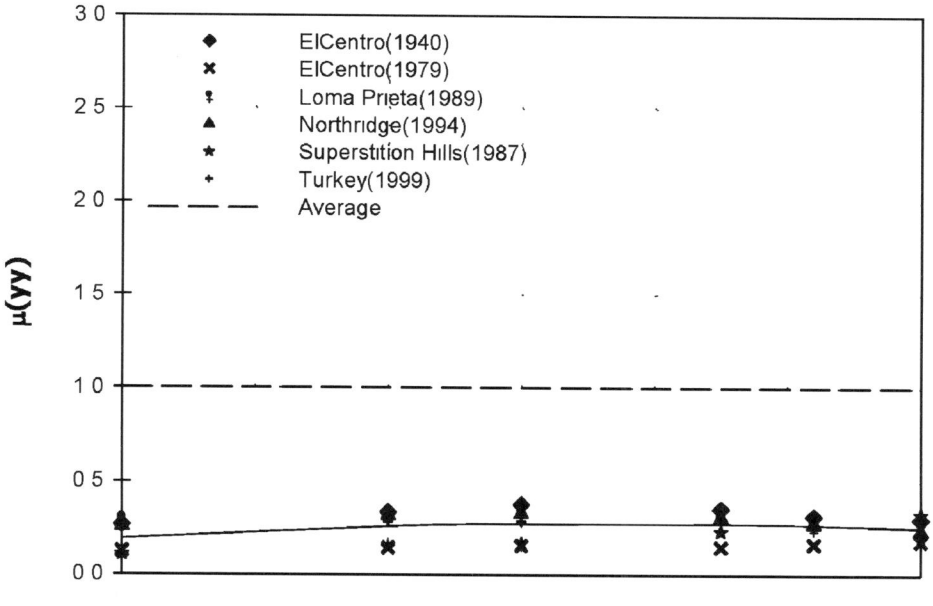

(a)

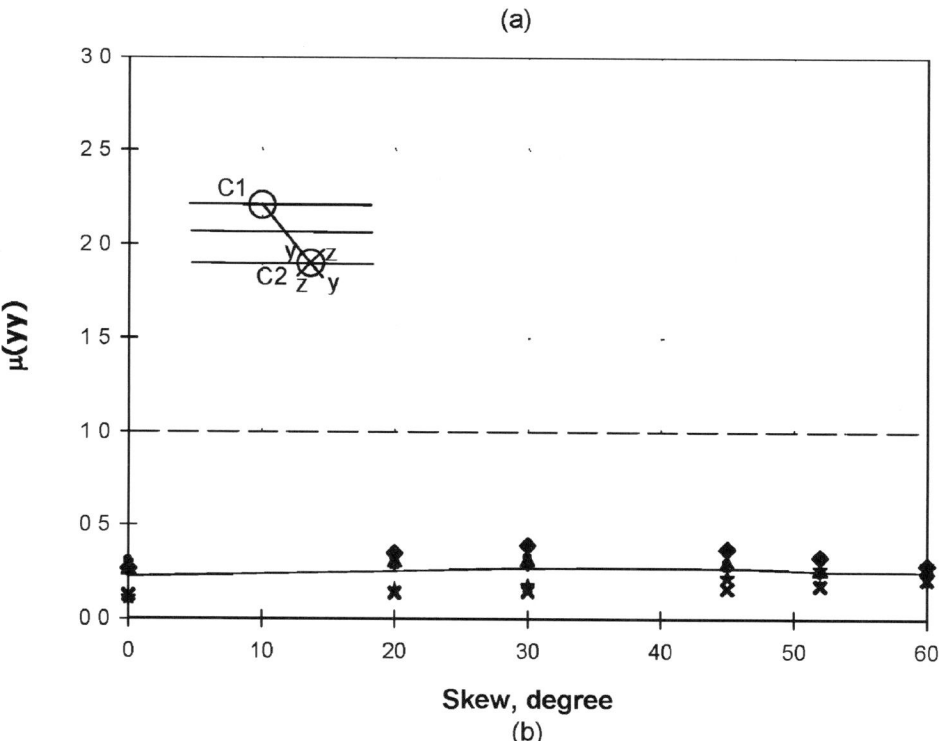

Skew, degree

(b)

Figure 3-19. Curvature Ductility in Y-Direction of (a) C1 and (b) C2 (0.3g)

164

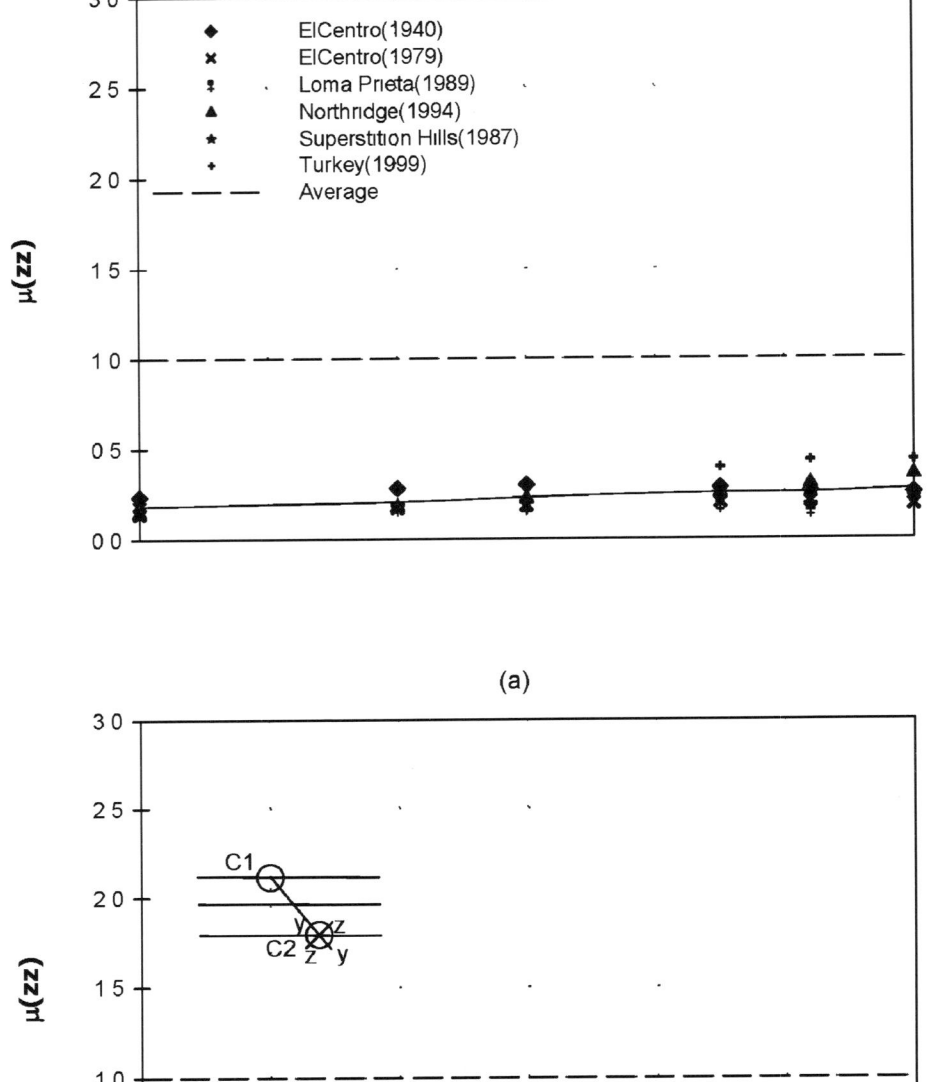

(a)

(b)

Figure 3-20. Curvature Ductility in Z-Direction of (a) C1 and (b) C2 (0.3g)

Skew, degree

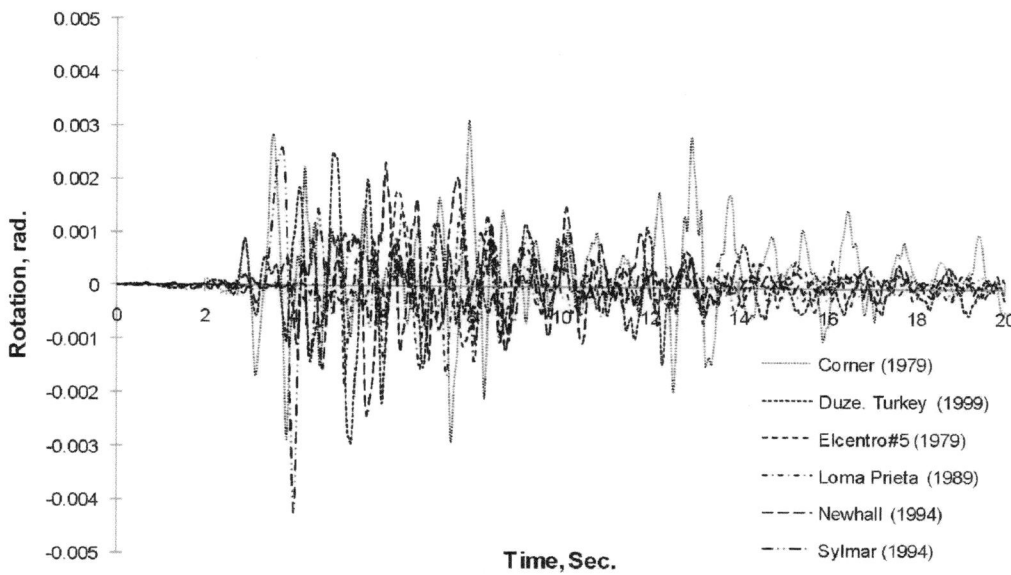

Figure 3-21. Sample of Deck Rotation at the Abutment

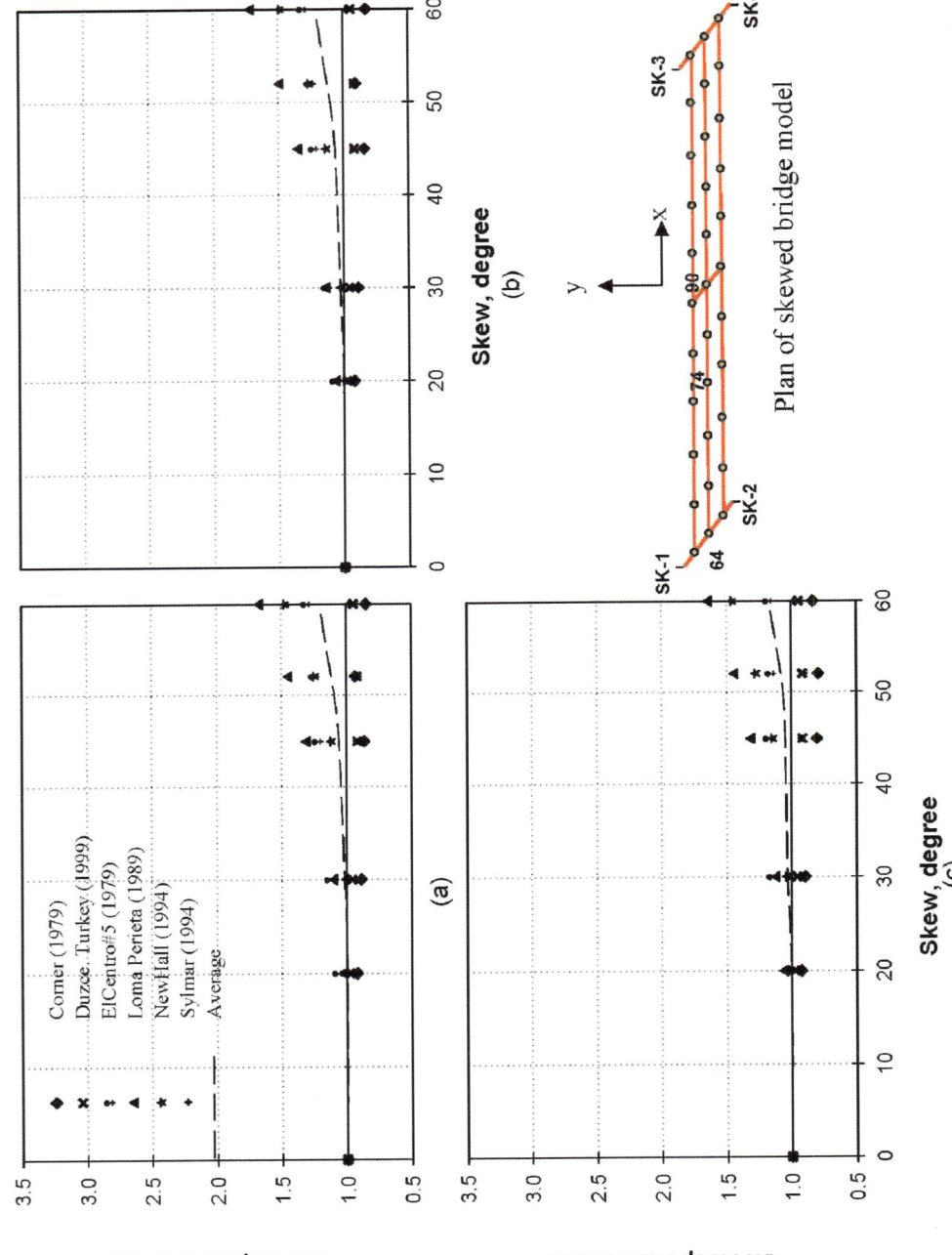

Figure 3-22. Displacement Response Index in X-Direction for nodes (a) 90, (b) 74, and (c) 64 (0.6g)

166

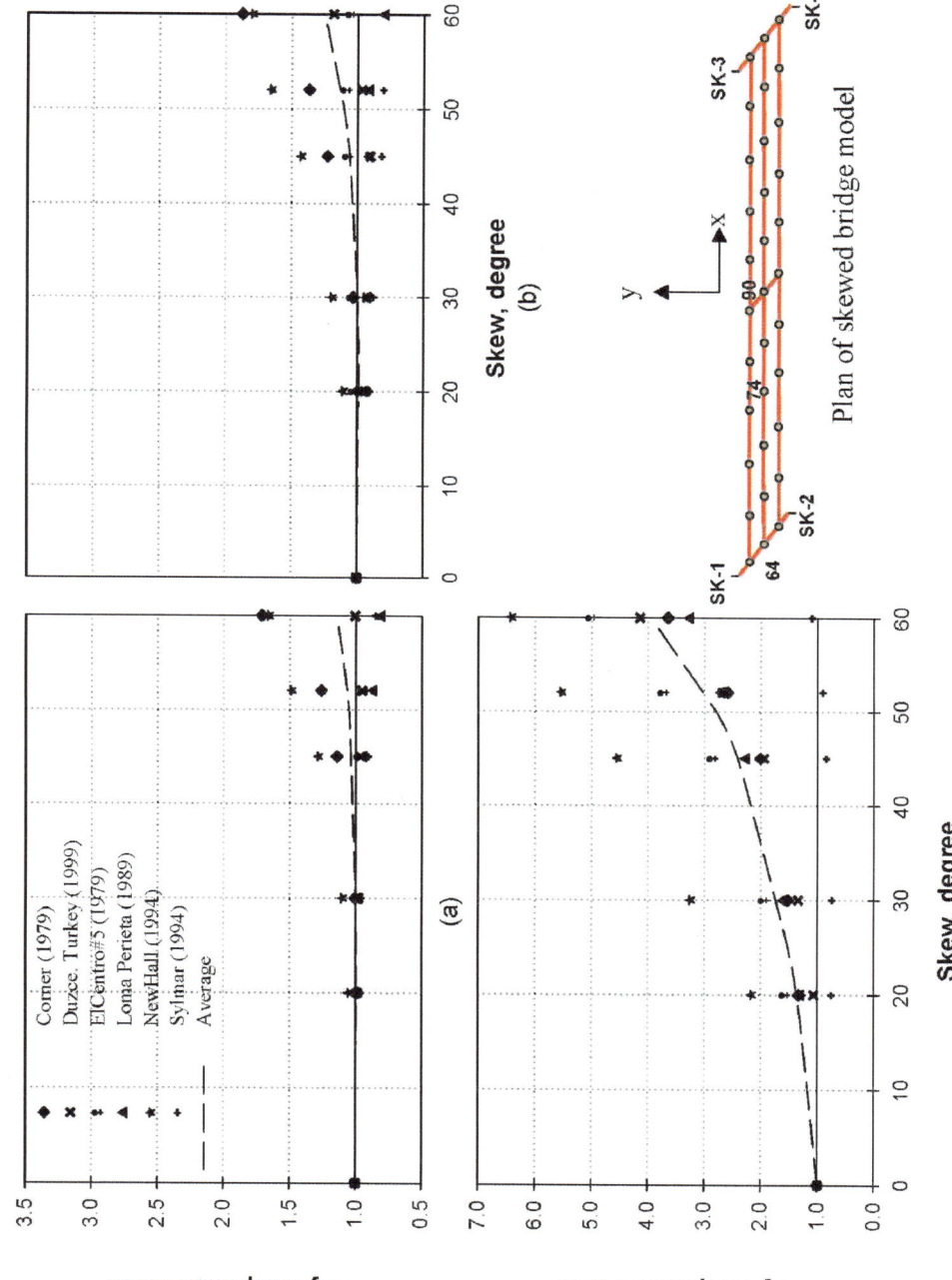

Figure 3-23. Displacement Response Index in Y-Direction for nodes (a) 90, (b) 74, and (c) 64 (0.6g)

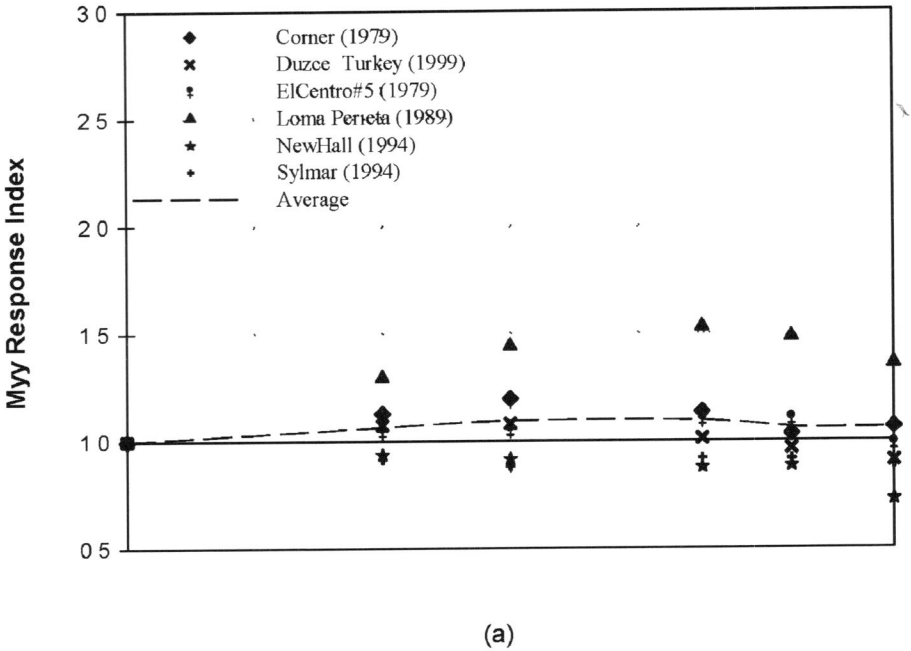

(a)

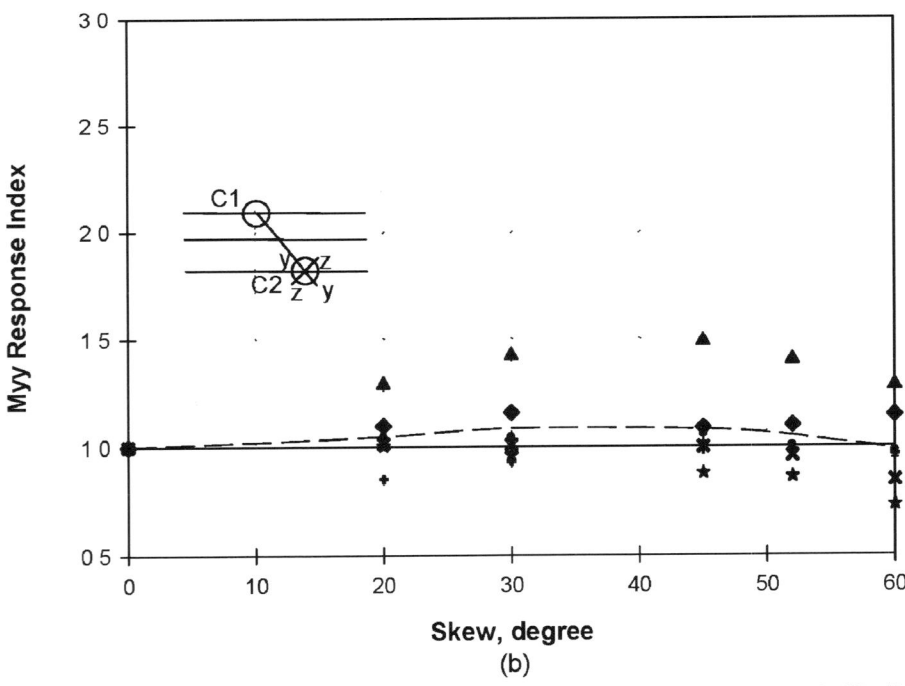

(b)

Figure 3-24. Moment Response Index in Y-Direction of (a) C1, and (b) C2 (0.6g)

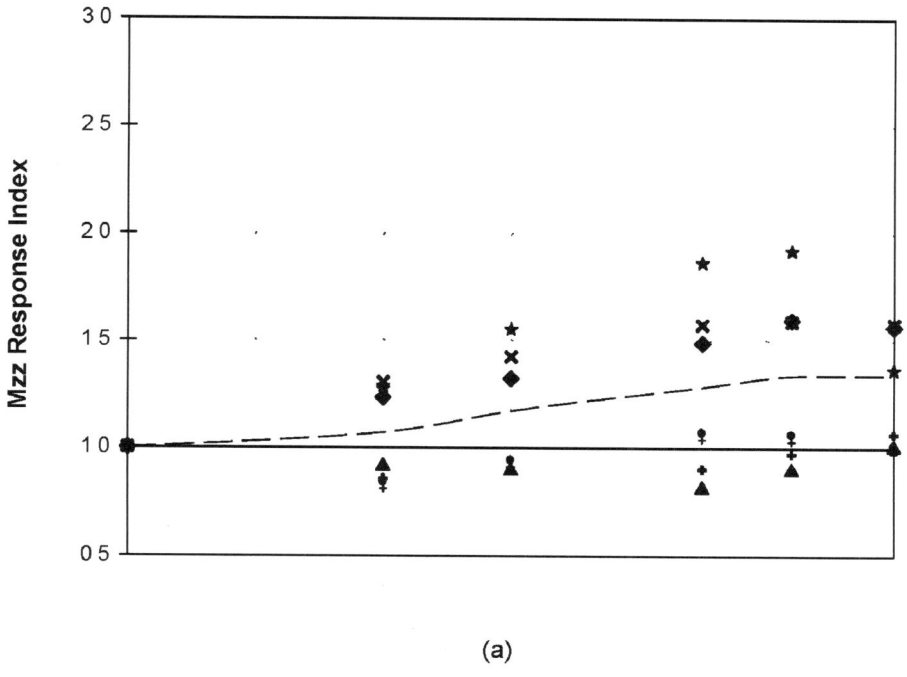

(a)

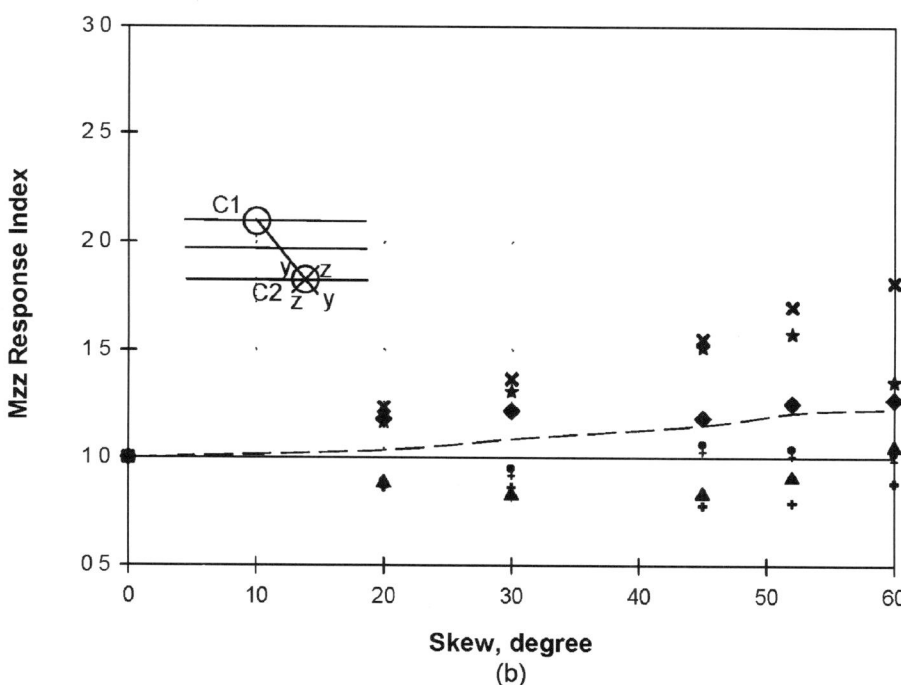

(b)

Figure 3-25. Moment Response Index in Z-Direction of (a) C1, and (b) C2 (0.6g)

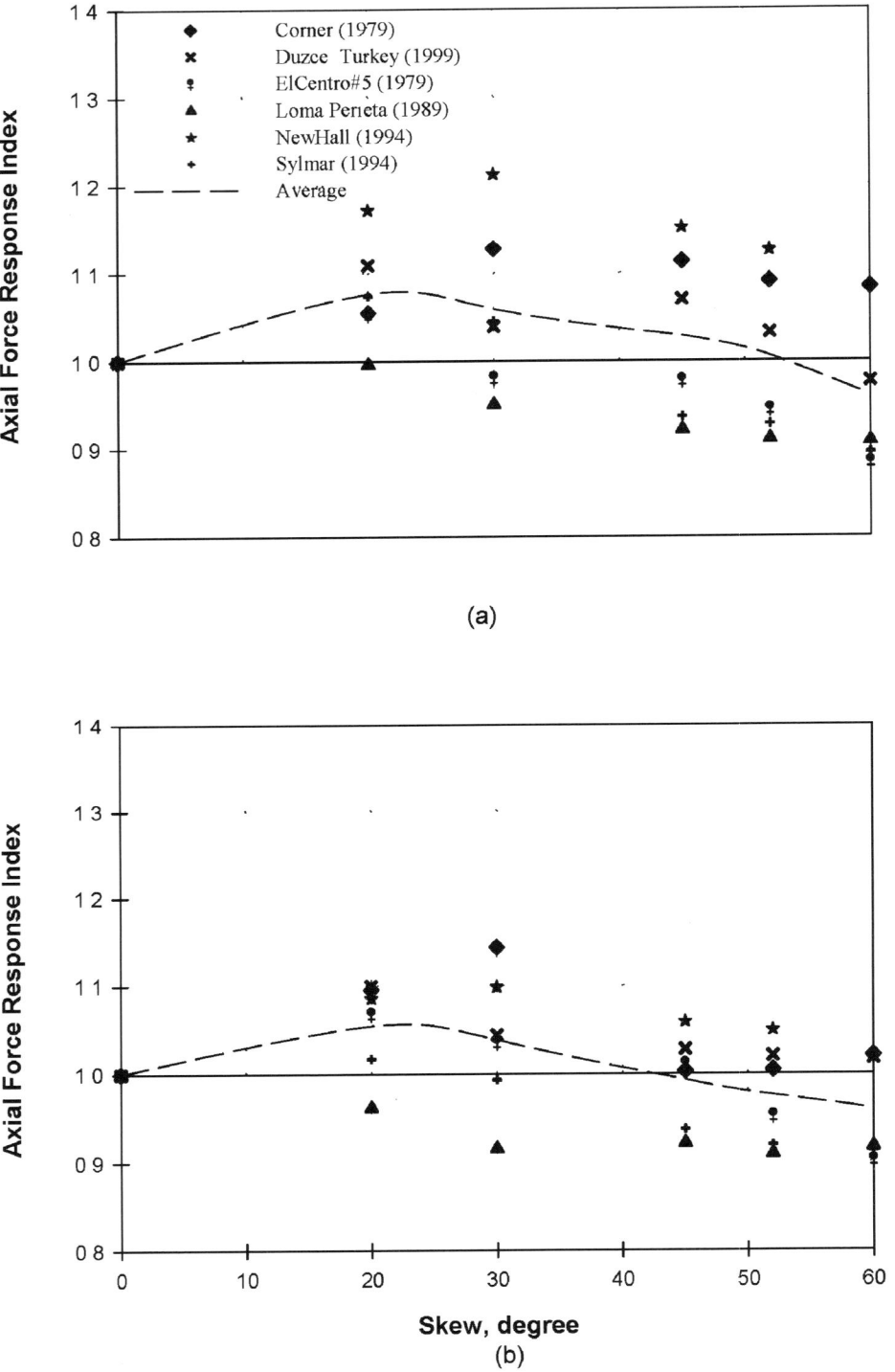

(a)

(b)

Figure 3-26. Axial Force Response Index in (a) C1, and (b) C2 (0.6g)

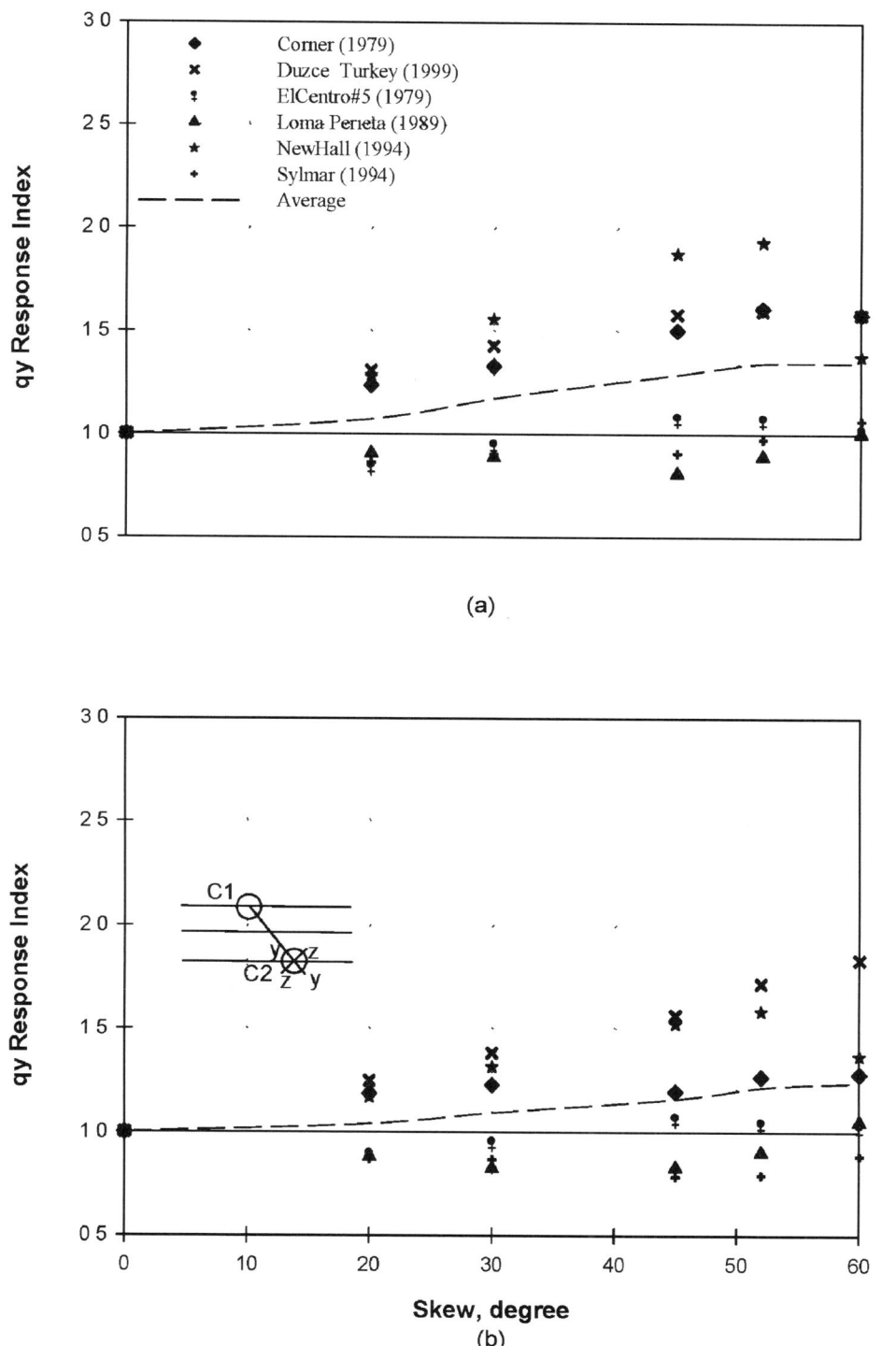

(a)

(b)

Figure 3-27. Shear Force Response Index in Y-Direction of (a) C1, and (b) C2 (0.6g)

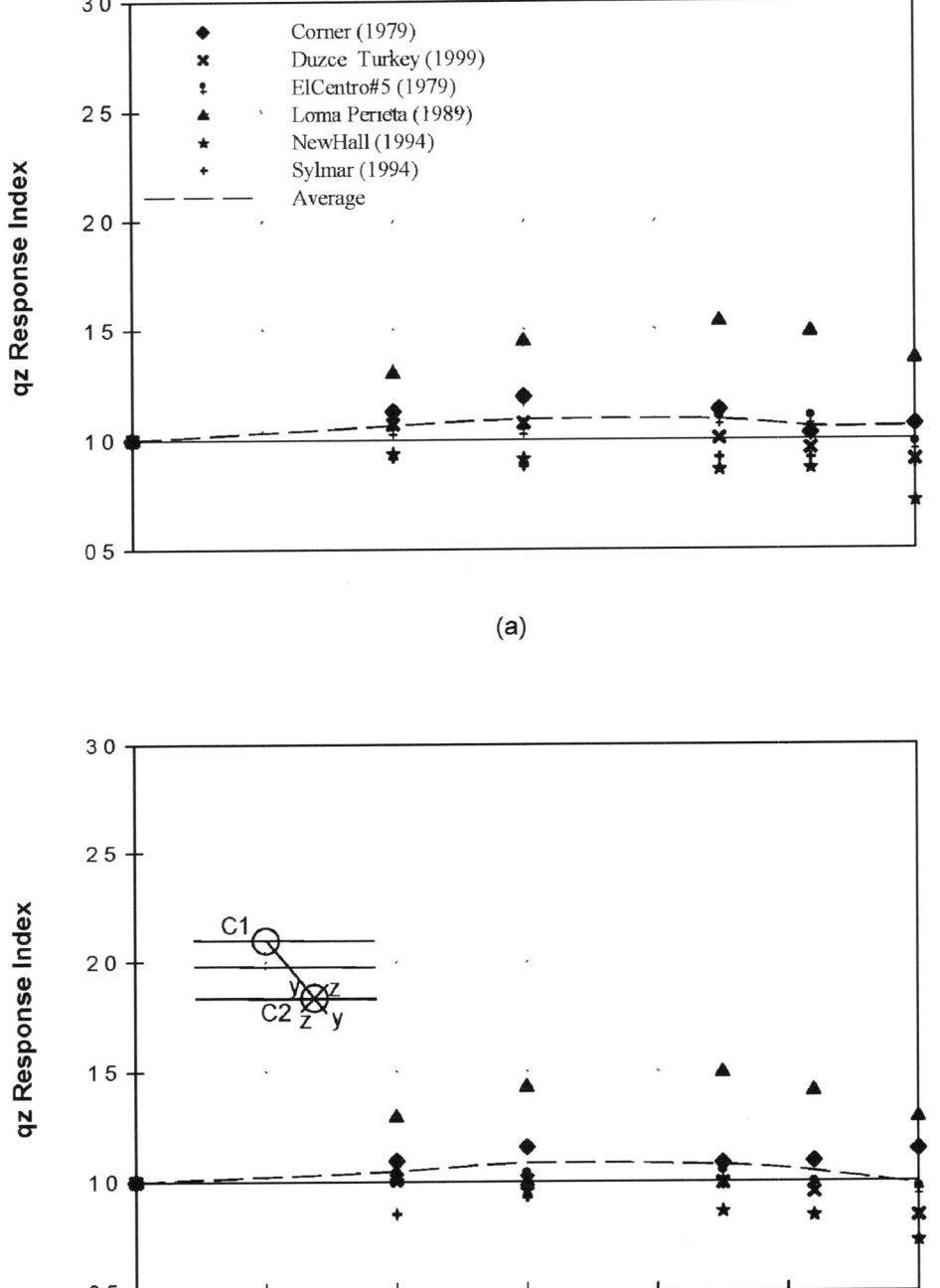

(a)

(b)

Figure 3-28. Shear Force Response Index in Z-Direction of (a) C1, and (b) C2 (0.6g)

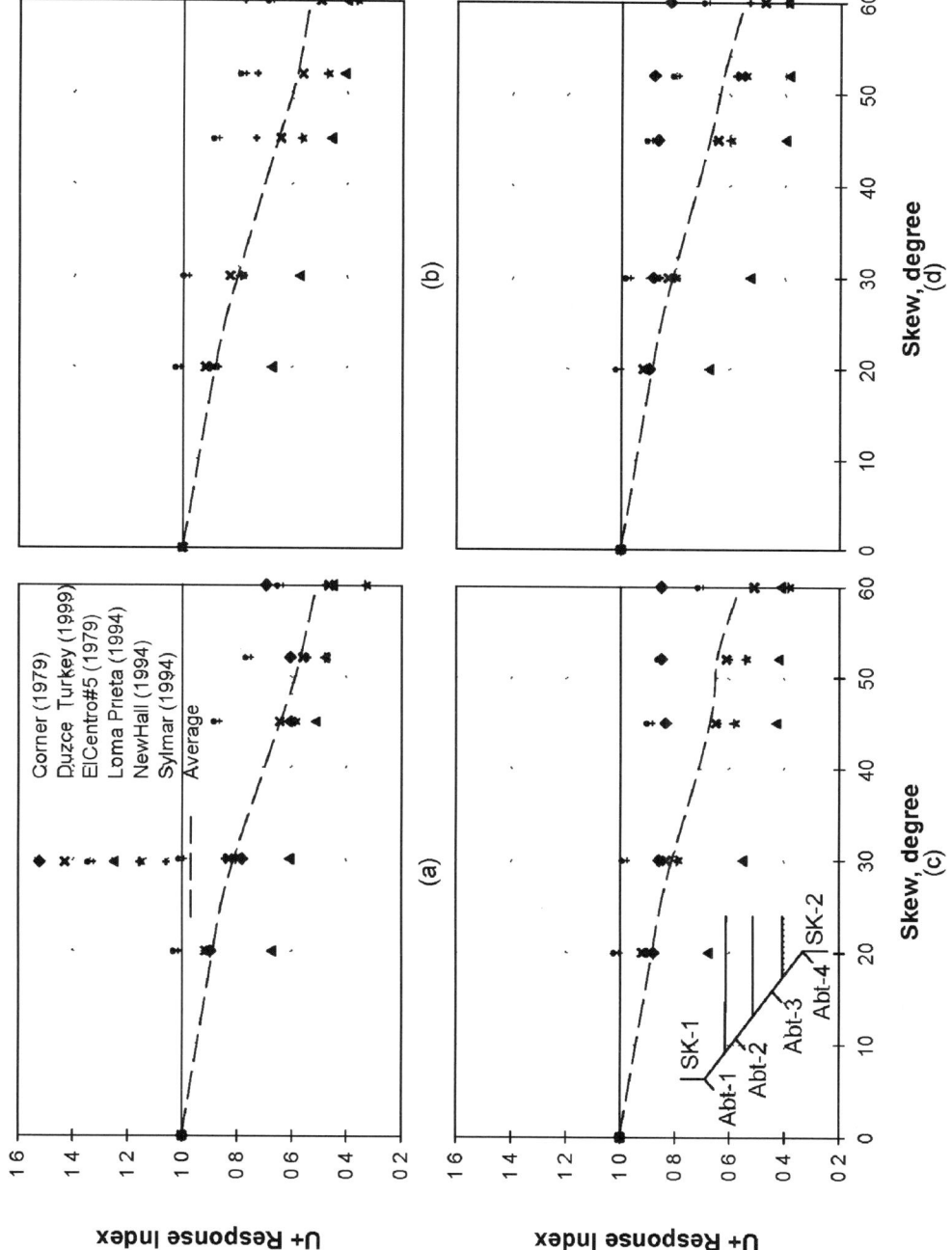

Figure 3-29. Displacement Response Index "into" springs (a) Abt-1, (b) Abt-2, (c) Abt-3, and (d) Abt-4 (0.6g)

173

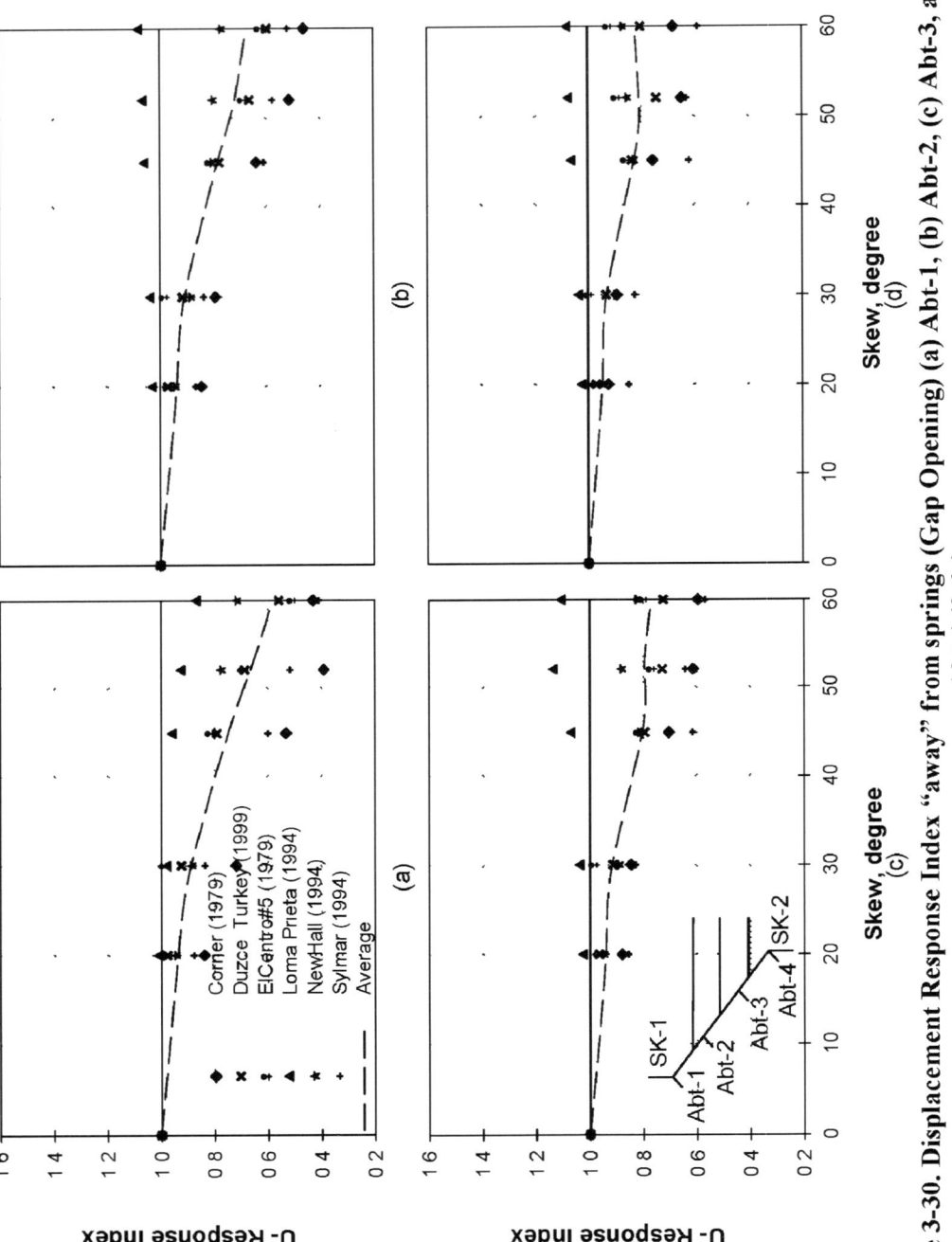

Figure 3-30. Displacement Response Index "away" from springs (Gap Opening) (a) Abt-1, (b) Abt-2, (c) Abt-3, and (d) Abt-4 (0.6g)

174

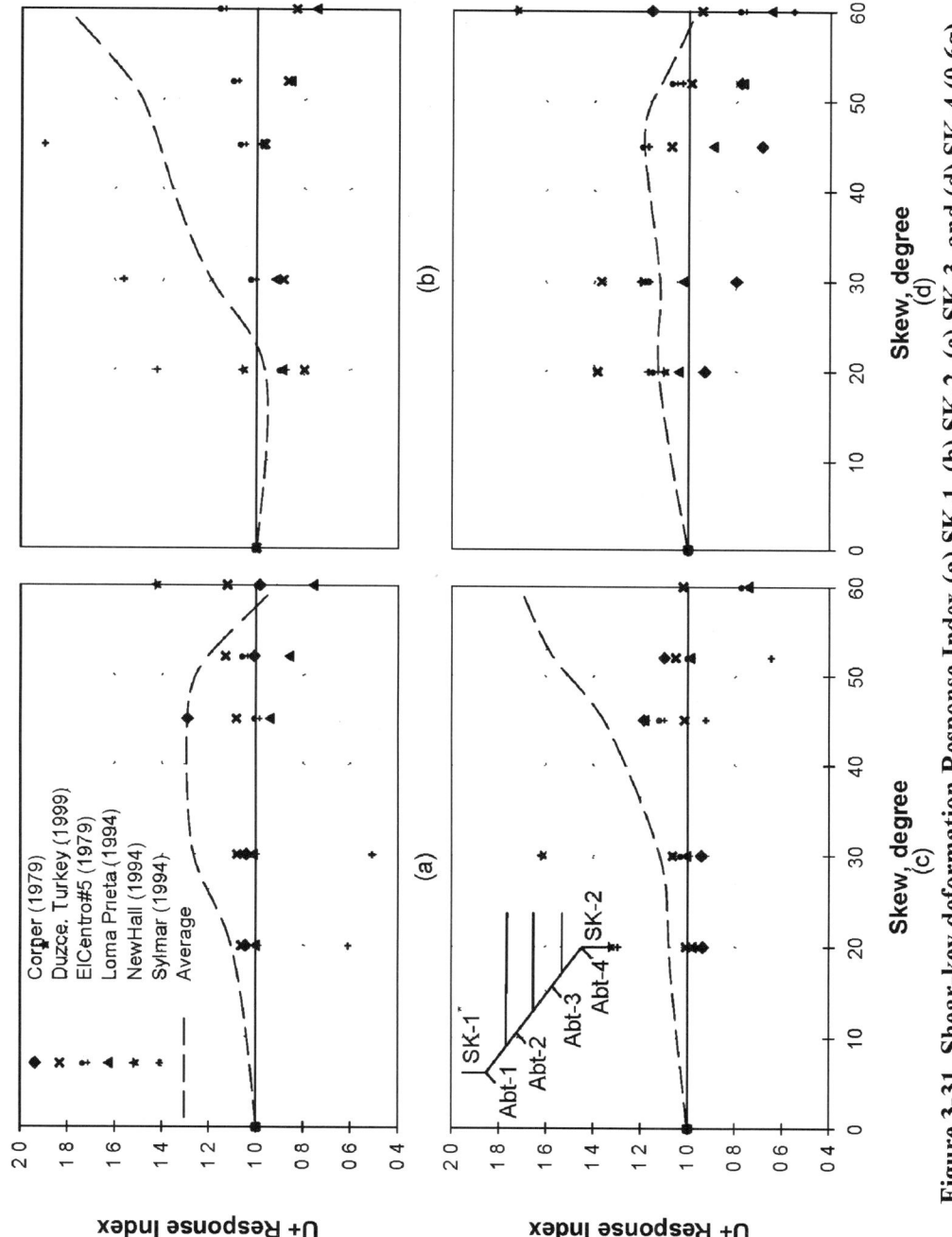

Figure 3-31. Shear-key deformation Response Index (a) SK-1, (b) SK-2, (c) SK-3, and (d) SK-4 (0.6g)

*SK-3 and SK-4 are at the other abutment

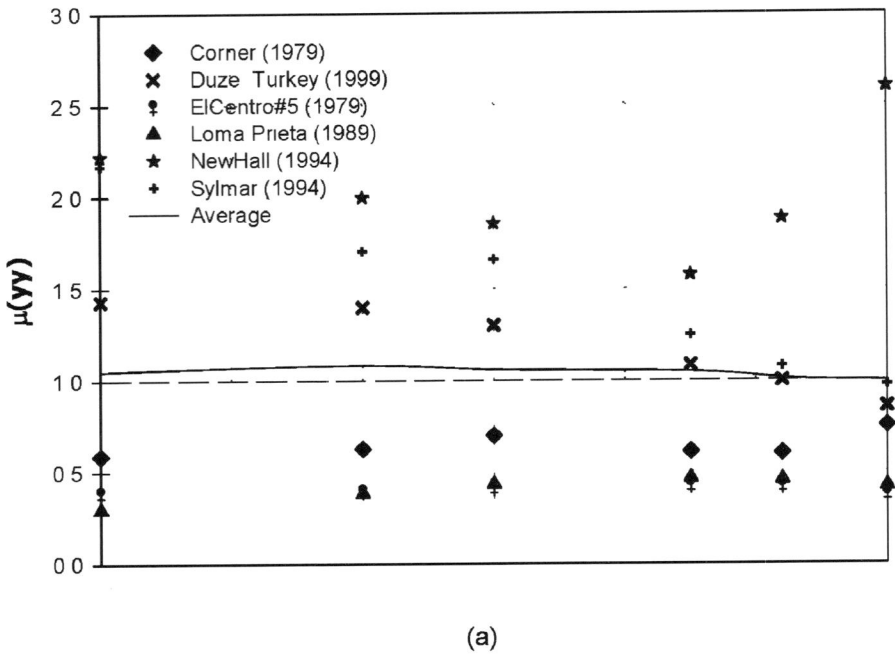

(a)

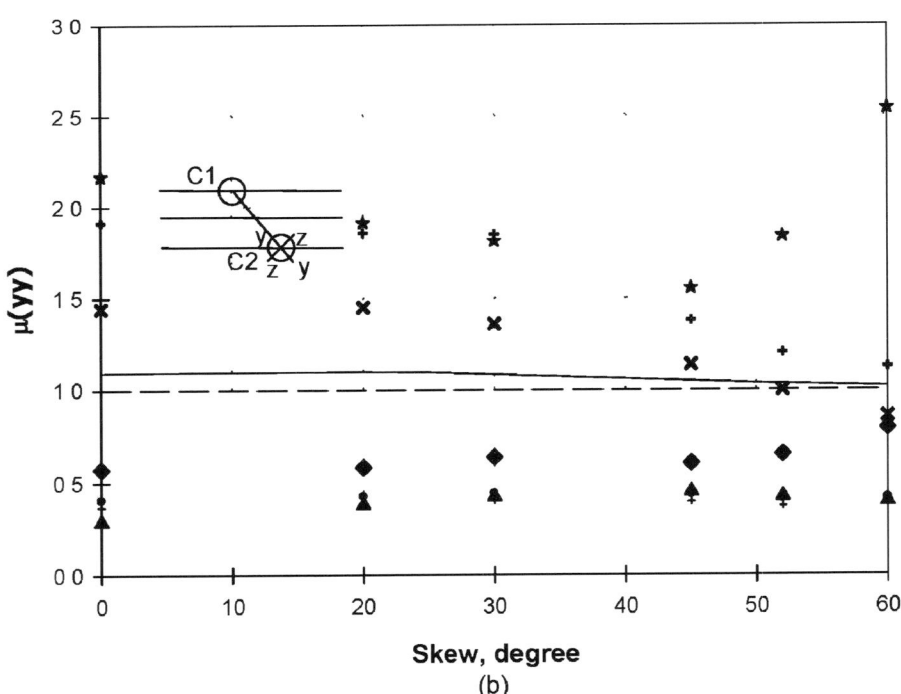

(b)

Figure 3-32. Curvature Ductility in Y-Direction of (a) C1 and (b) C2 (0.6g)

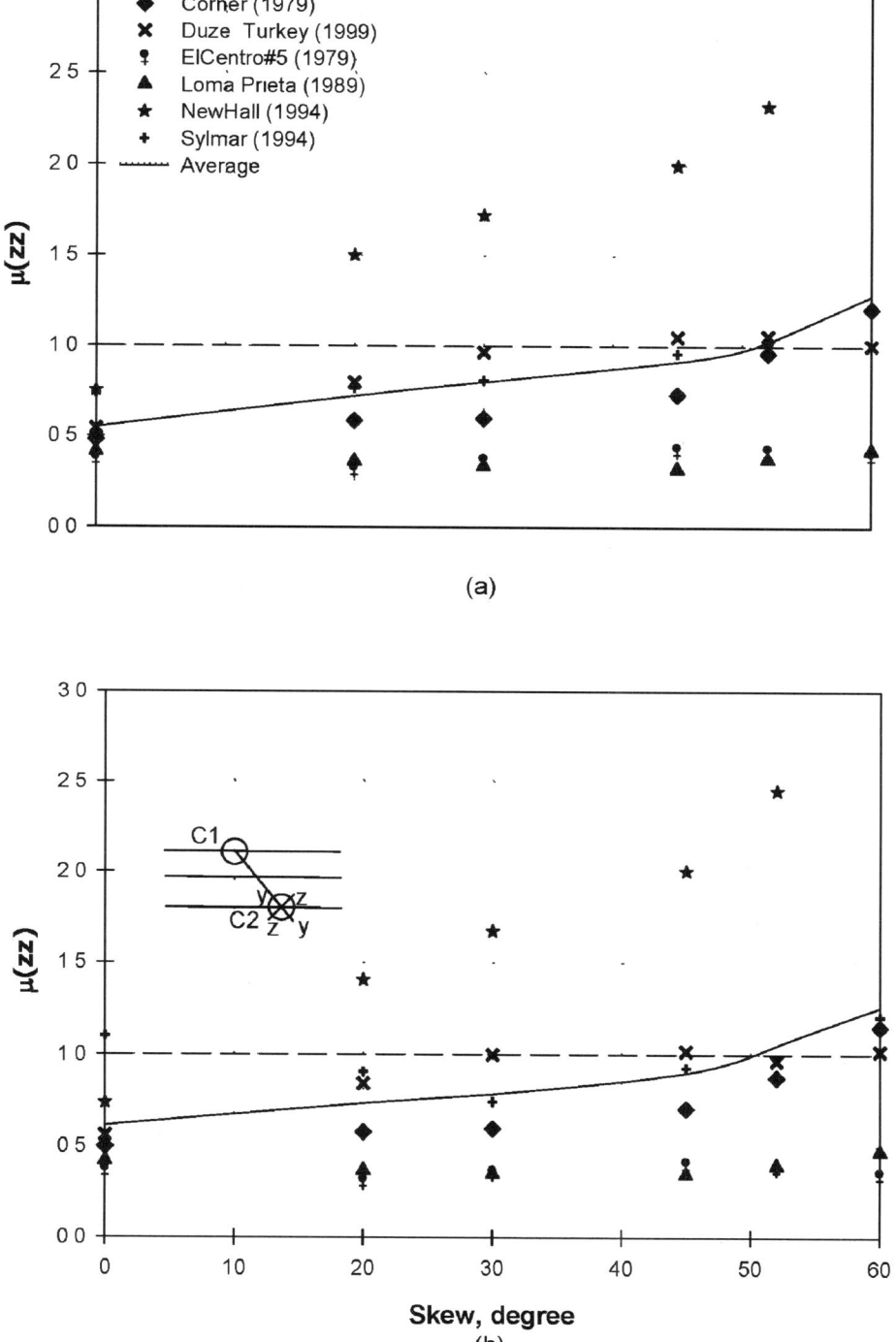

(a)

(b)

Figure 3-33. Curvature Ductility in Z-Direction of (a) C1 and (b) C2 (0.6g)

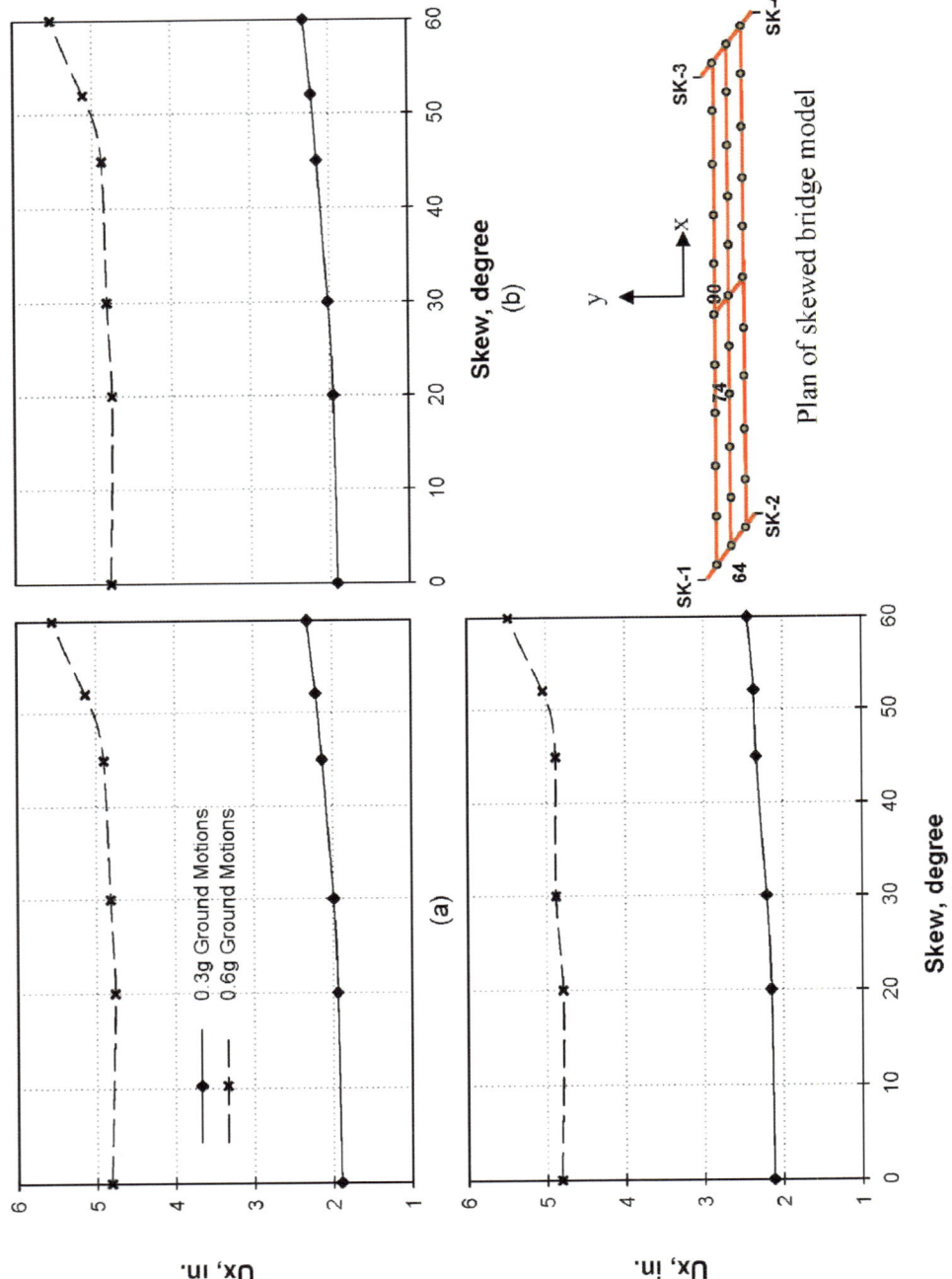

Figure 3-34. Average Displacement in X-Direction for nodes (a) 90, (b) 74, and (c) 64 (0.3g vs 0.6g)

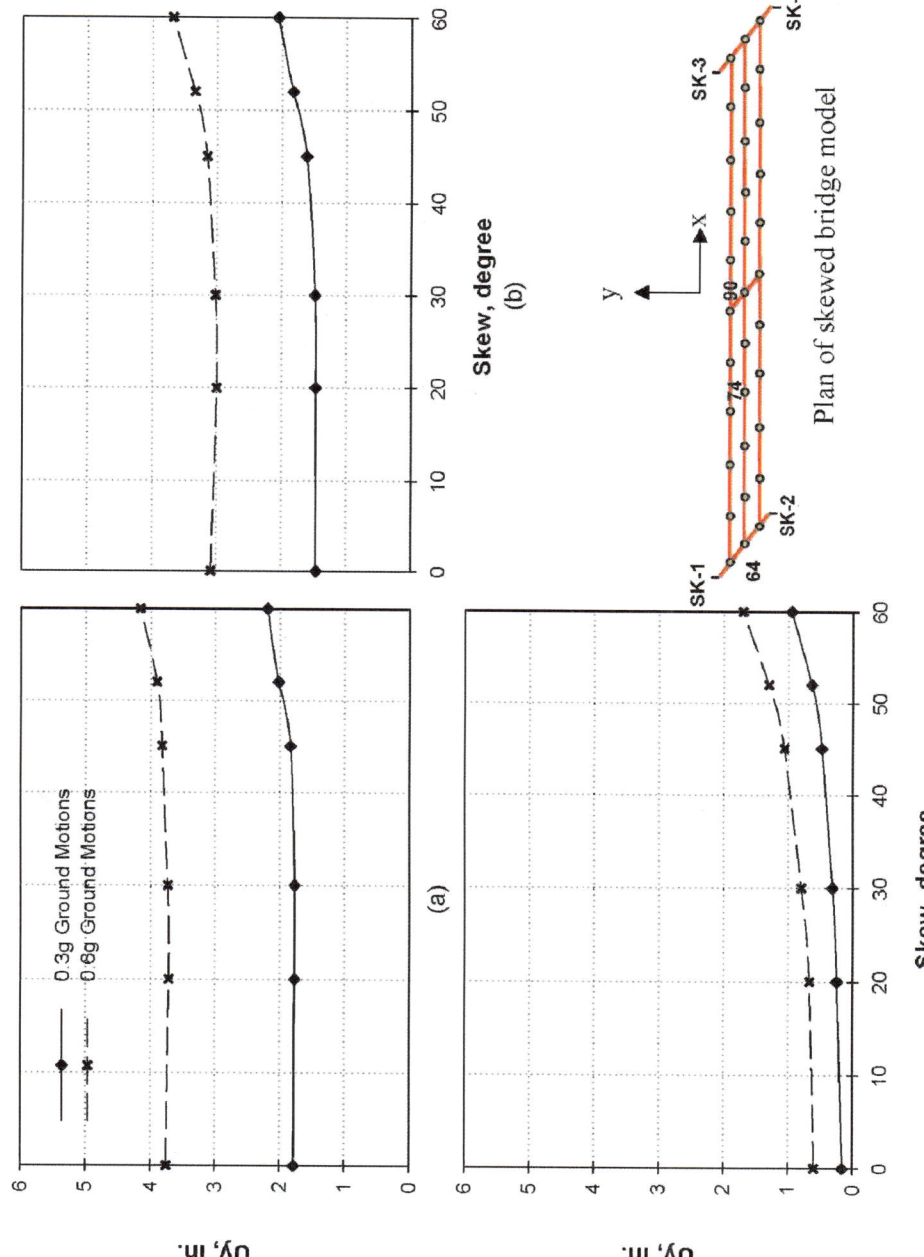

Figure 3-35. Average Displacement in Y-Direction for nodes (a) 90, (b) 74, and (c) 64 (0.3g vs 0.6g)

180

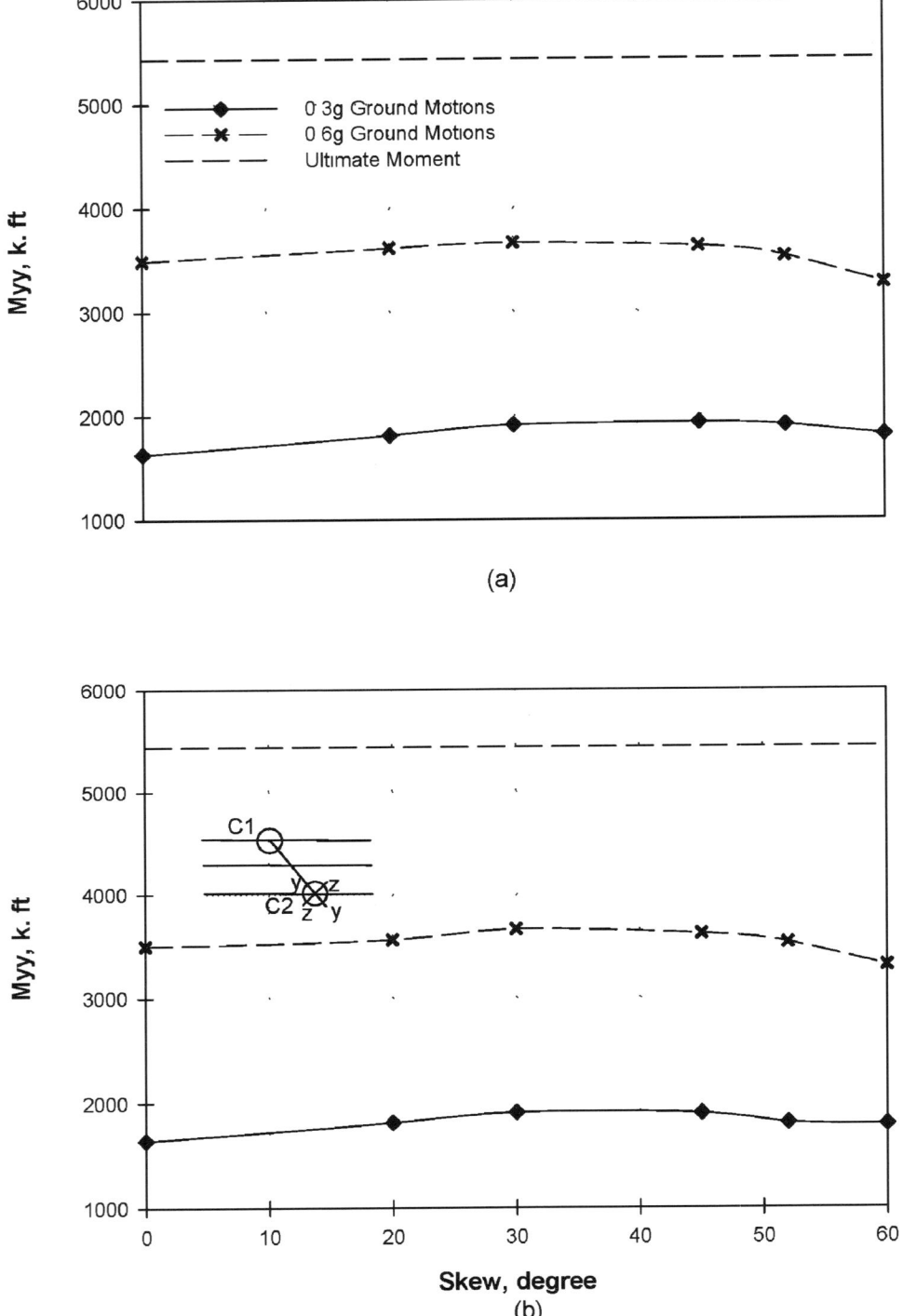

(a)

(b)

Figure 3-36. Average Moment in Y-Direction of (a) C1, and (b) C2 (0.3g vs 0.6g)

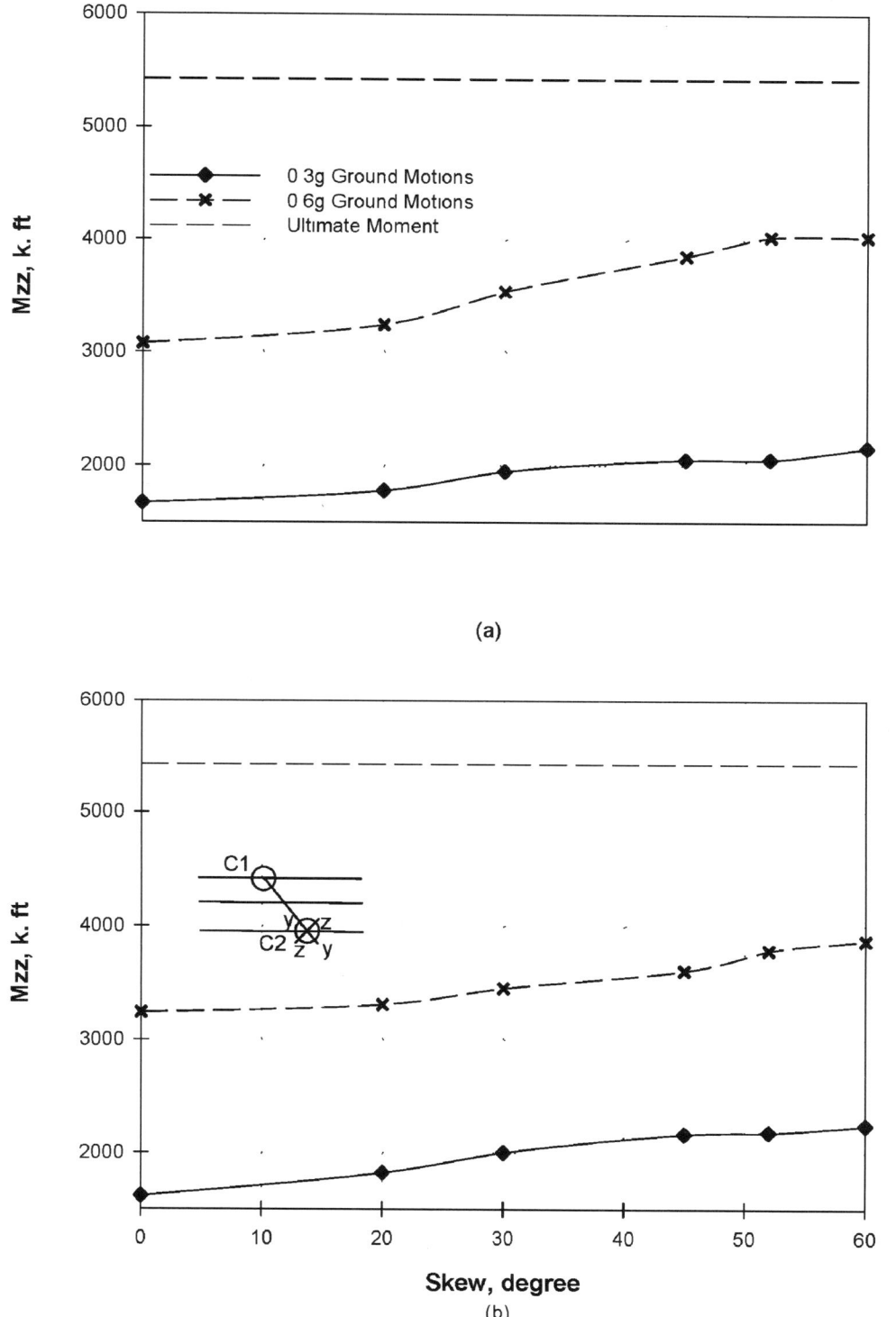

(a)

(b)

Figure 3-37. Average Moment in Z-Direction of (a) C1, and (b) C2 (0.3g vs 0.6g)

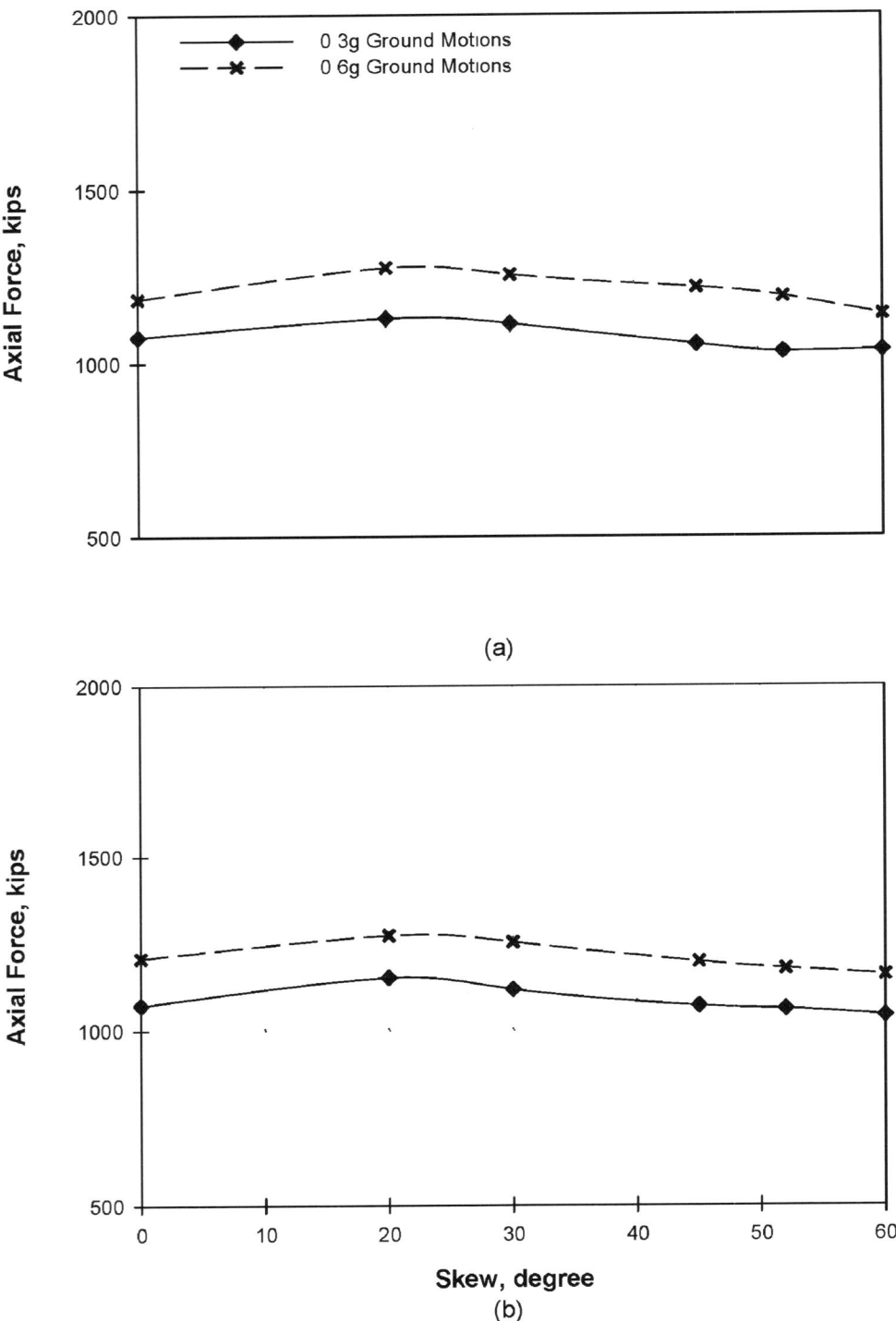

(a)

(b)

Figure 3-38. Average Axial Force in (a) C1, and (b) C2 (0.3g vs 0.6g)

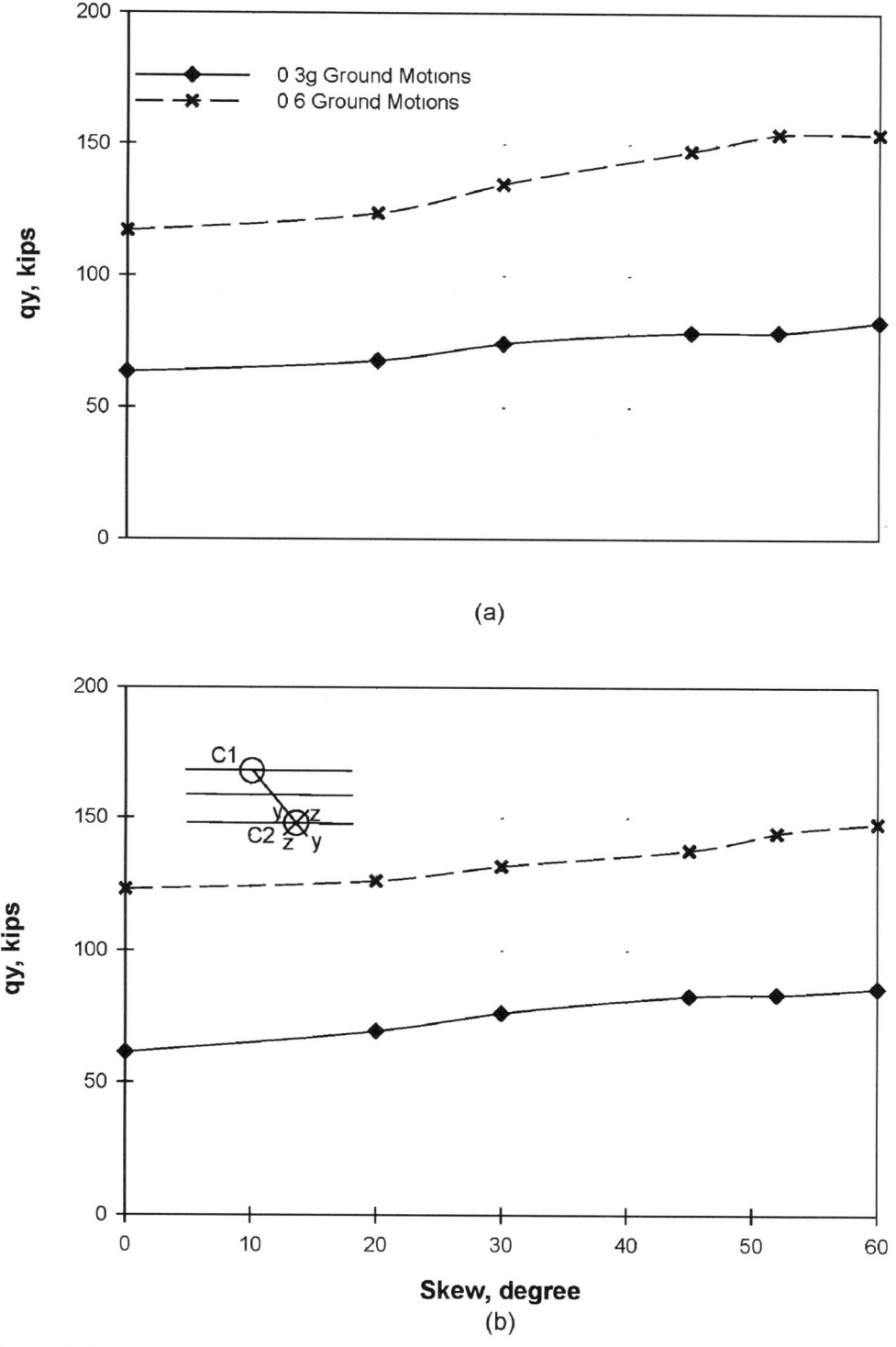

(a)

(b)

Figure 3-39. Average Shear Force in Y-Direction of (a) C1, and (b) C2 (0.3g vs 0.6g)

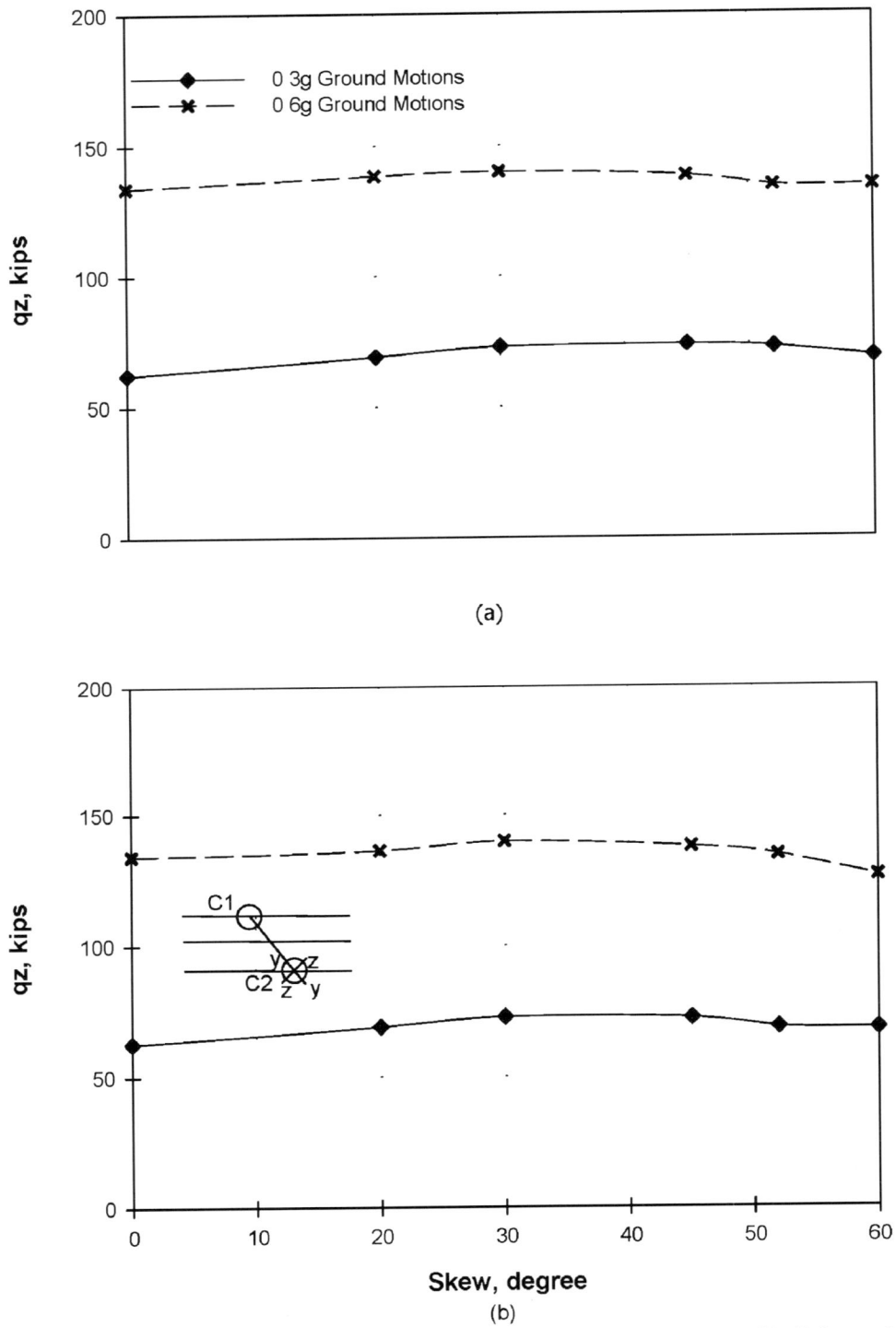

(a)

(b)

Figure 3-40. Average Shear Force in Z-Direction of (a) C1, and (b) C2 (0.3g vs 0.6g)

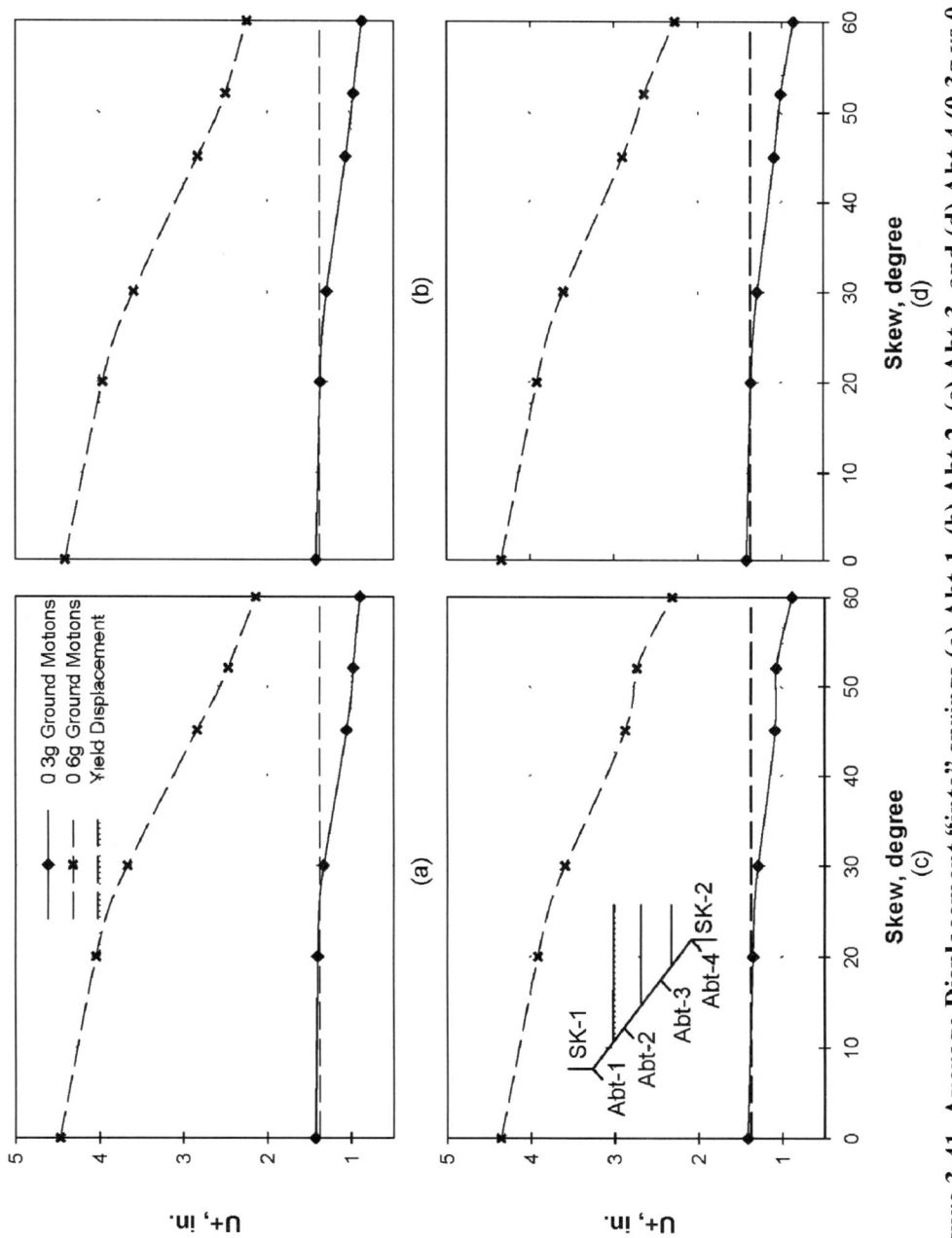

Figure 3-41. Average Displacement "into" springs (a) Abt-1, (b) Abt-2, (c) Abt-3, and (d) Abt-4 (0.3g vs 0.6g)

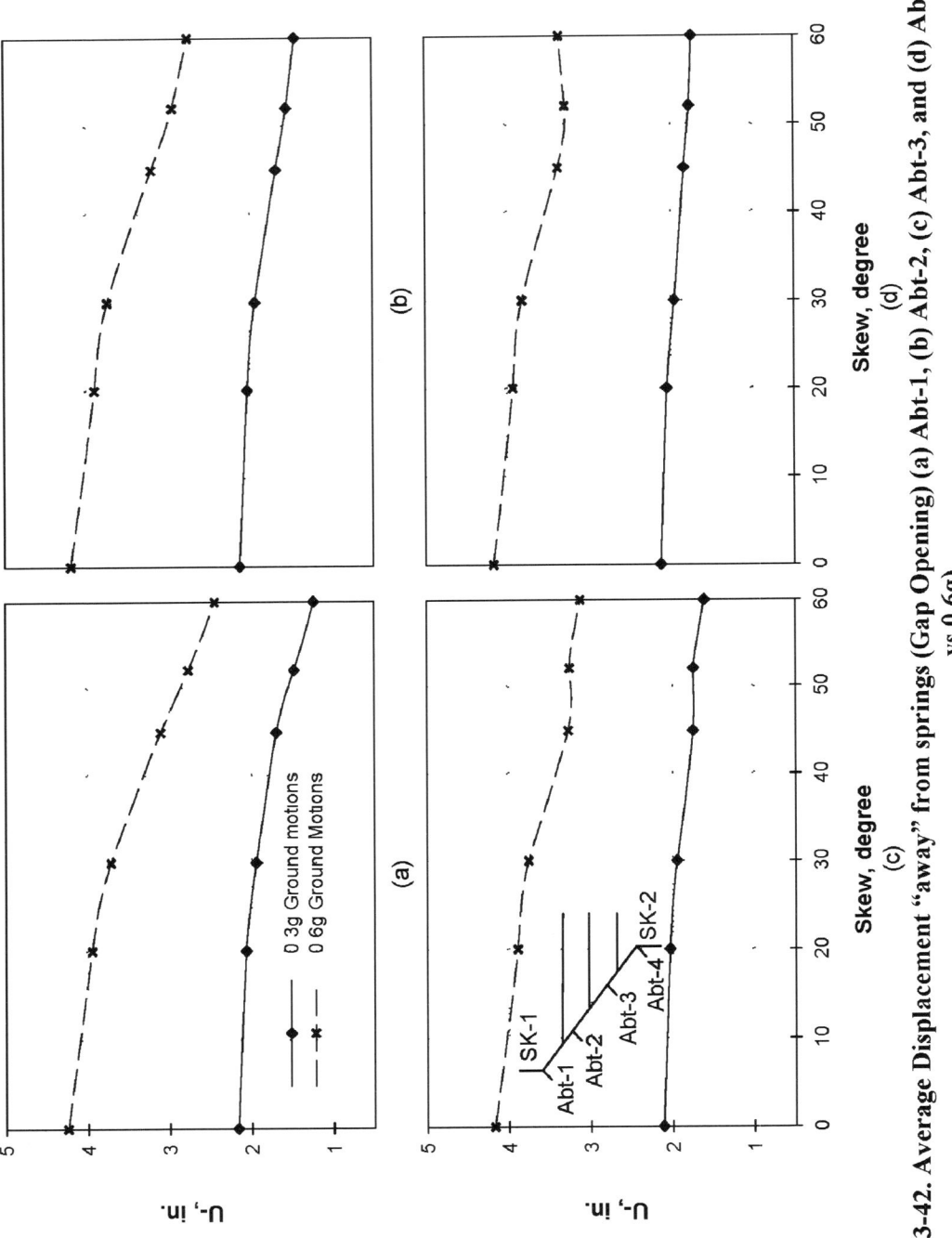

Figure 3-42. Average Displacement "away" from springs (Gap Opening) (a) Abt-1, (b) Abt-2, (c) Abt-3, and (d) Abt-4 (0.3g vs 0.6g)

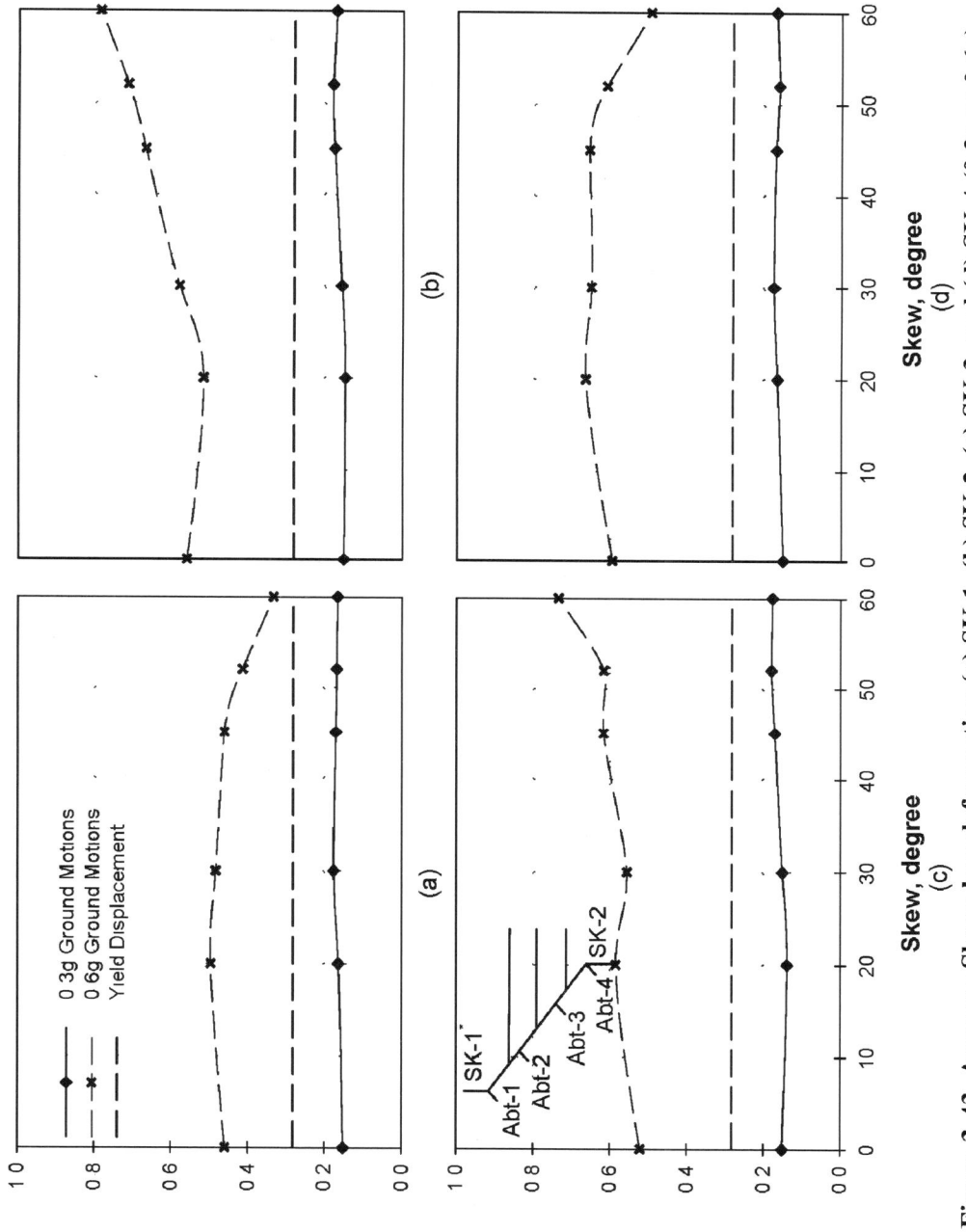

Figure 3-43. Average Shear-key deformation (a) SK-1, (b) SK-2, (c) SK-3, and (d) SK-4 (0.3g vs 0.6g)

*SK-3 and SK-4 are at the other abutment

187

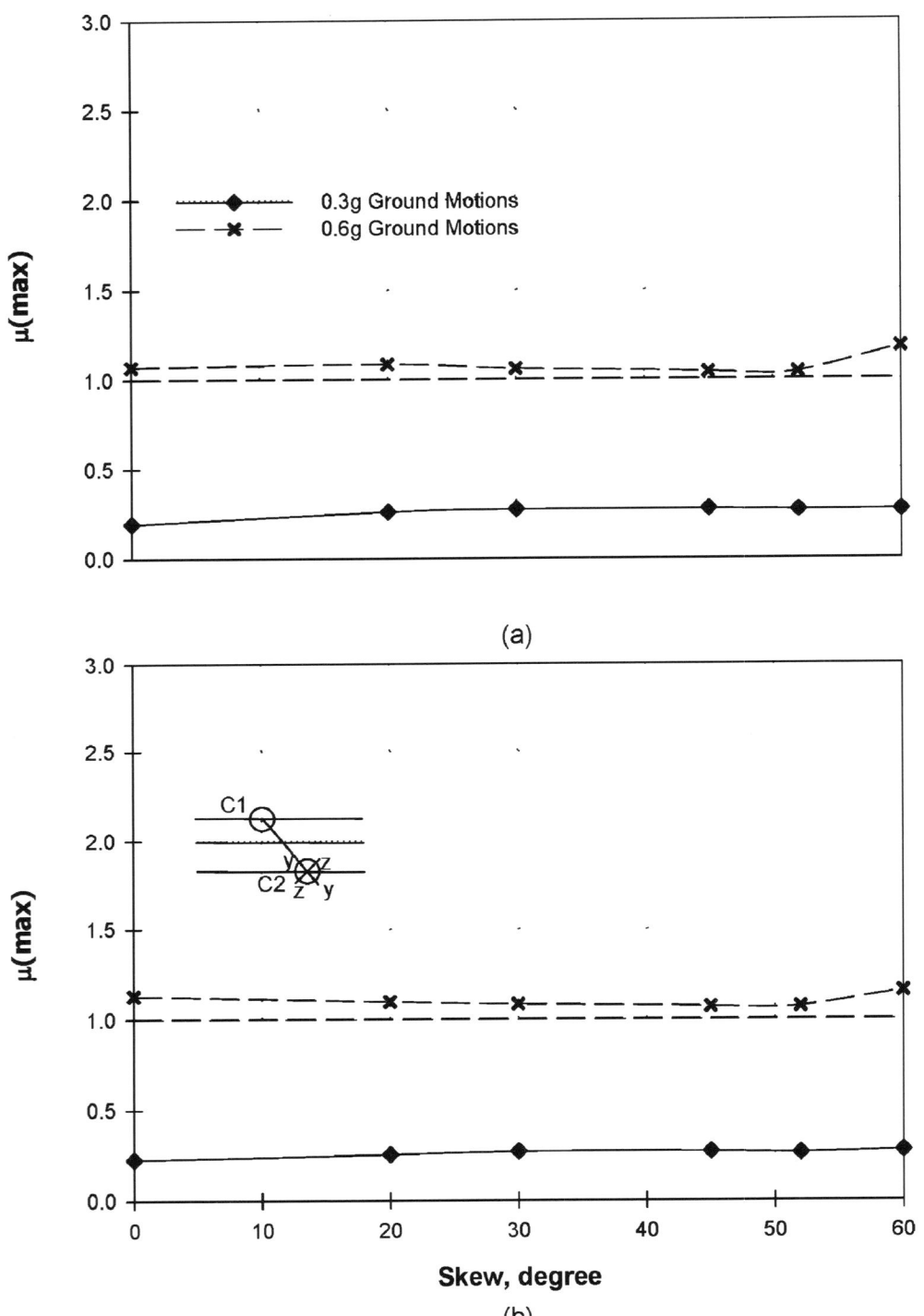

(a)

(b)

Figure 3-44. Maximum Average Curvature Ductility of (a) C1 and (b) C2 (0.3g vs 0.6g)

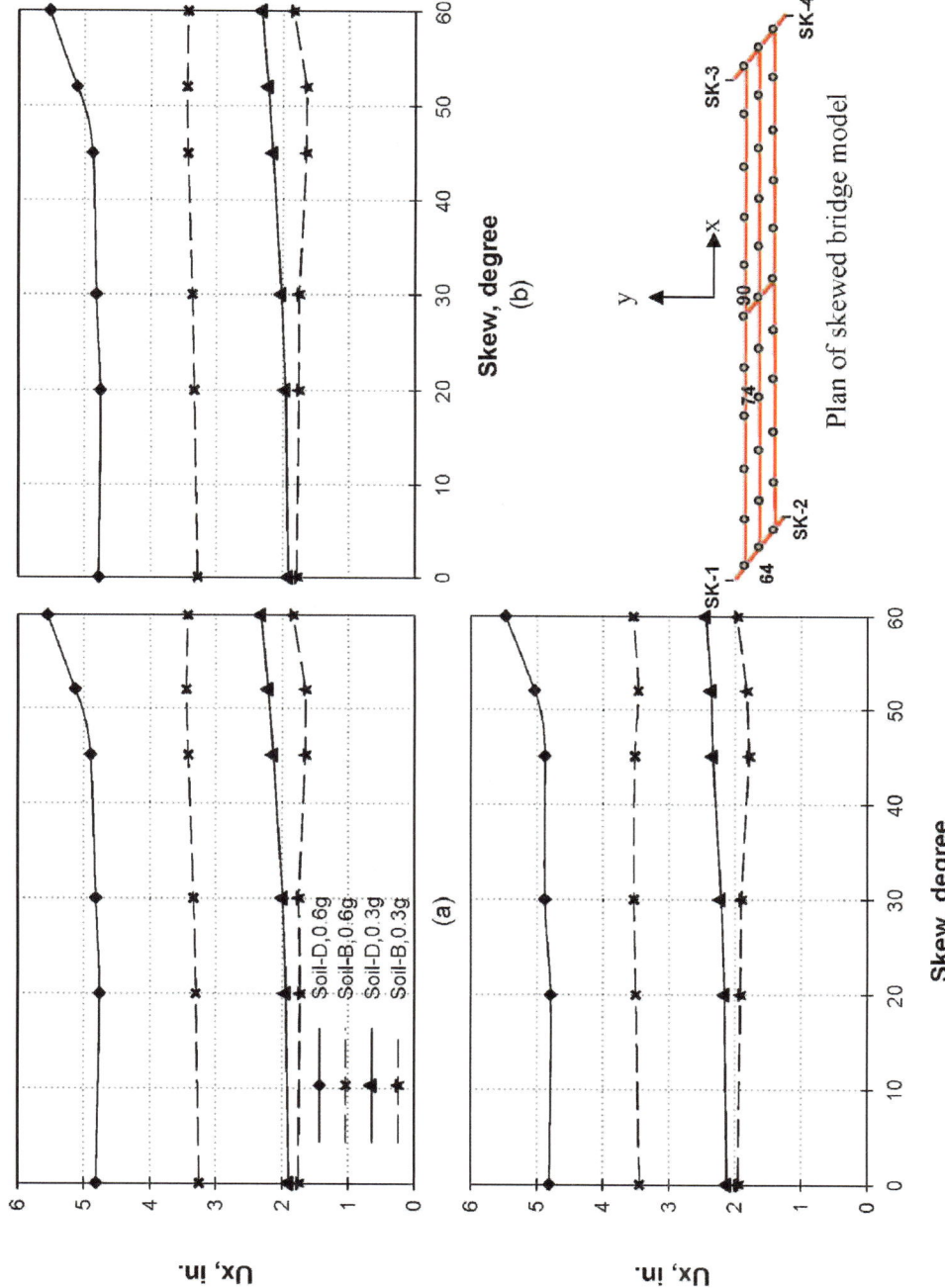

Figure 3-45. Average Displacement in X-direction for nodes (a) 90, (b) 74, and (c) 64

189

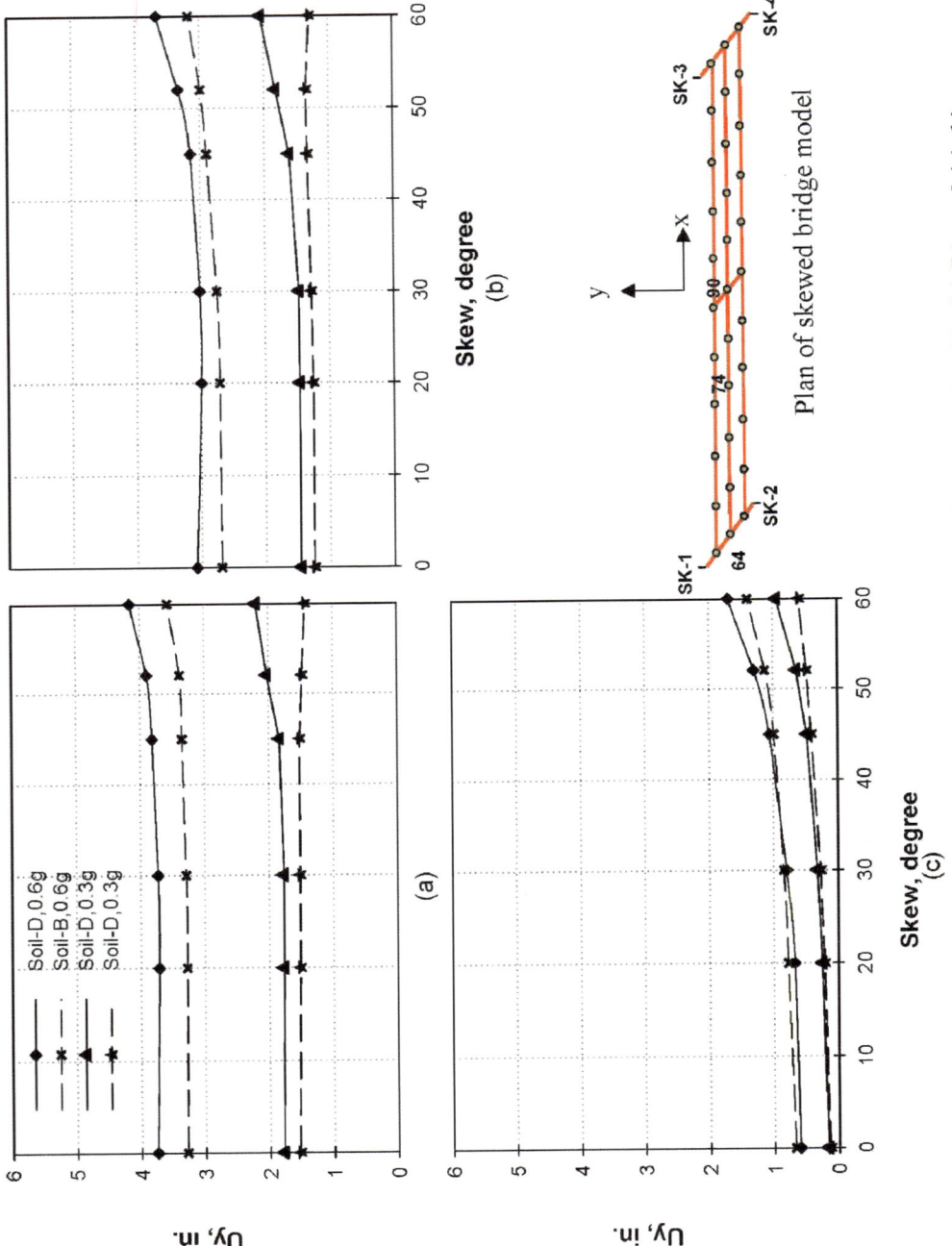

Figure 3-46. Average Displacement in Y-direction for nodes (a) 90, (b) 74, and (c) 64

191

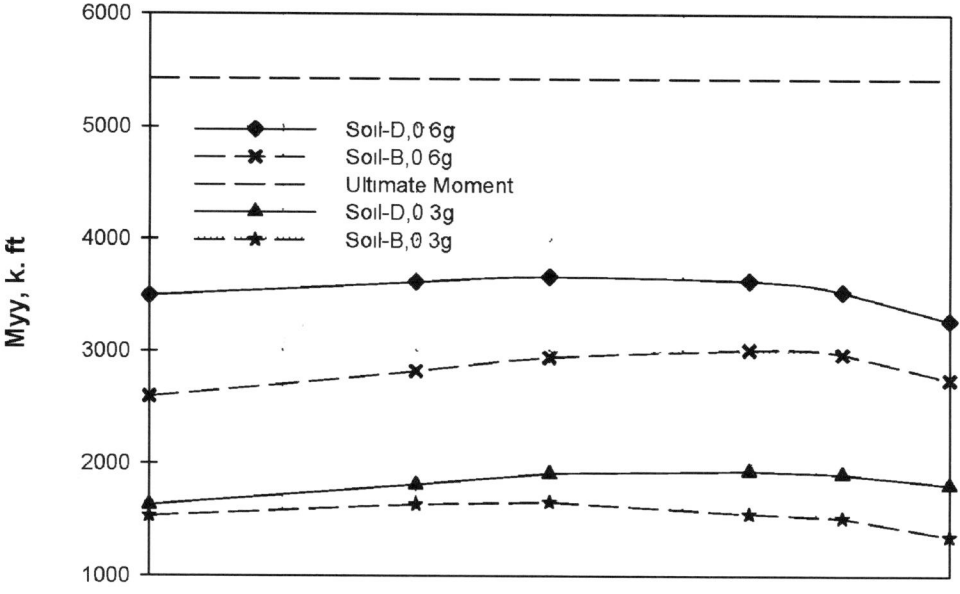

(a)

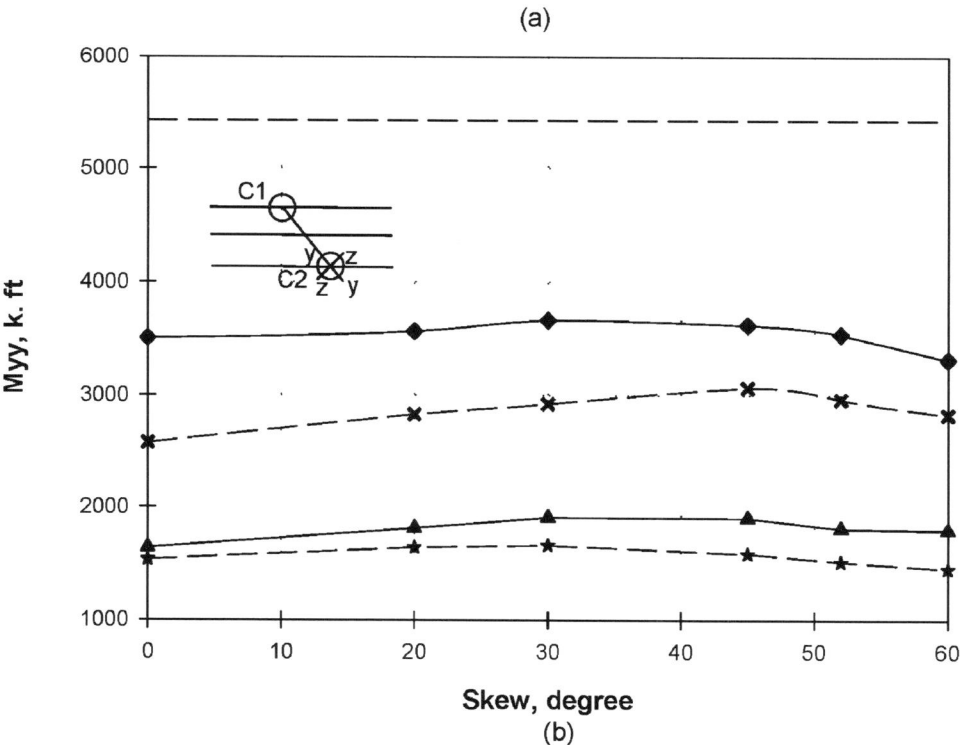

Skew, degree
(b)

Figure 3-47. Average Moment in Y-Directions of (a) C1, and (b) C2

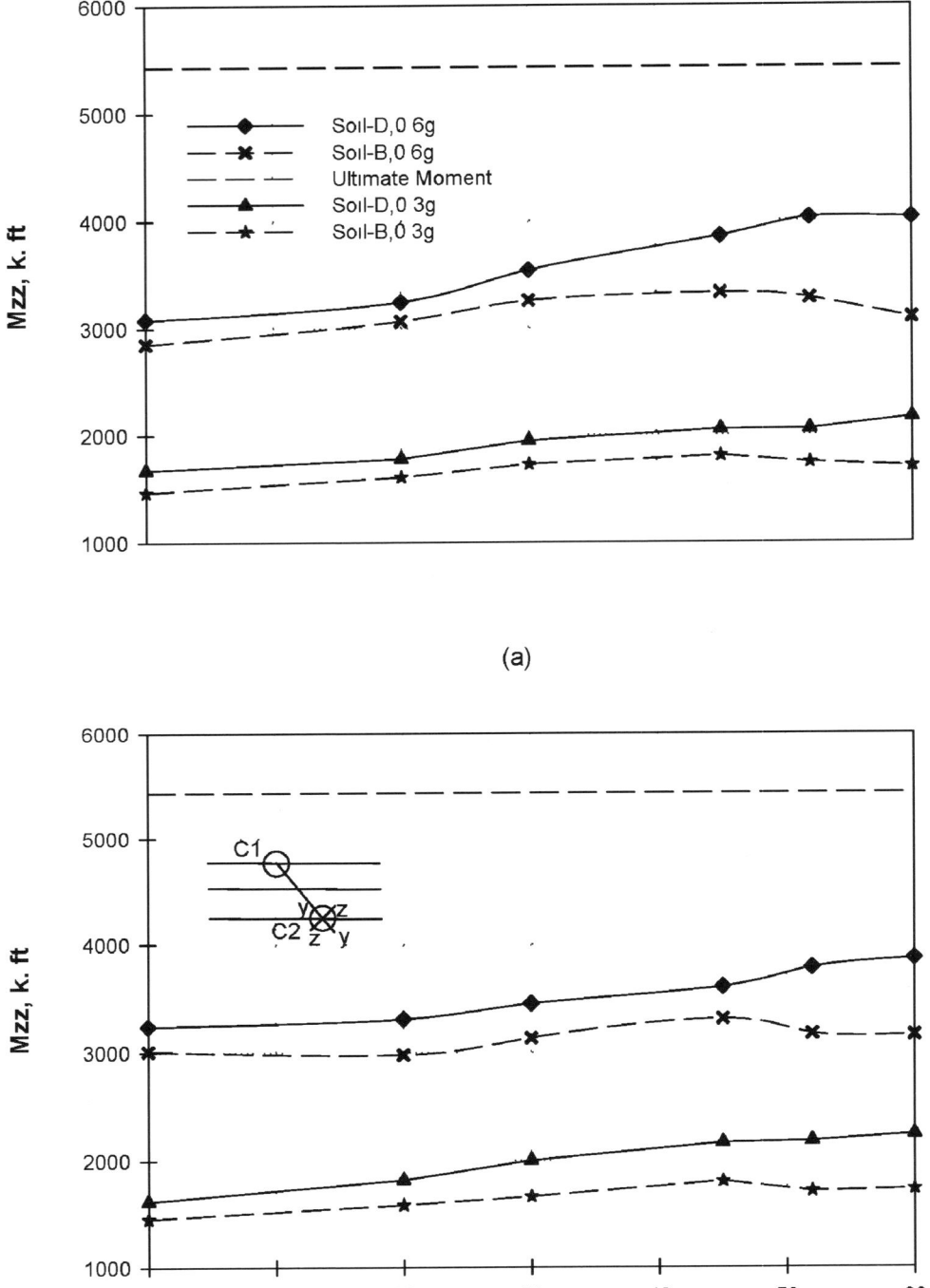

(a)

(b)

Figure 3-48. Average Moment in Z-Directions of (a) C1, and (b) C2

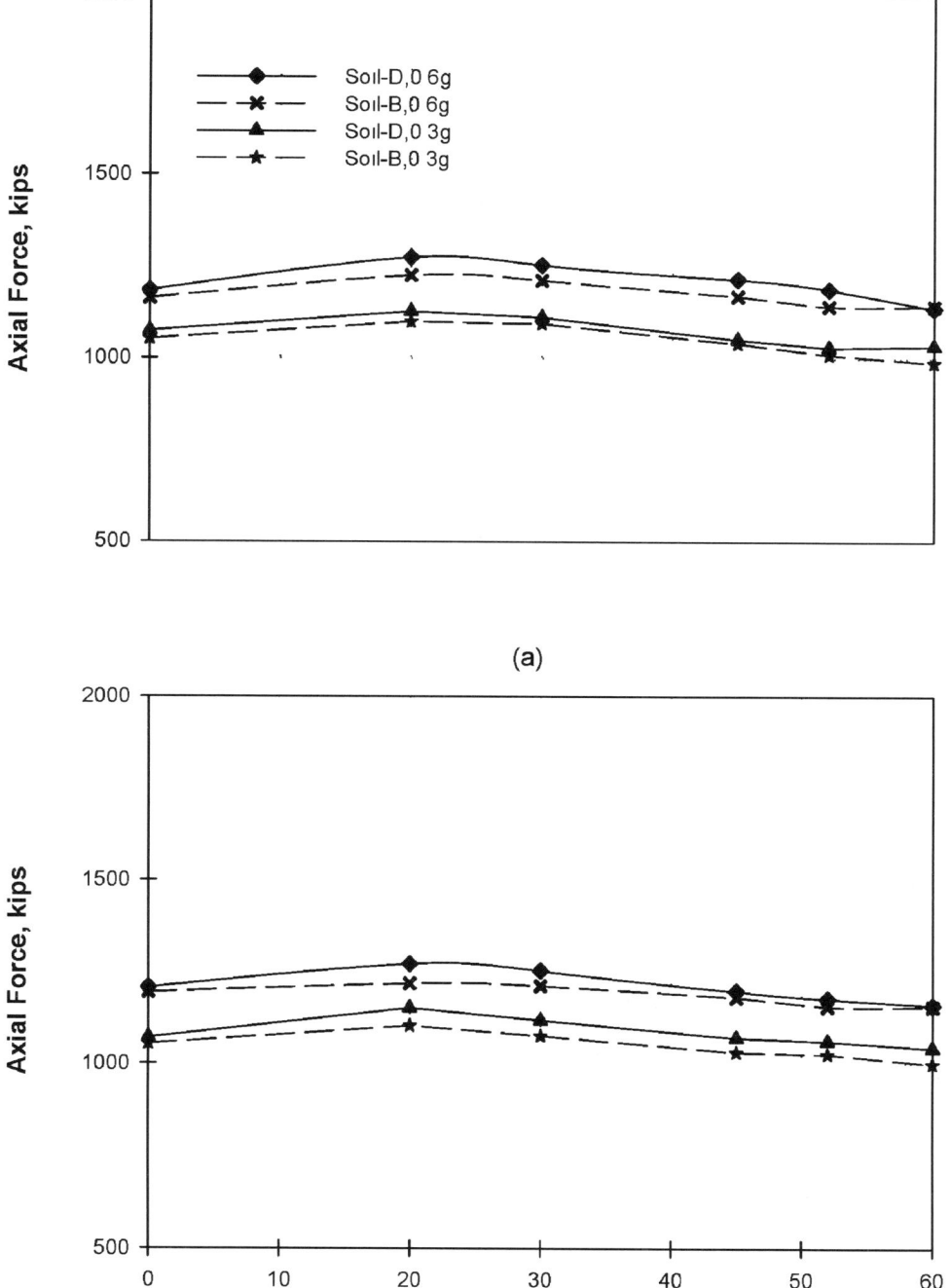

(a)

(b)

Figure 3-49. Average Axial Force in (a) C1, and (b) C2

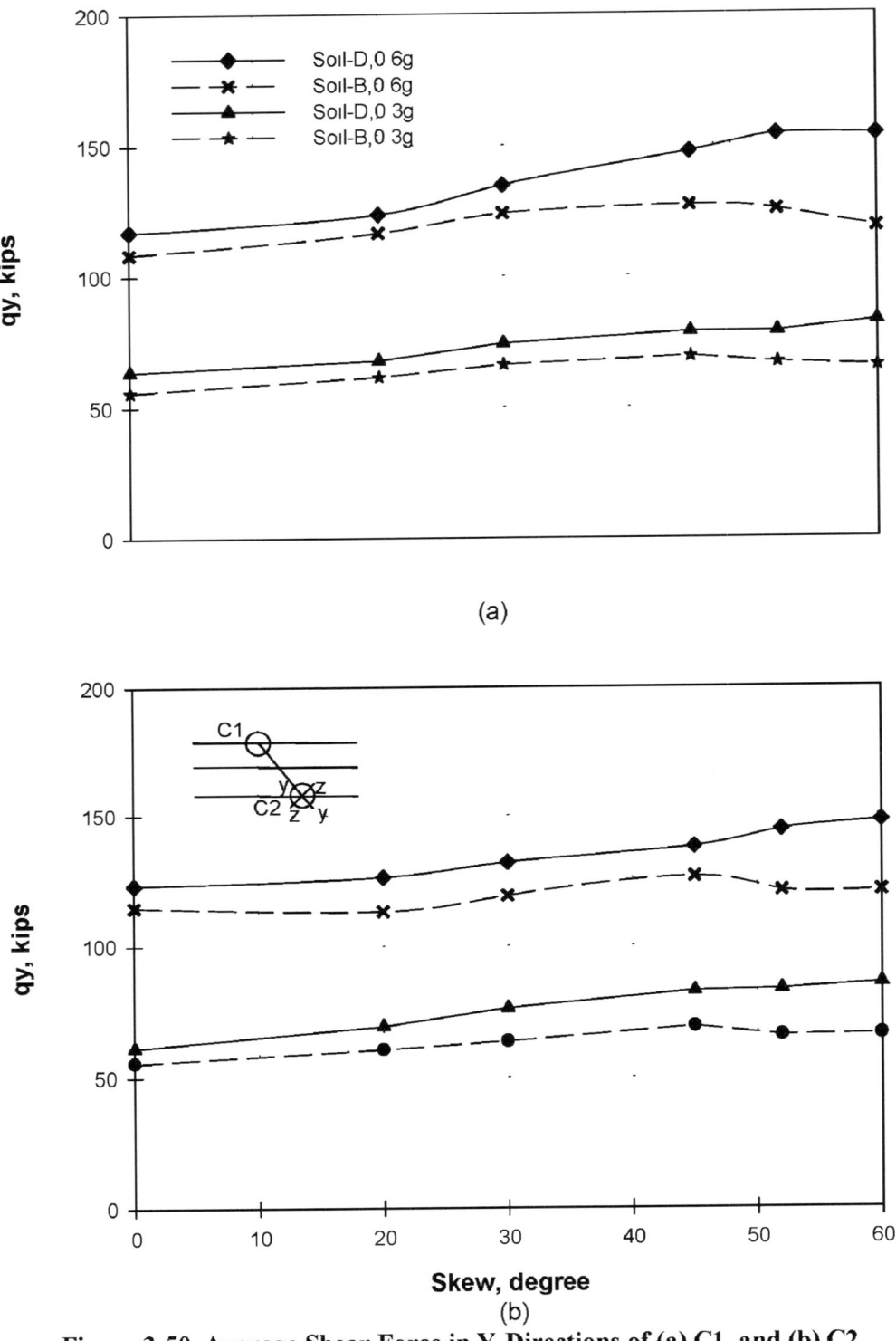

(a)

(b)

Skew, degree

Figure 3-50. Average Shear Force in Y-Directions of (a) C1, and (b) C2

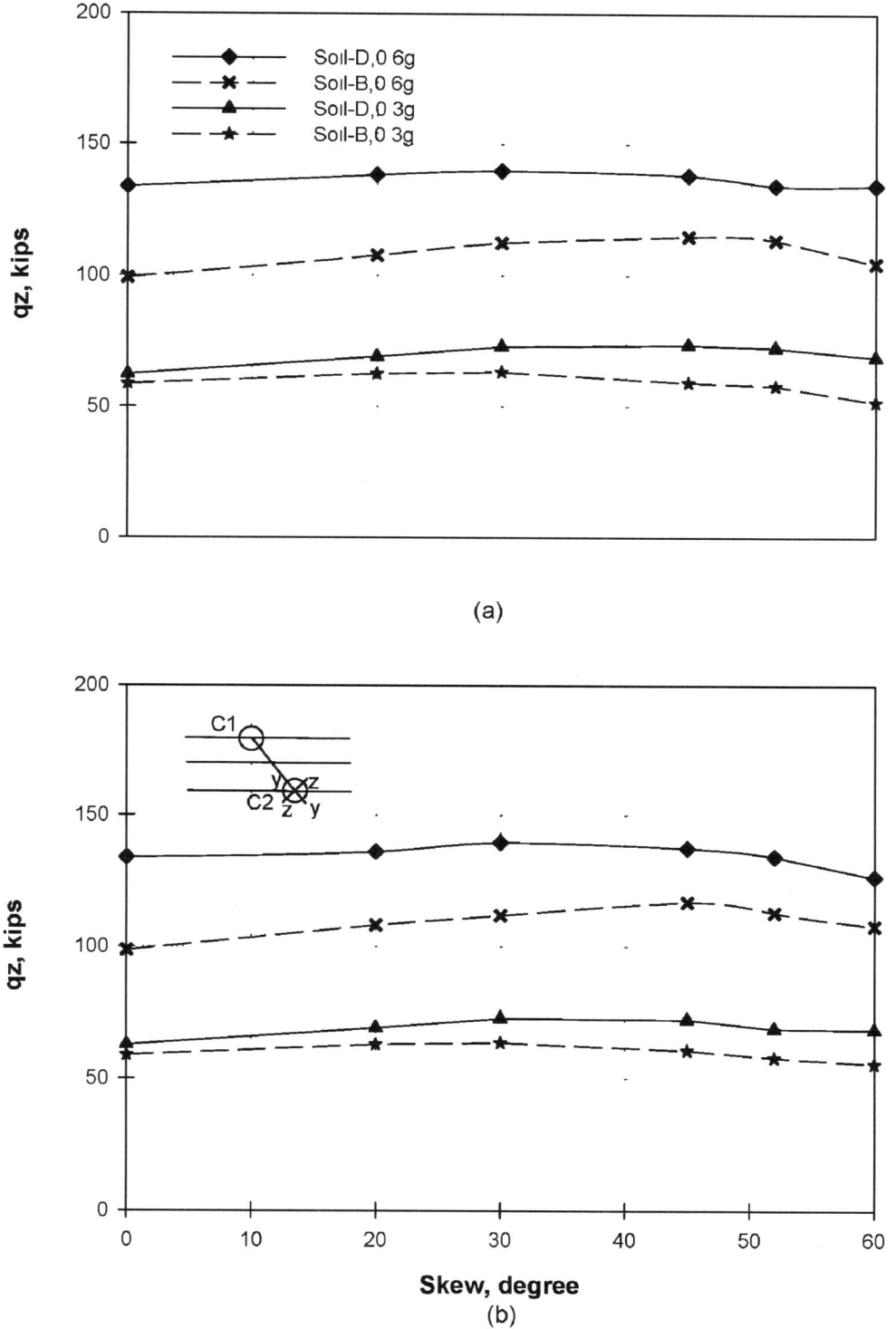

(a)

(b)

Figure 3-51. Average Shear Force in Z-Directions of (a) C1, and (b) C2

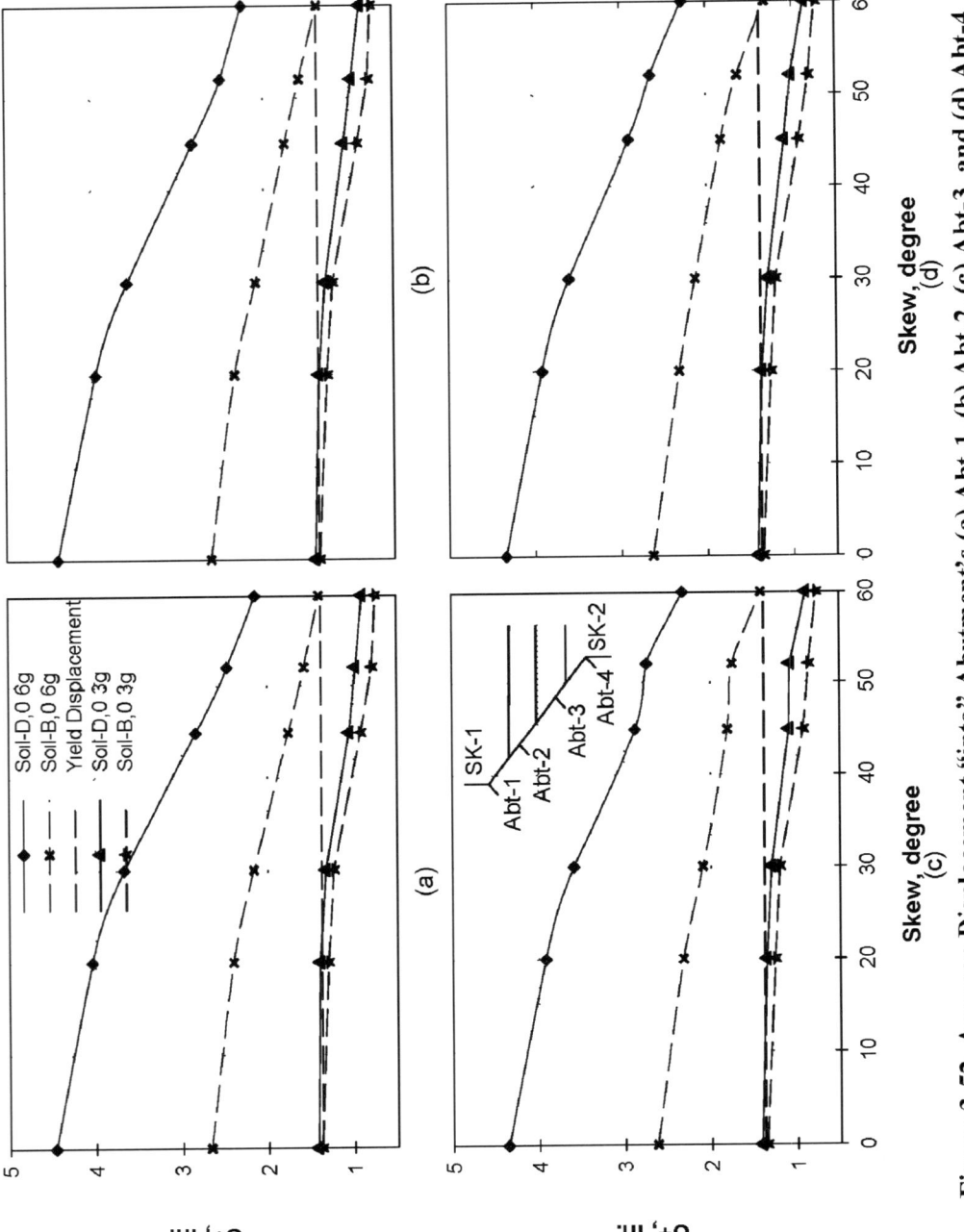

Figure 3-52. Average Displacement "into" Abutment's (a) Abt-1, (b) Abt-2, (c) Abt-3, and (d) Abt-4

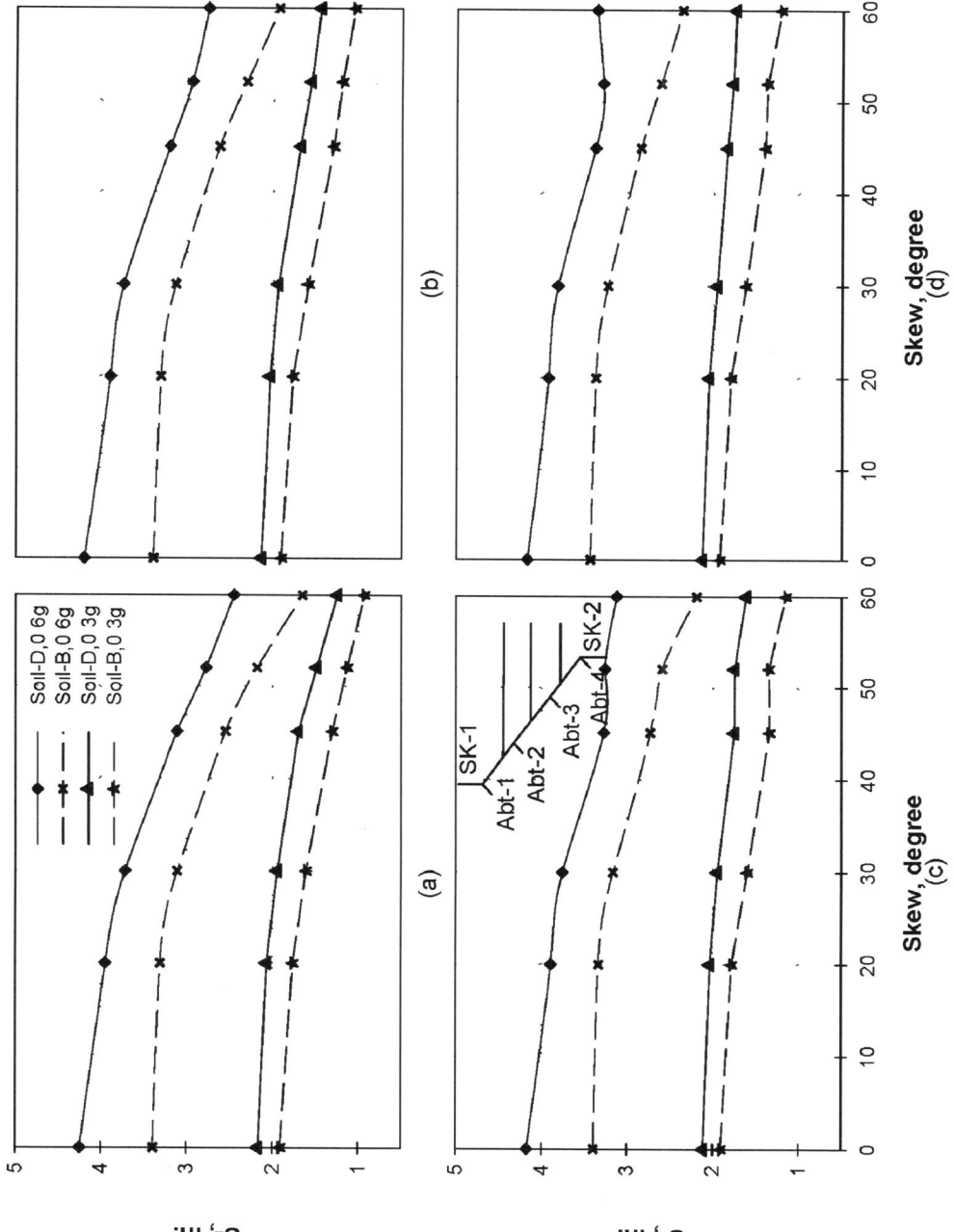

Figure 3-53. Average Displacement "away" from Abutment's (a) Abt-1, (b) Abt-2, (c) Abt-3, and (d) Abt-4

197

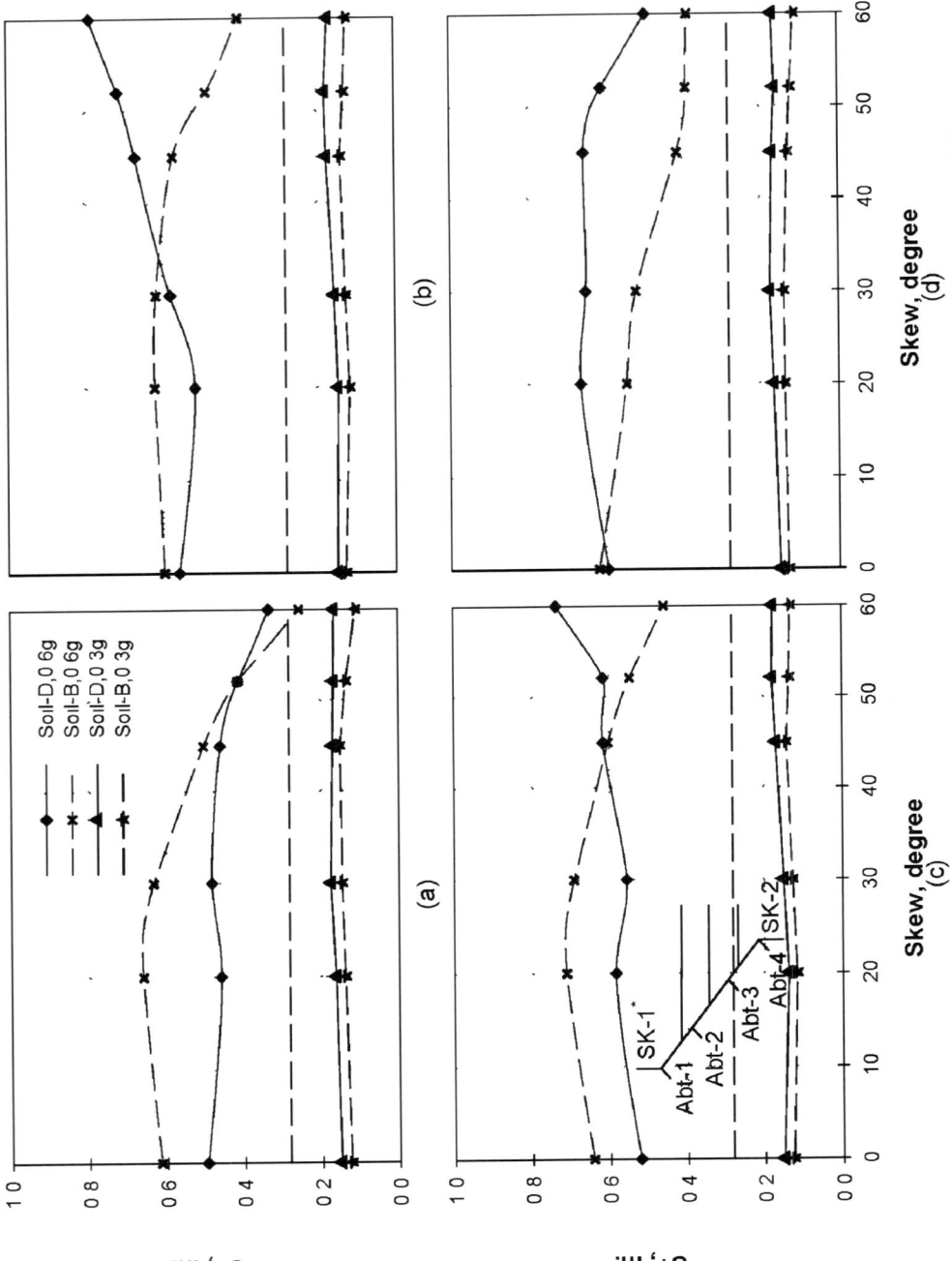

Figure 3-54. Average Shear-key deformation (a) SK-1, (b) SK-2, (c) SK-3, and (d) SK-4

*SK-3 and SK-4 are at the other abutment

198

199

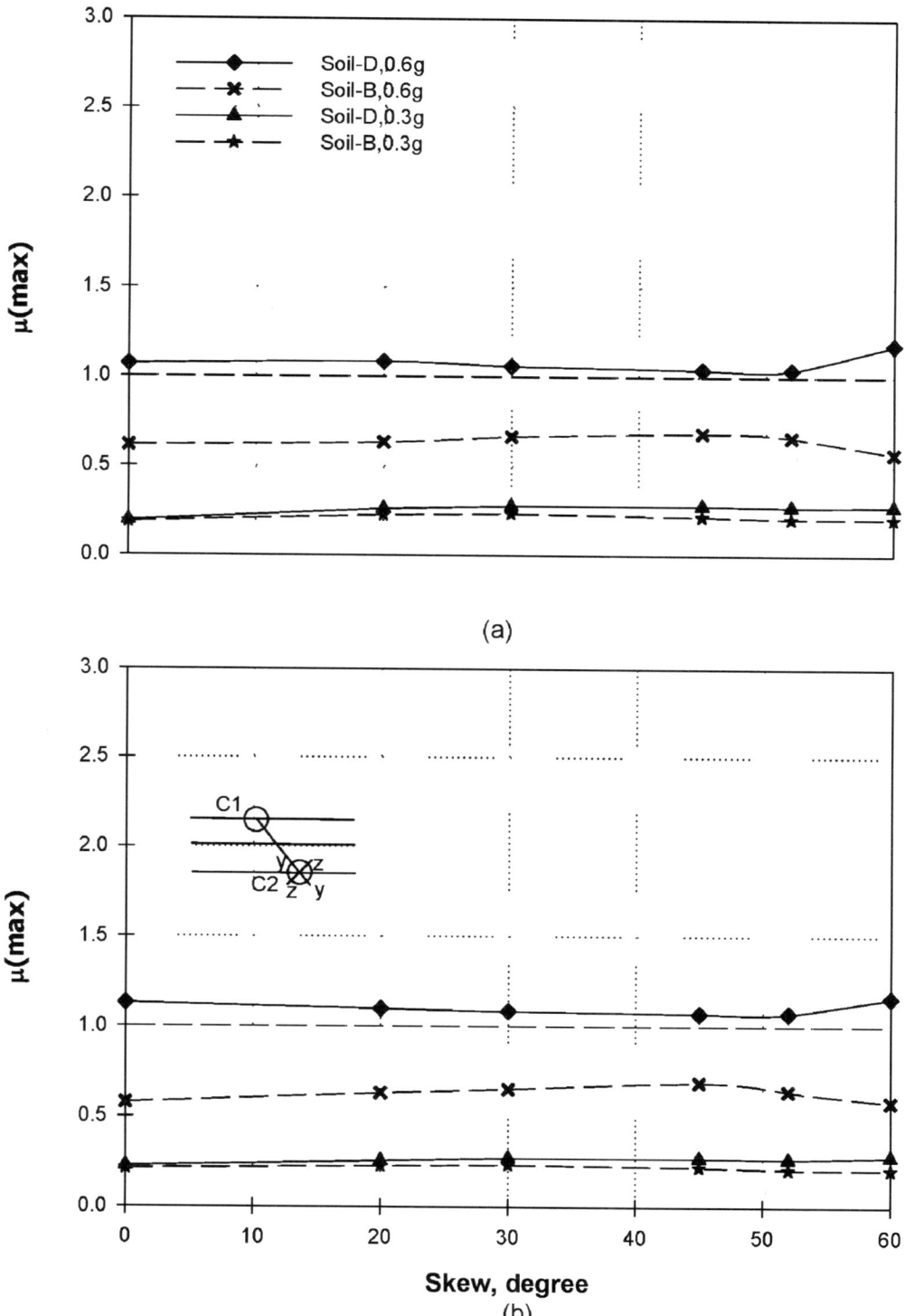

(a)

(b)

Skew, degree

Figure 3-55. Maximum Average Ductility Ratio of (a) C1 and (b) C2

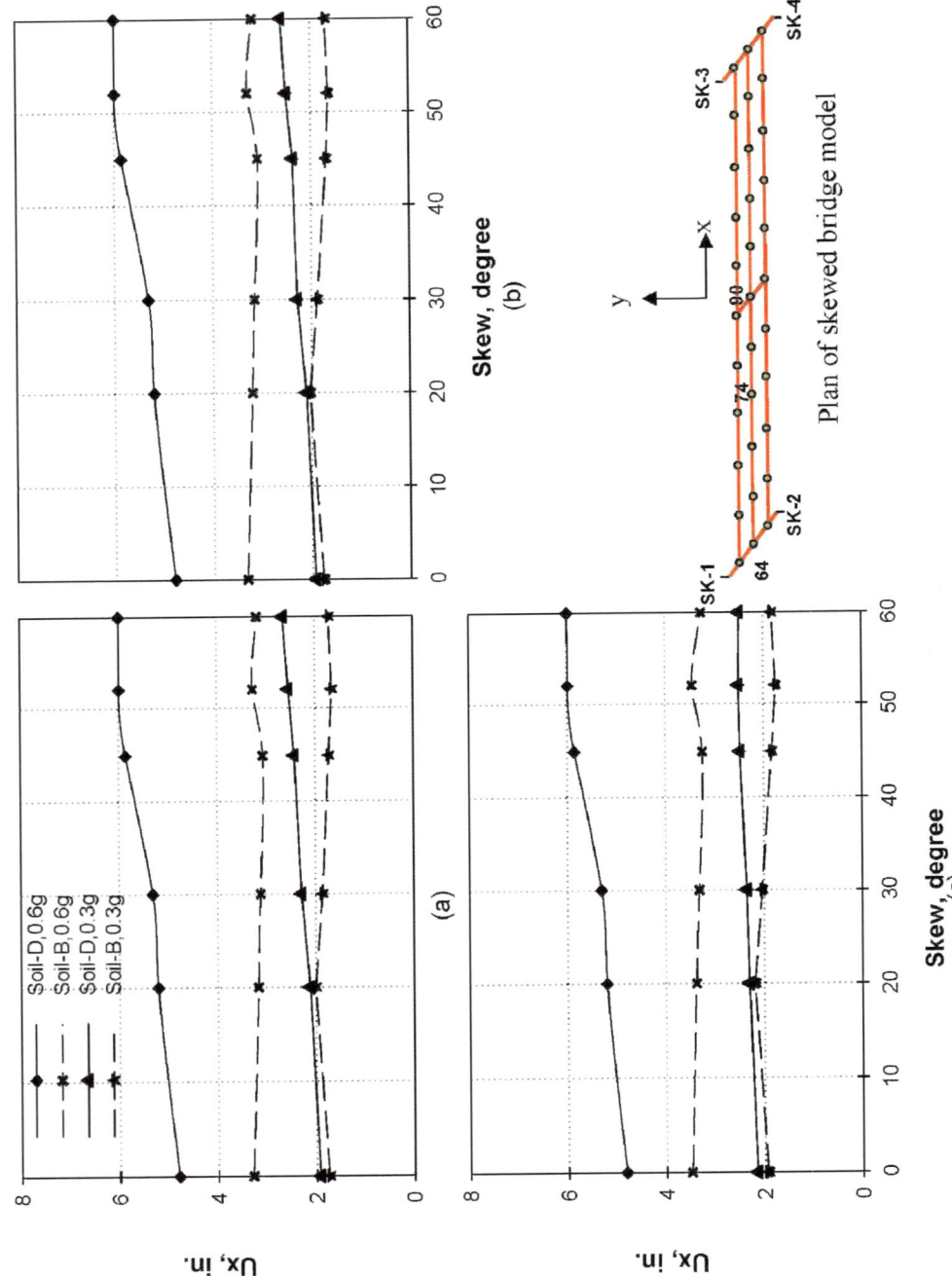

Figure 3-56. Average Displacement in X-direction for nodes (a) 90, (b) 74, and (c) 64 (W/O)

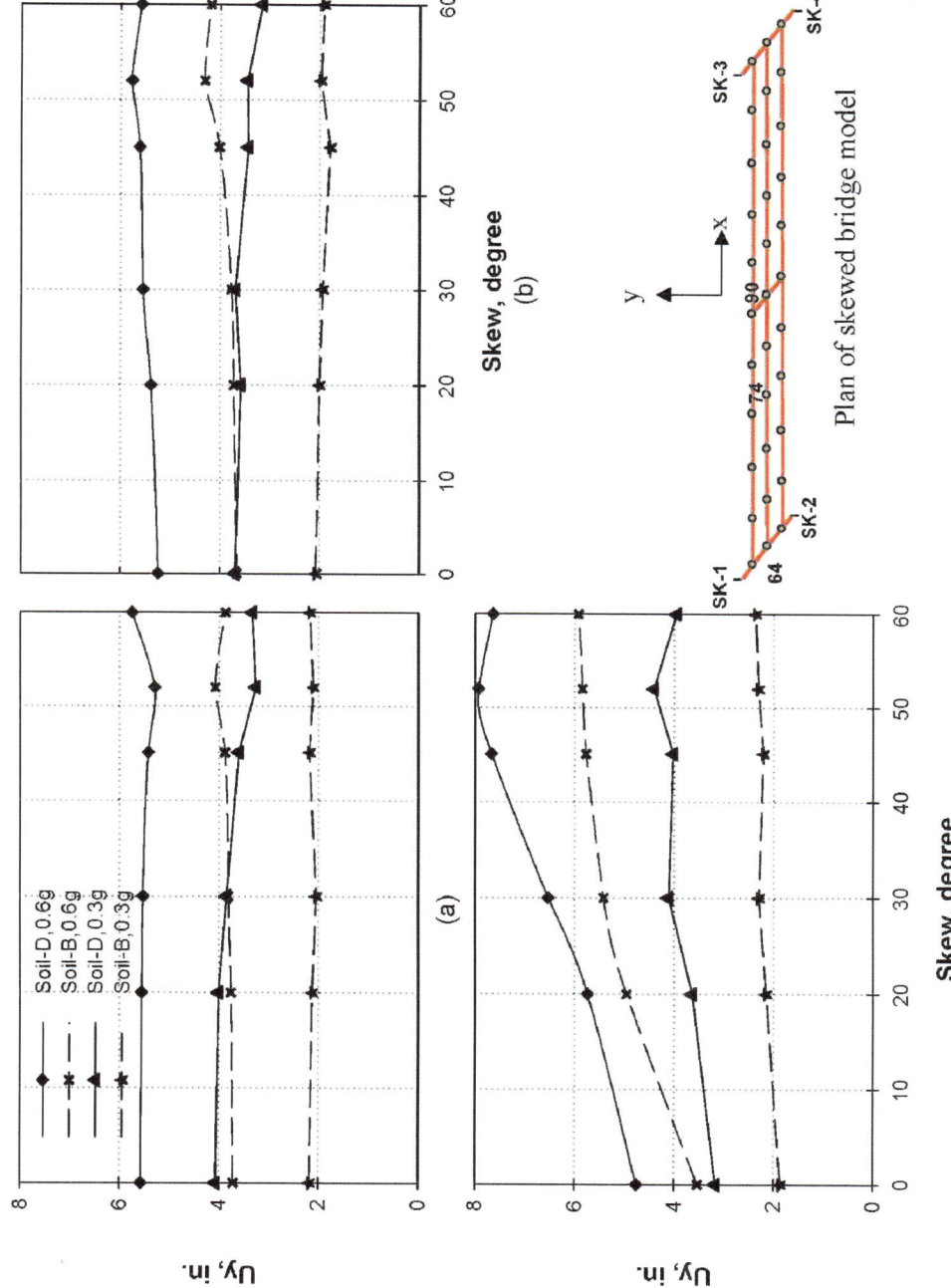

Figure 3-57. Average Displacement in Y-direction for nodes (a) 90, (b) 74, and (c) 64 (W/O)

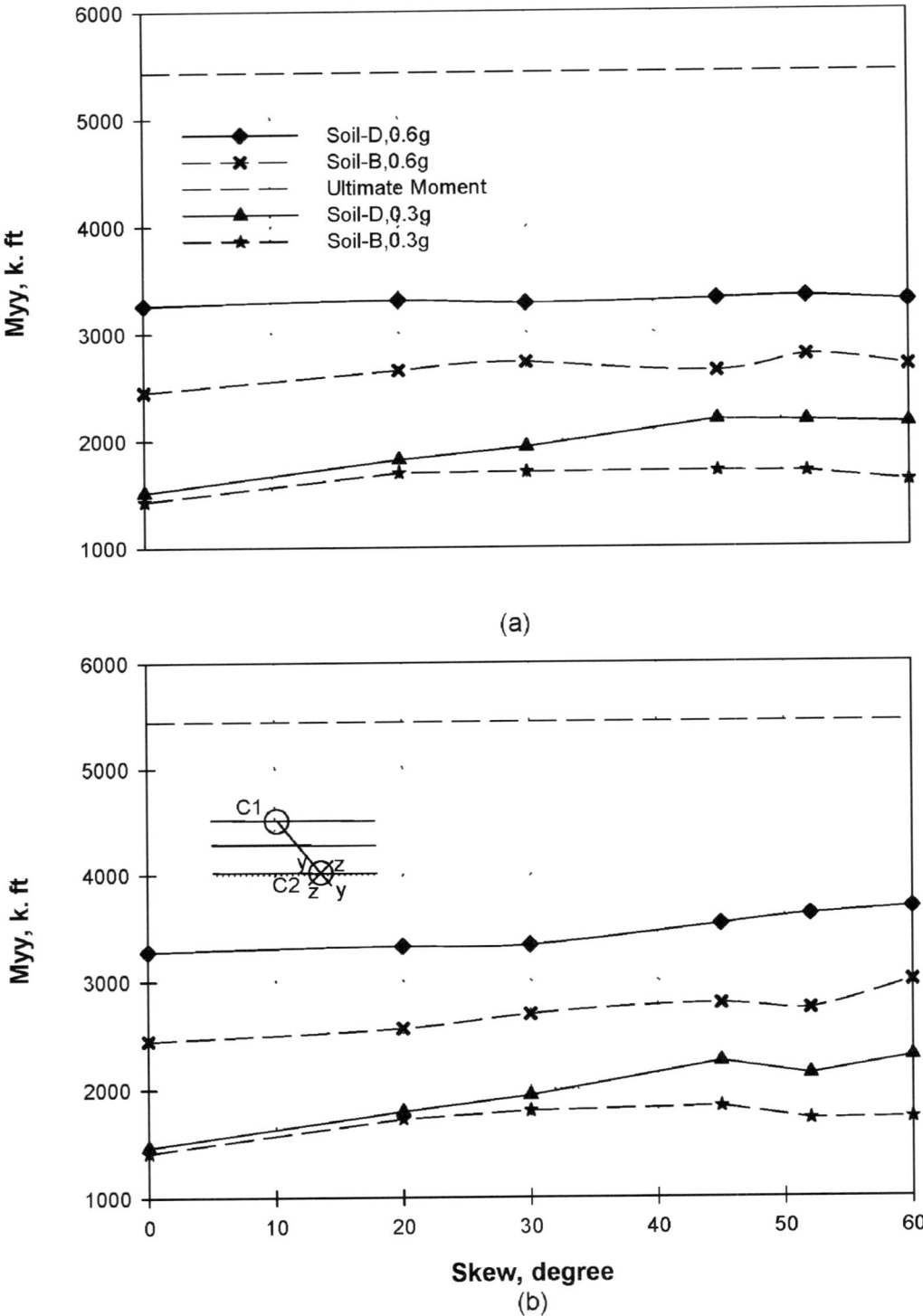

(a)

(b)

Figure 3-58. Average Moment in Y-Directions of (a) C1, and (b) C2 (W/O)

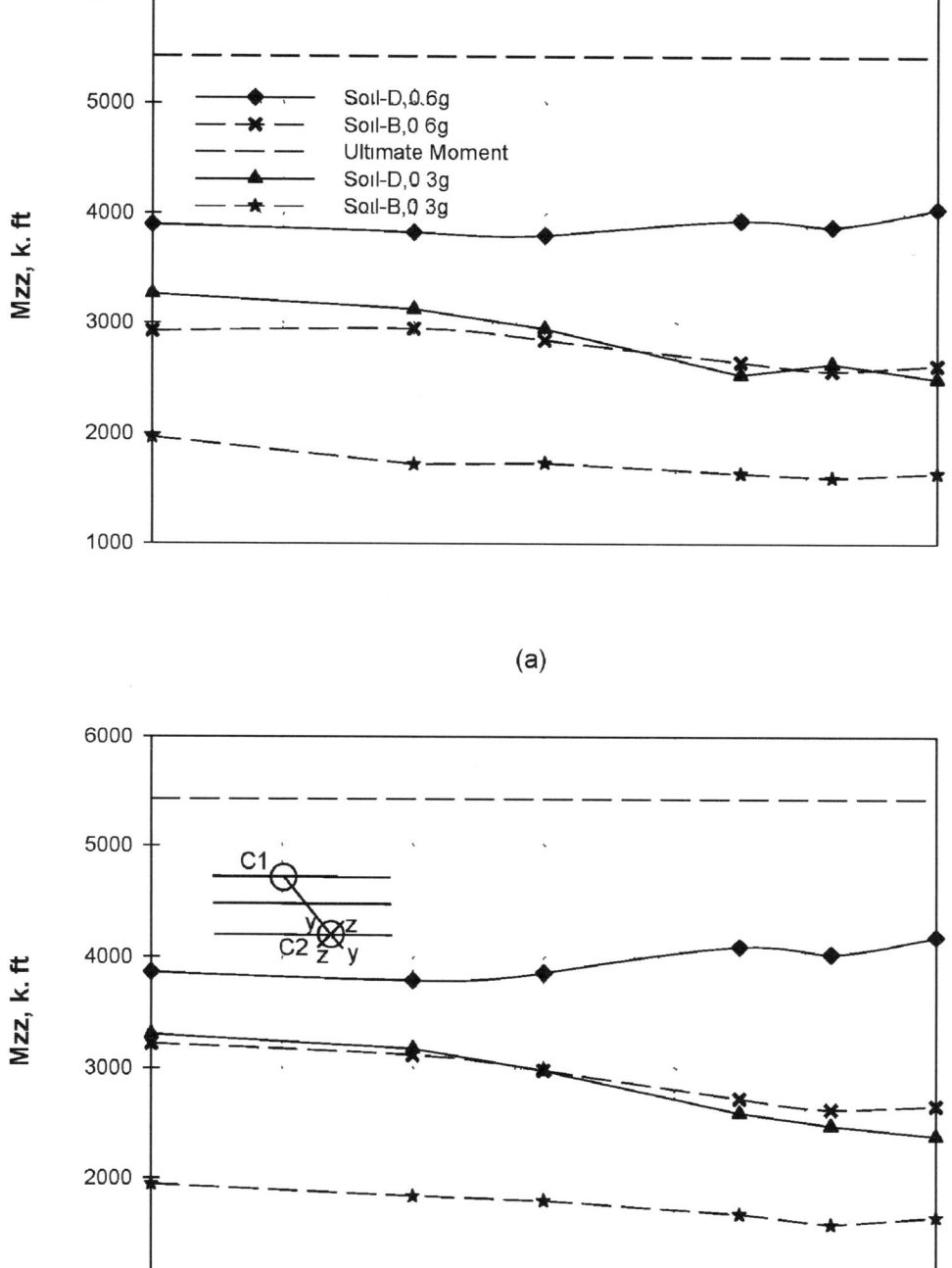

(a)

(b)

Figure 3-59. Average Moment in Z-Directions of (a) C1, and (b) C2 (W/O)

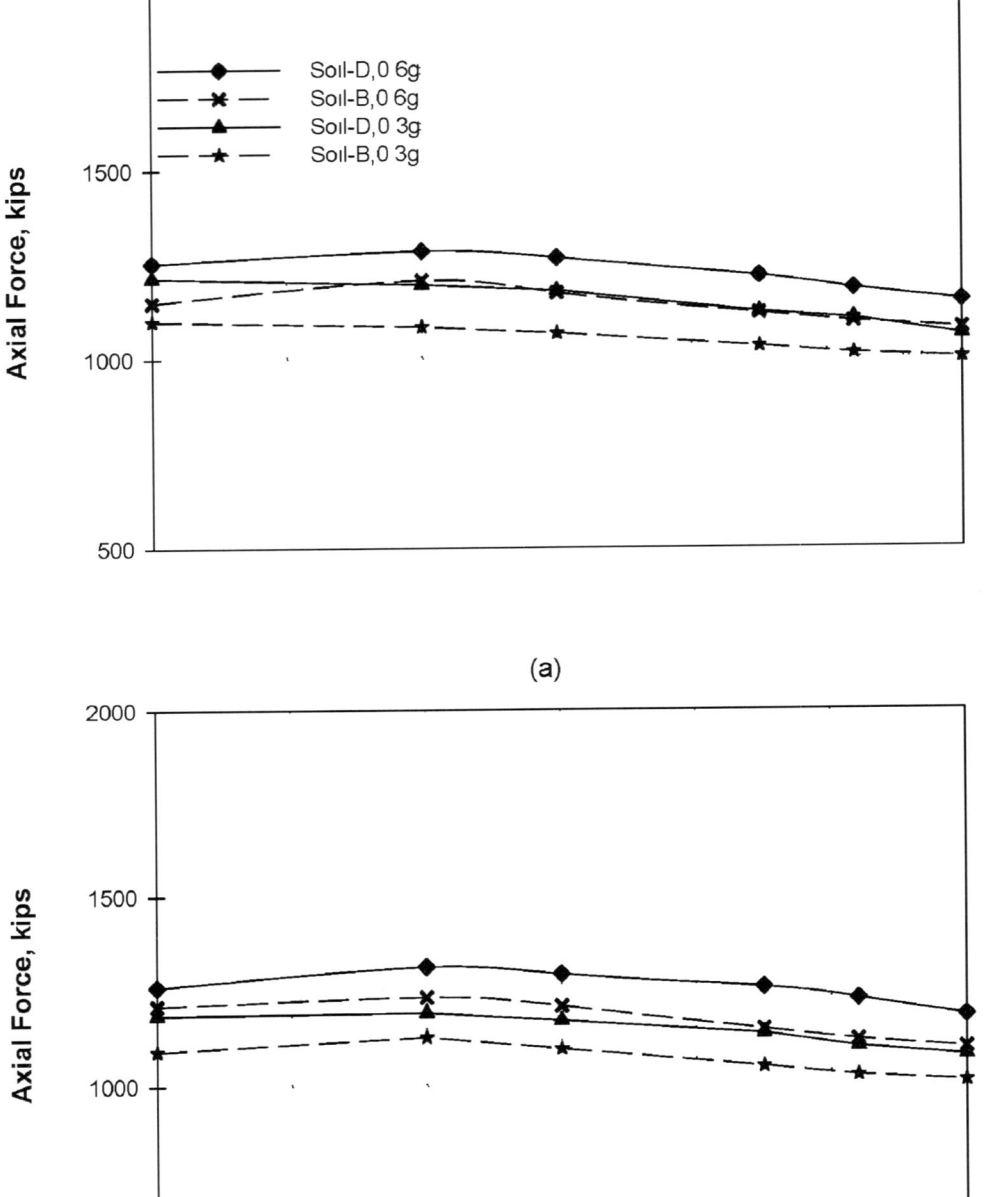

(a)

(b)

Figure 3-60. Average Axial Force in (a) C1, and (b) C2 (W/O)

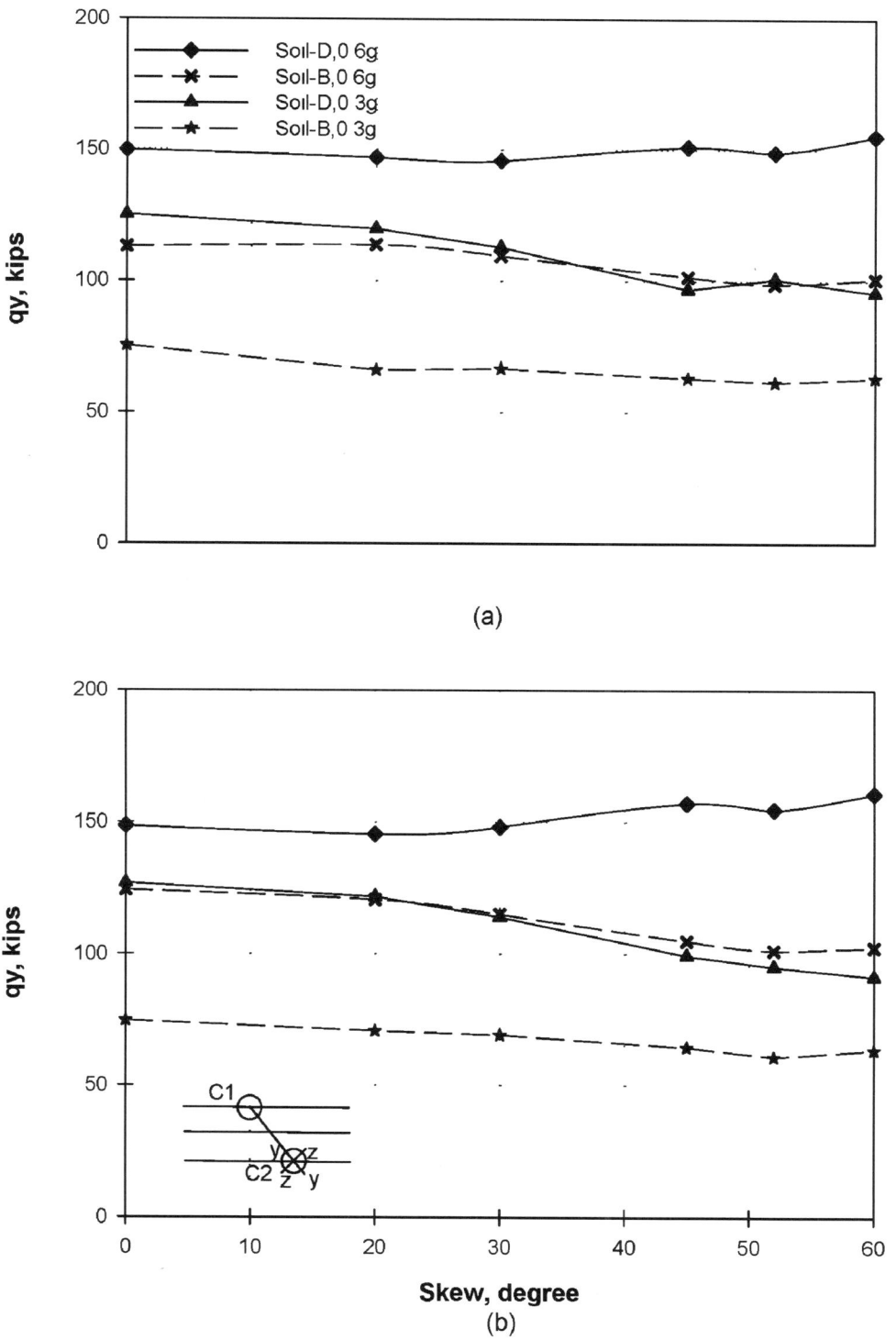

(a)

(b)

Figure 3-61. Average Shear Force in Y-Directions of (a) C1, and (b) C2 (W/O)

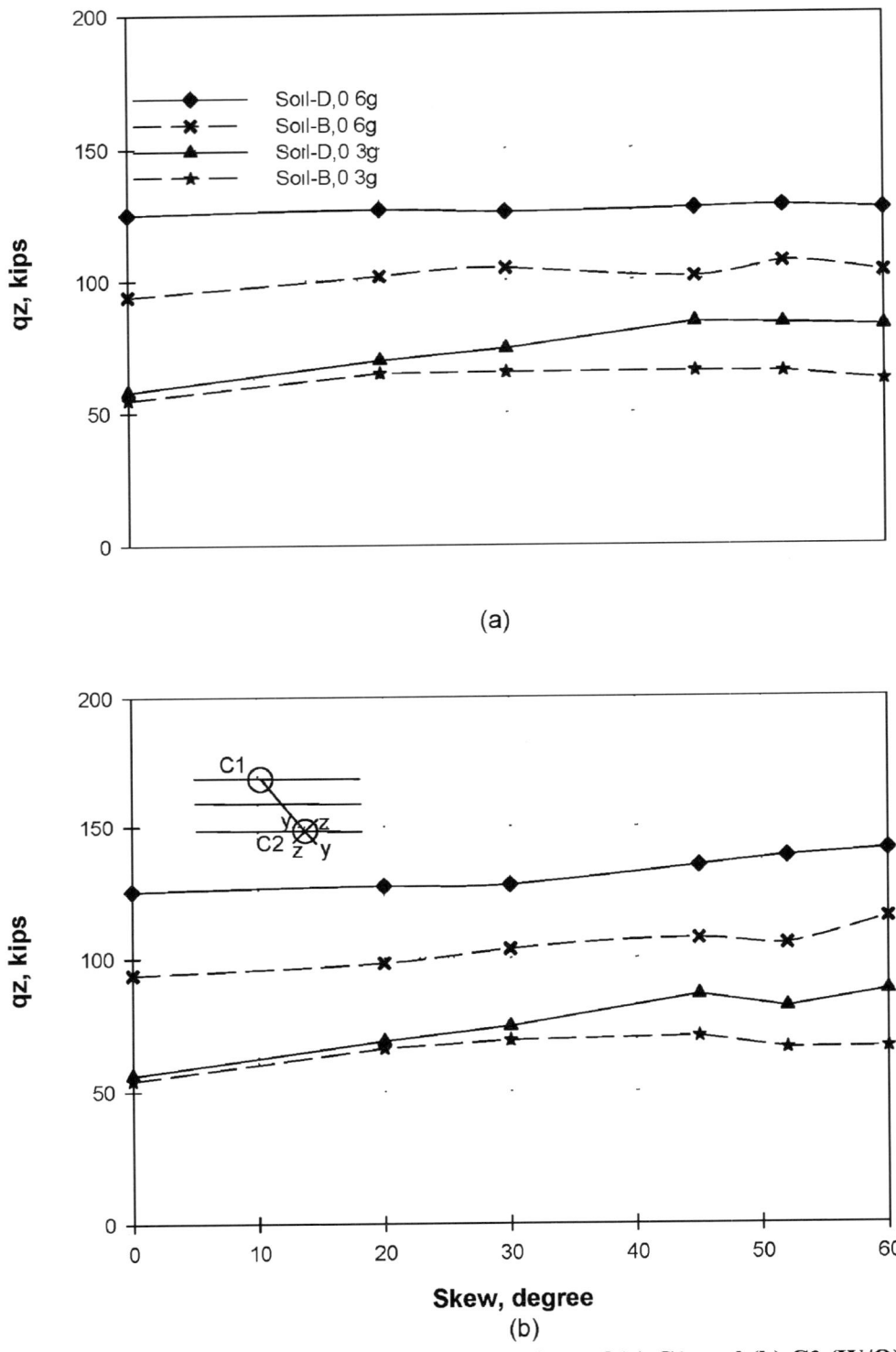

(a)

(b)

Figure 3-62. Average Shear Force in Z-Directions of (a) C1, and (b) C2 (W/O)

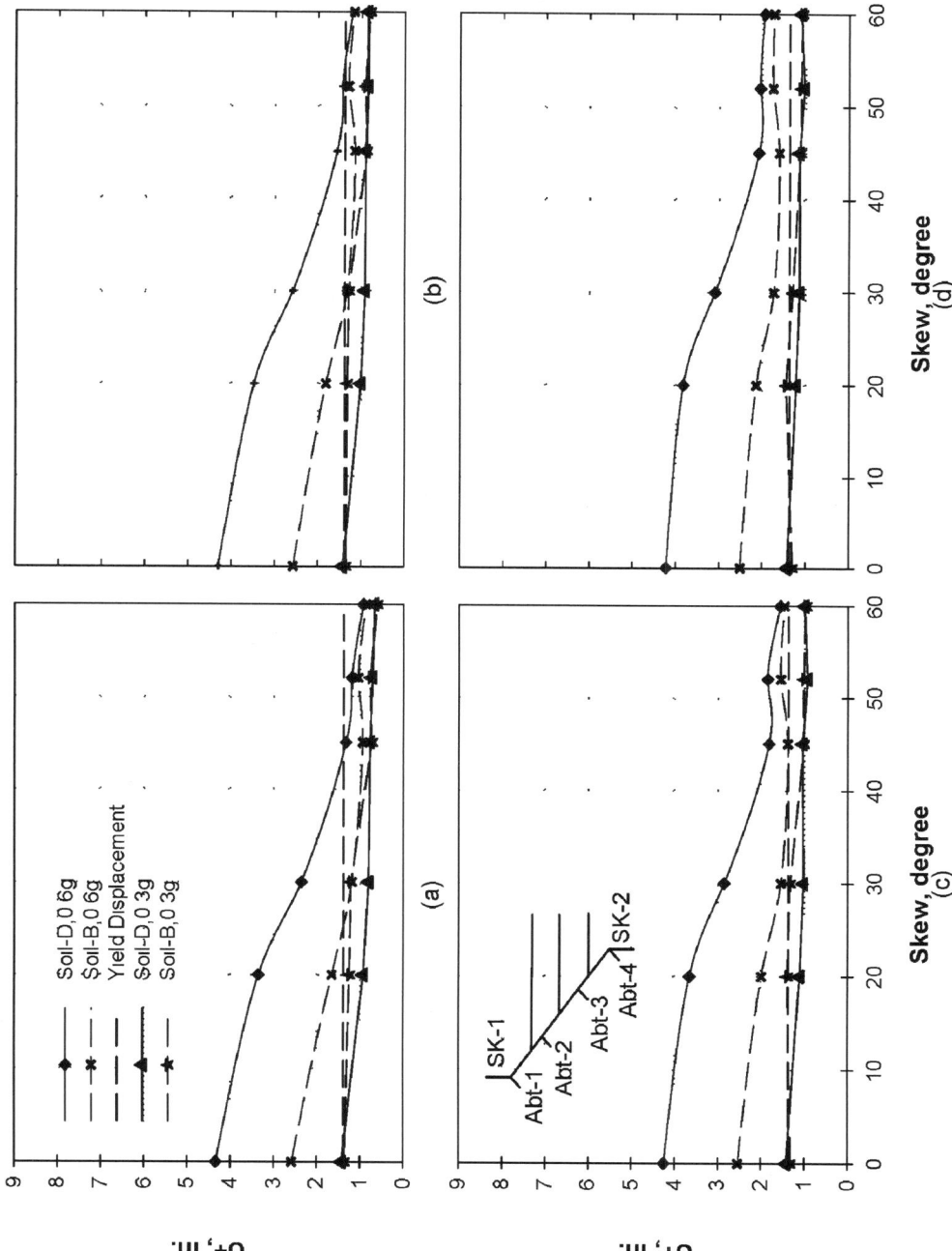

Figure 3-63. Average Displacement "into" Abutment's (a) Abt-1, (b) Abt-2, (c) Abt-3, and (d) Abt-4 (W/O)

207

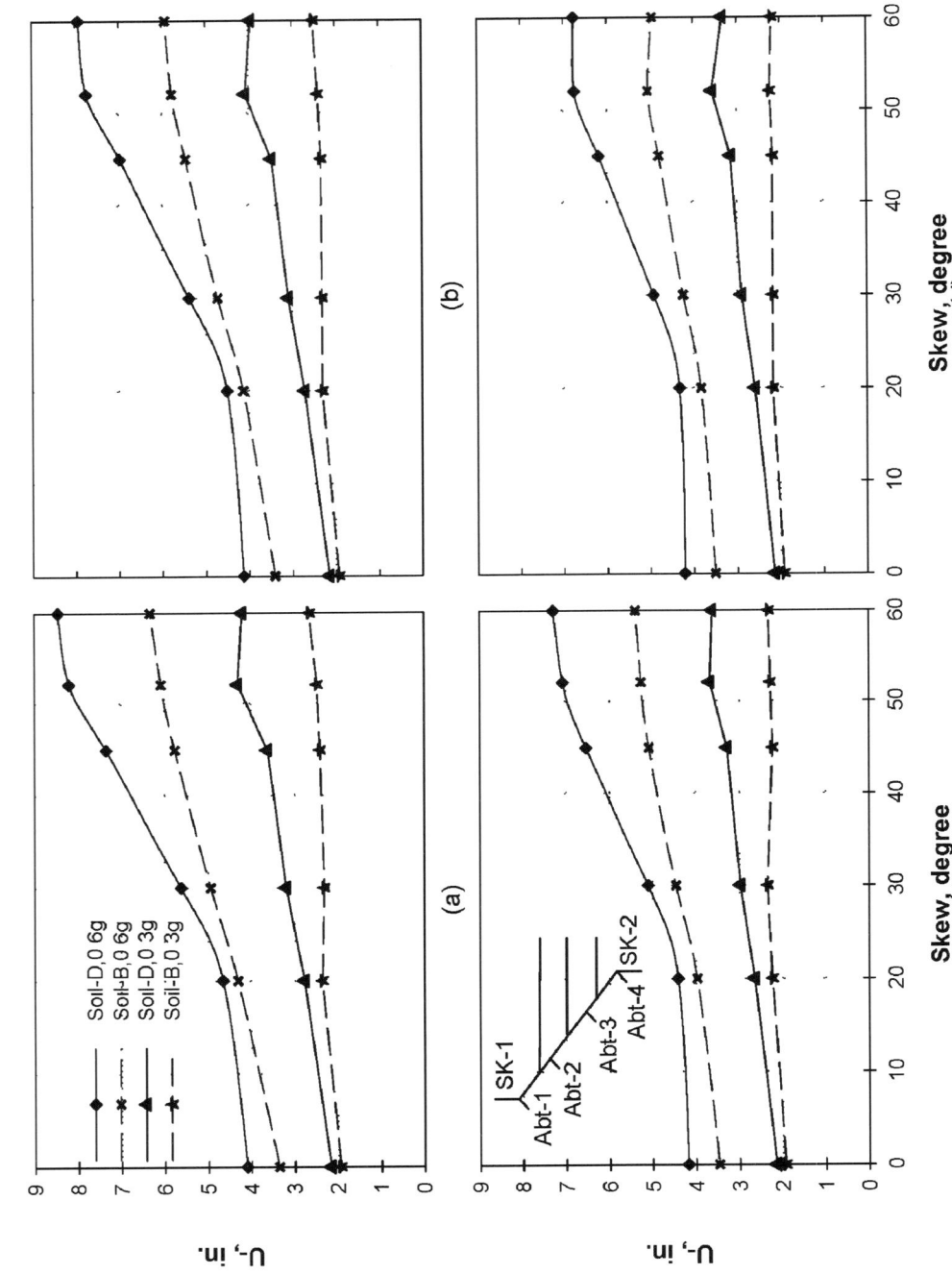

Figure 3-64. Average Displacement "away" from Abutment's (a) Abt-1, (b) Abt-2, (c) Abt-3, and (d) Abt-4 (W/O)

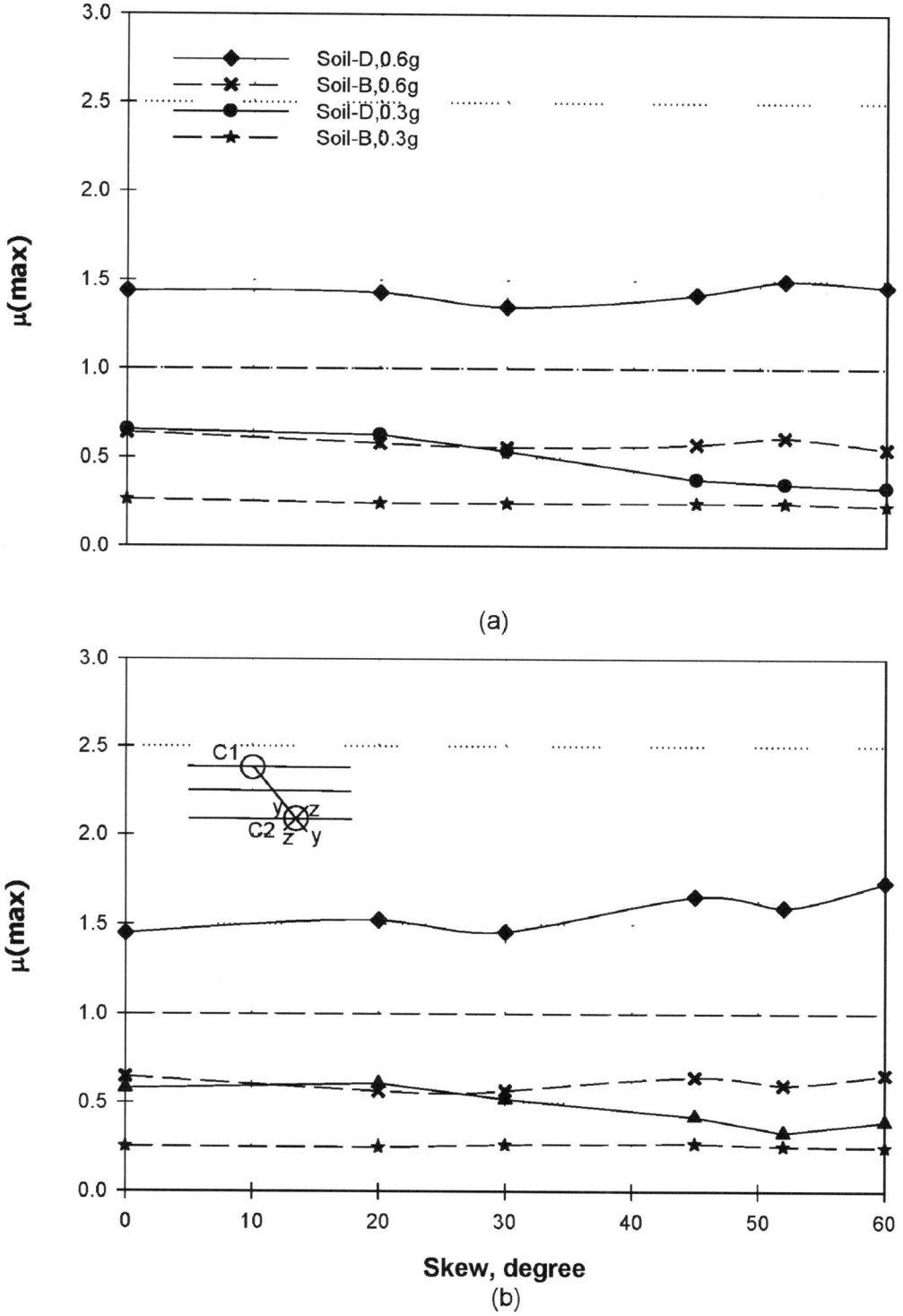

(a)

(b)

Figure 3-65. Maximum Average Ductility Ratio of (a) C1 and (b) C2 (W/O)

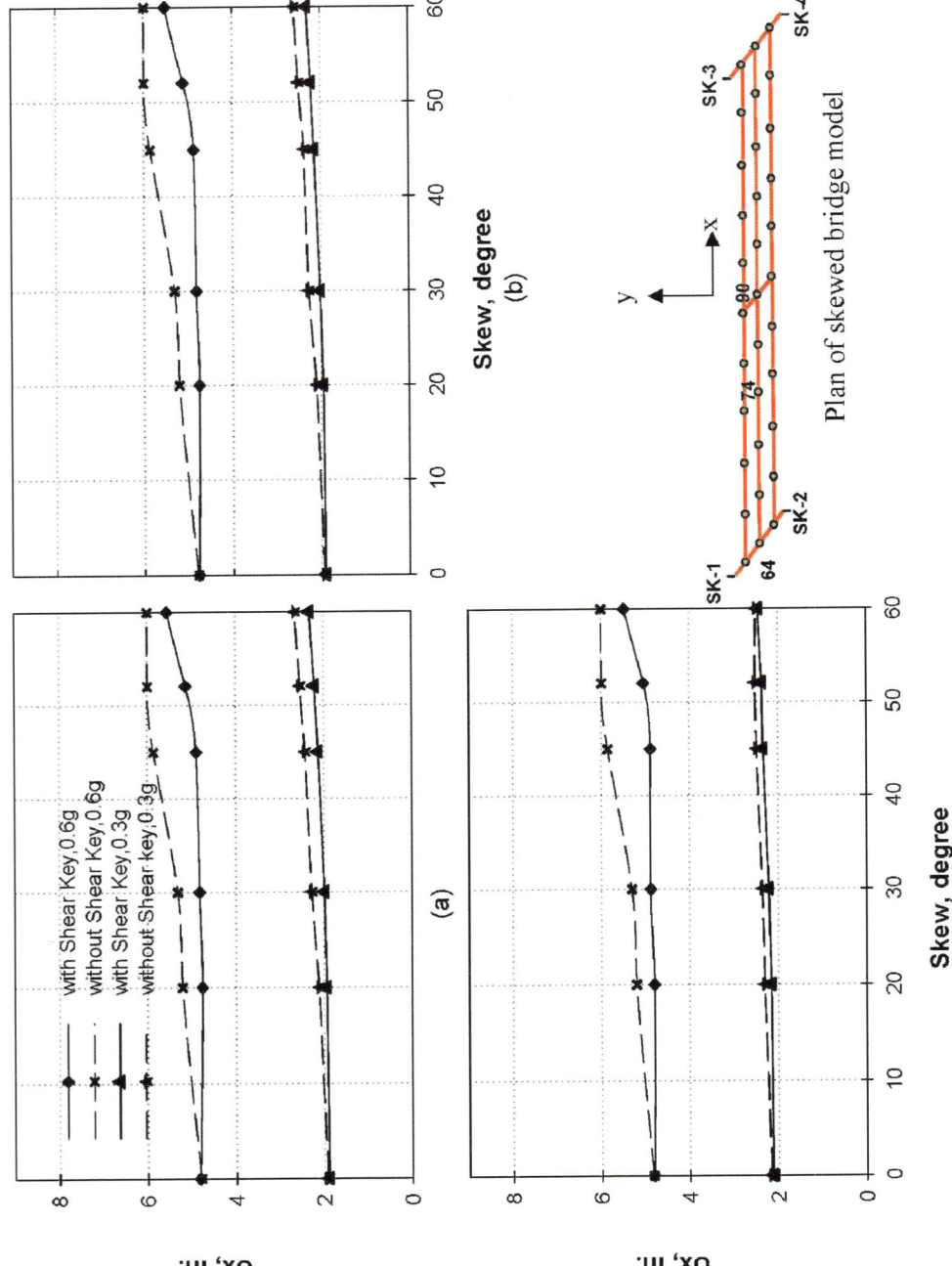

Figure 3-66. Average Displacement in X-direction for nodes (a) 90, (b) 74, and (c) 64 (Soil-D)

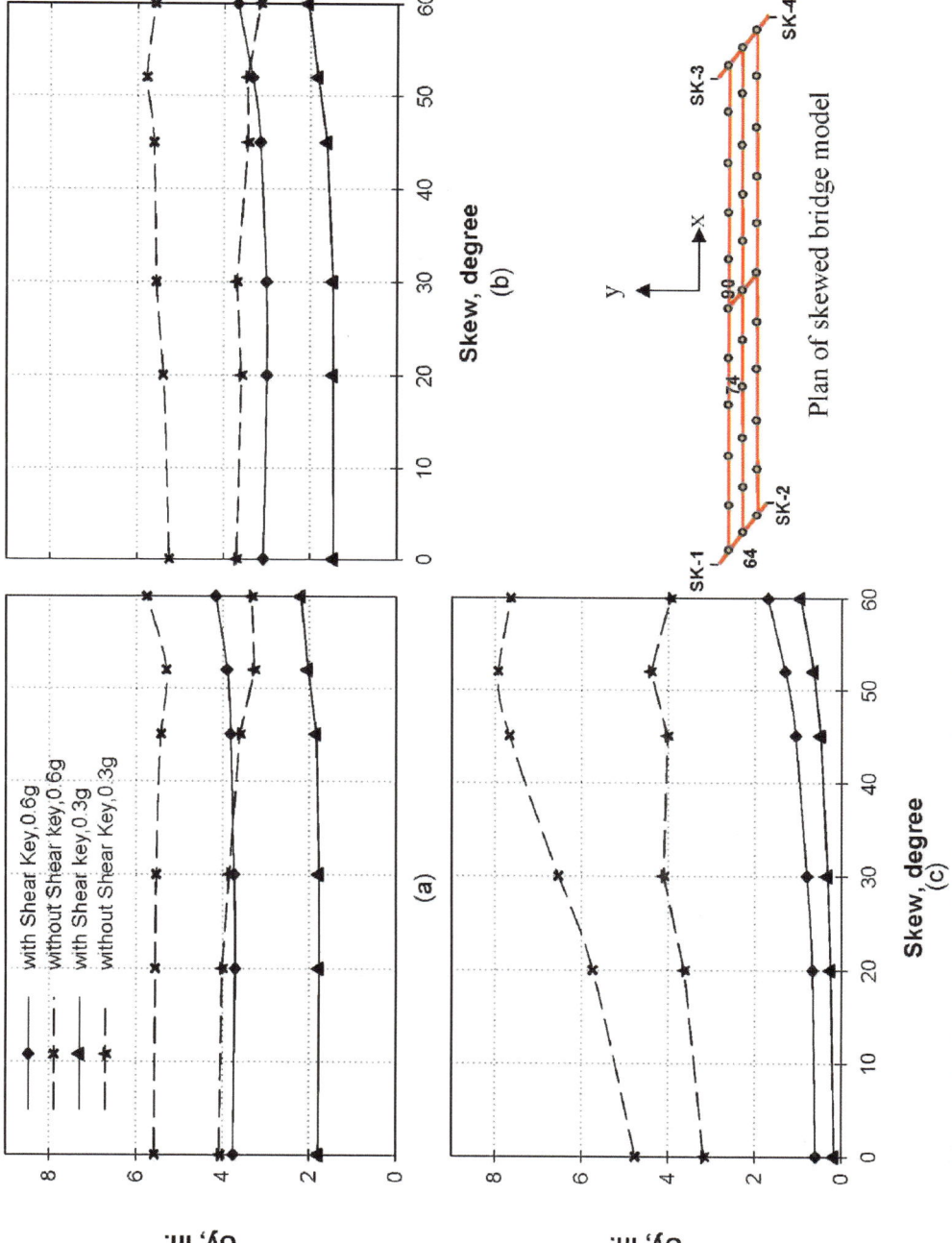

Figure 3-67. Average Displacement in Y-direction for nodes (a) 90, (b) 74, and (c) 64 (Soil-D)

211

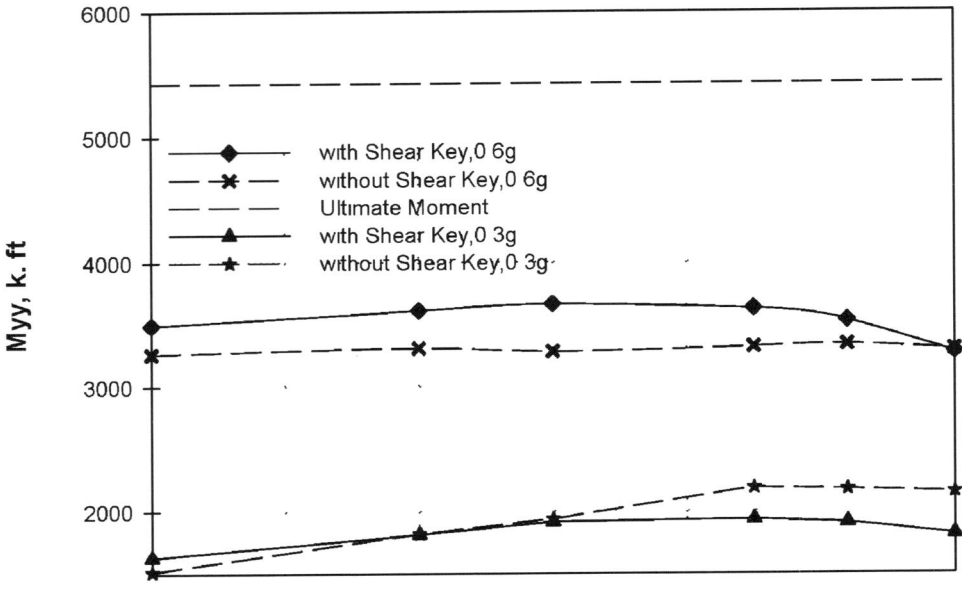

(a)

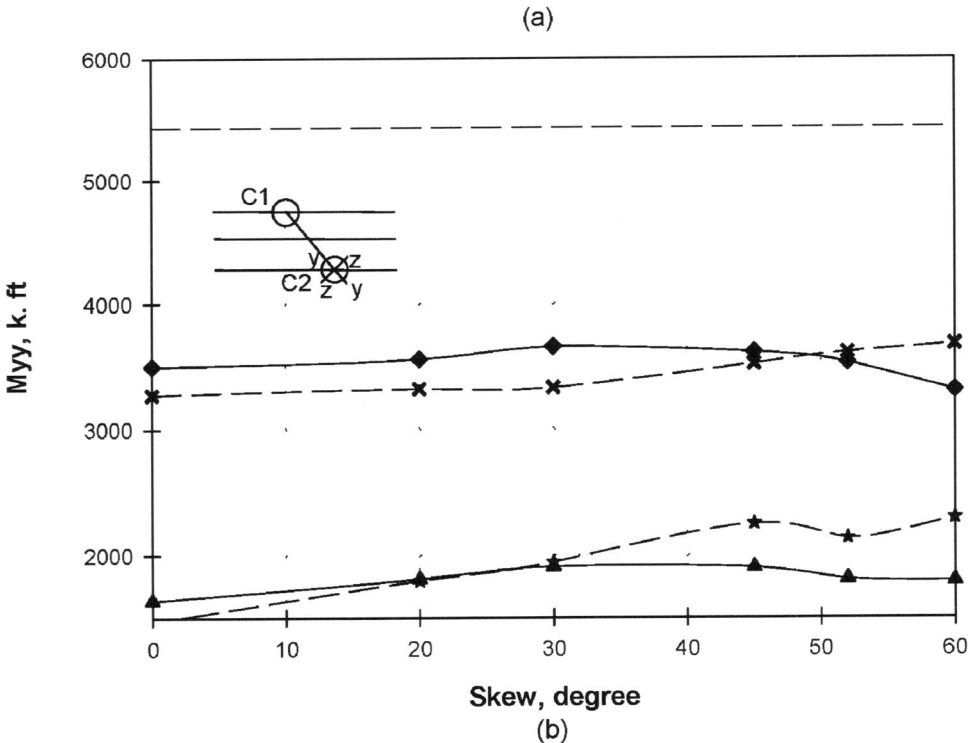

Skew, degree

(b)

Figure 3-68. Average Moment in Y-Directions of (a) C1, and (b) C2 (Soil-D)

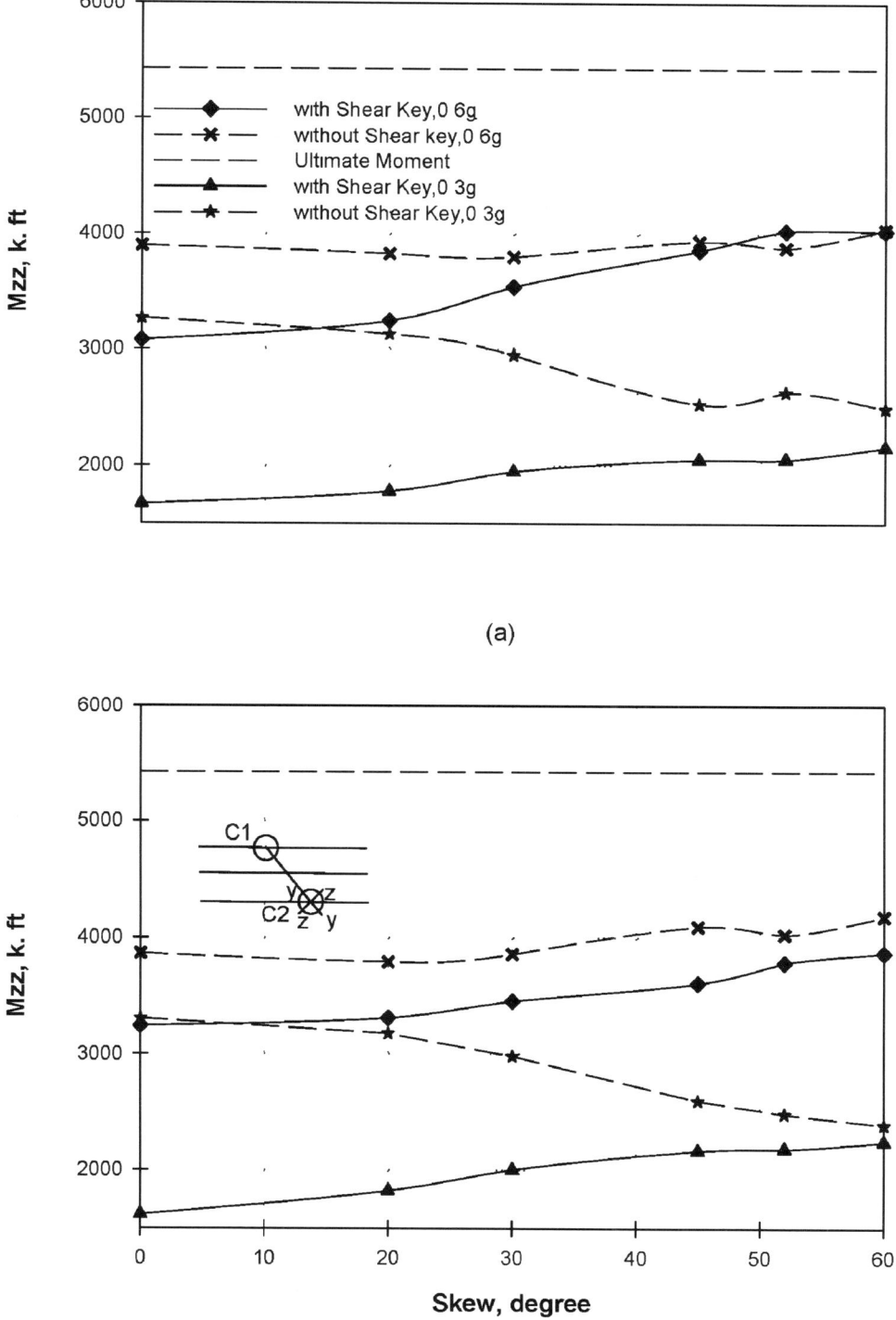

(a)

(b)

Figure 3-69. Average Moment in Z-Directions of (a) C1, and (b) C2 (Soil-D)

214

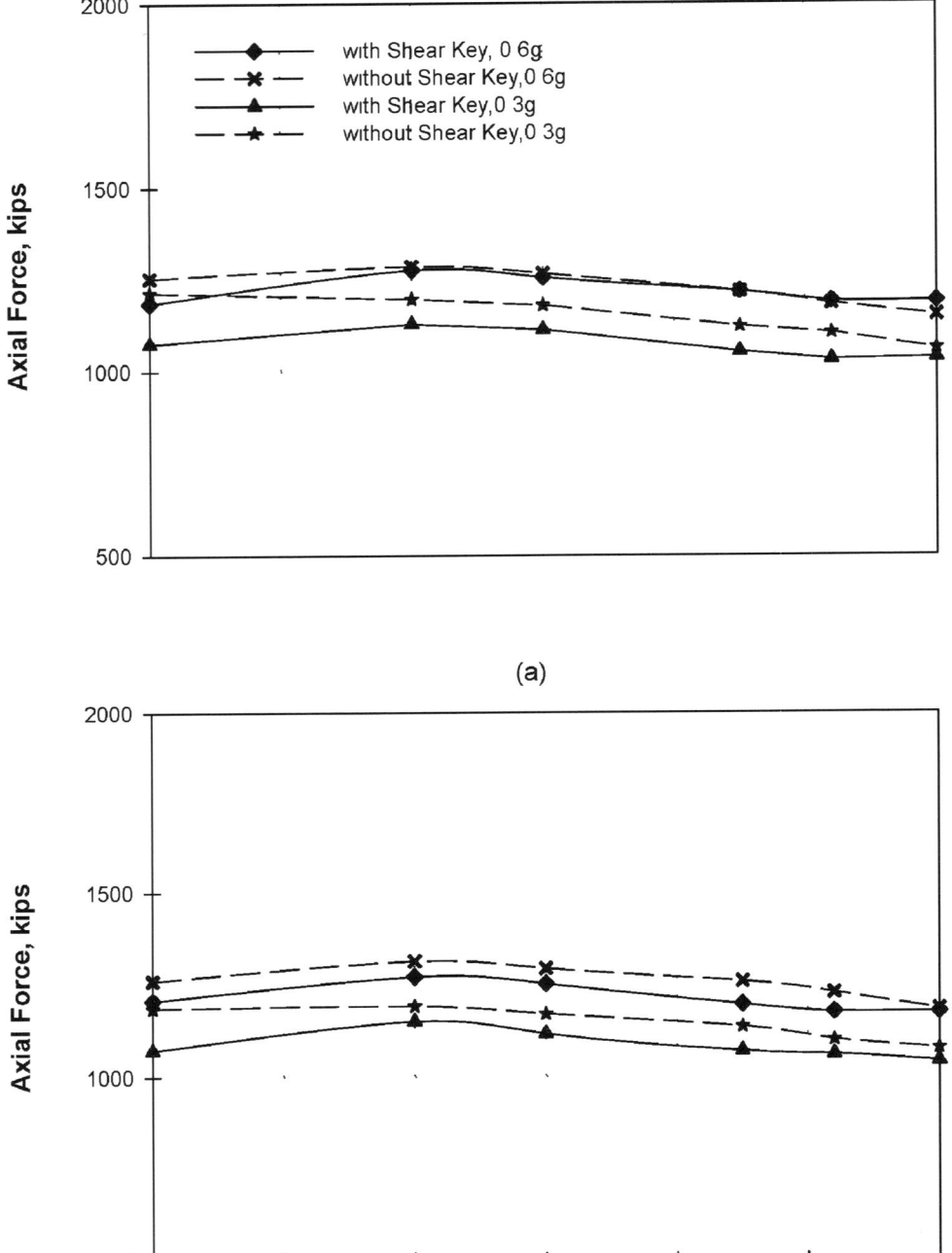

(a)

(b)

Figure 3-70. Average Axial Force in (a) C1, and (b) C2 (Soil-D)

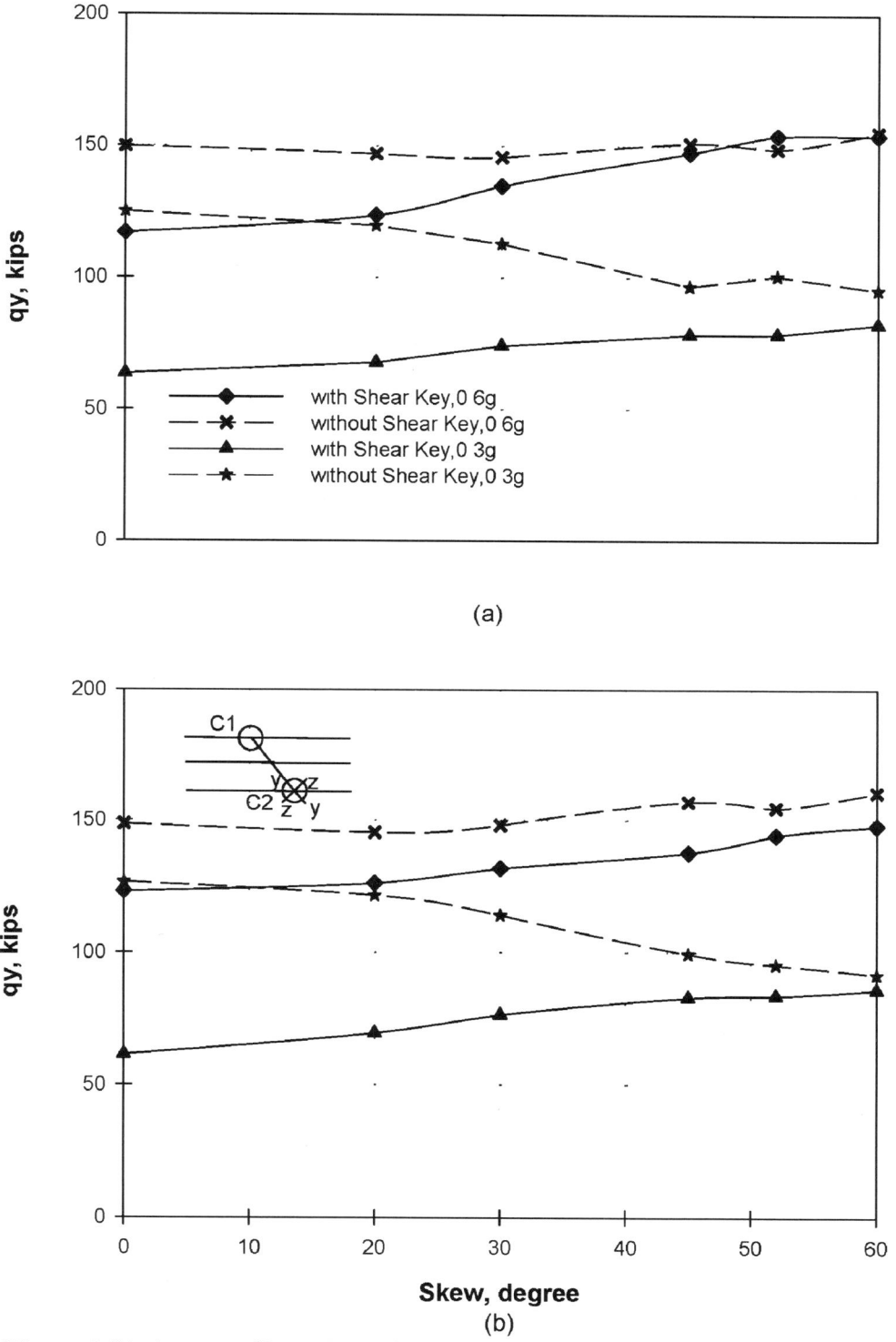

(a)

(b)

Figure 3-71. Average Shear Force in Y-Directions of (a) C1, and (b) C2 (Soil-D)

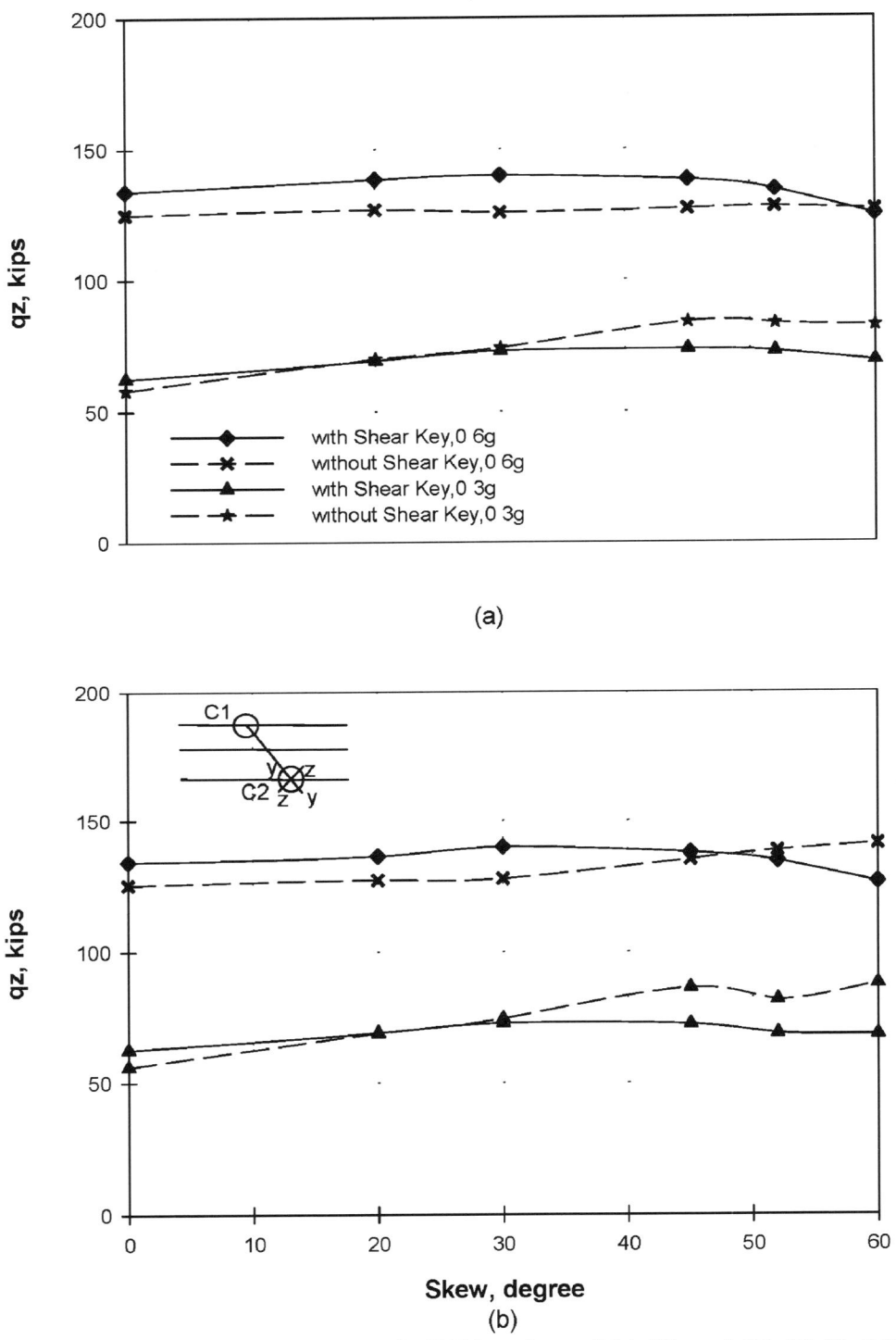

(a)

(b)

Figure 3-72. Average Shear Force in Z-Directions of (a) C1, and (b) C2 (Soil-D)

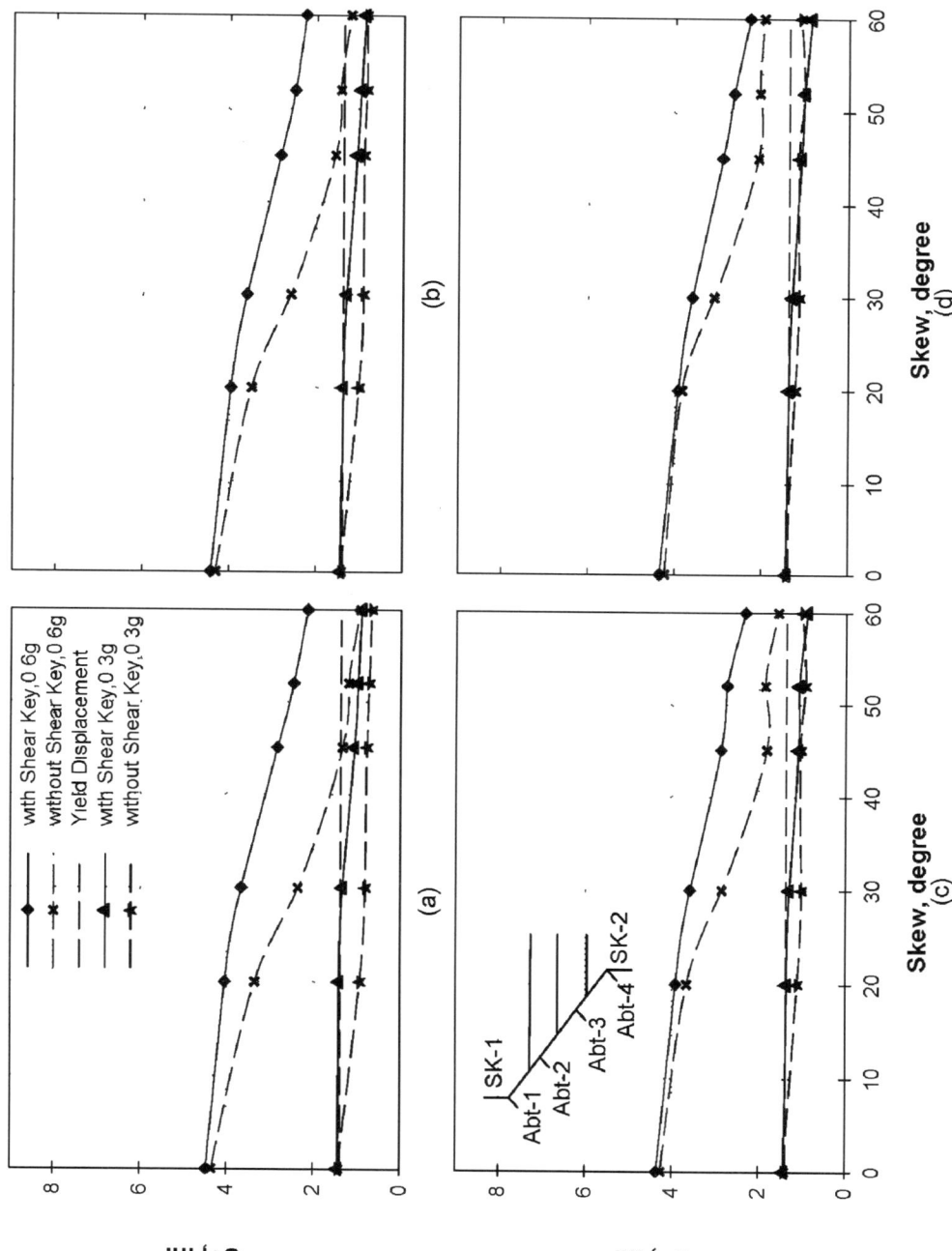

Figure 3-73. Average Displacement "into" Abutment's (a) Abt-1, (b) Abt-2, (c) Abt-3, and (d) Abt-4 (Soil-D)

217

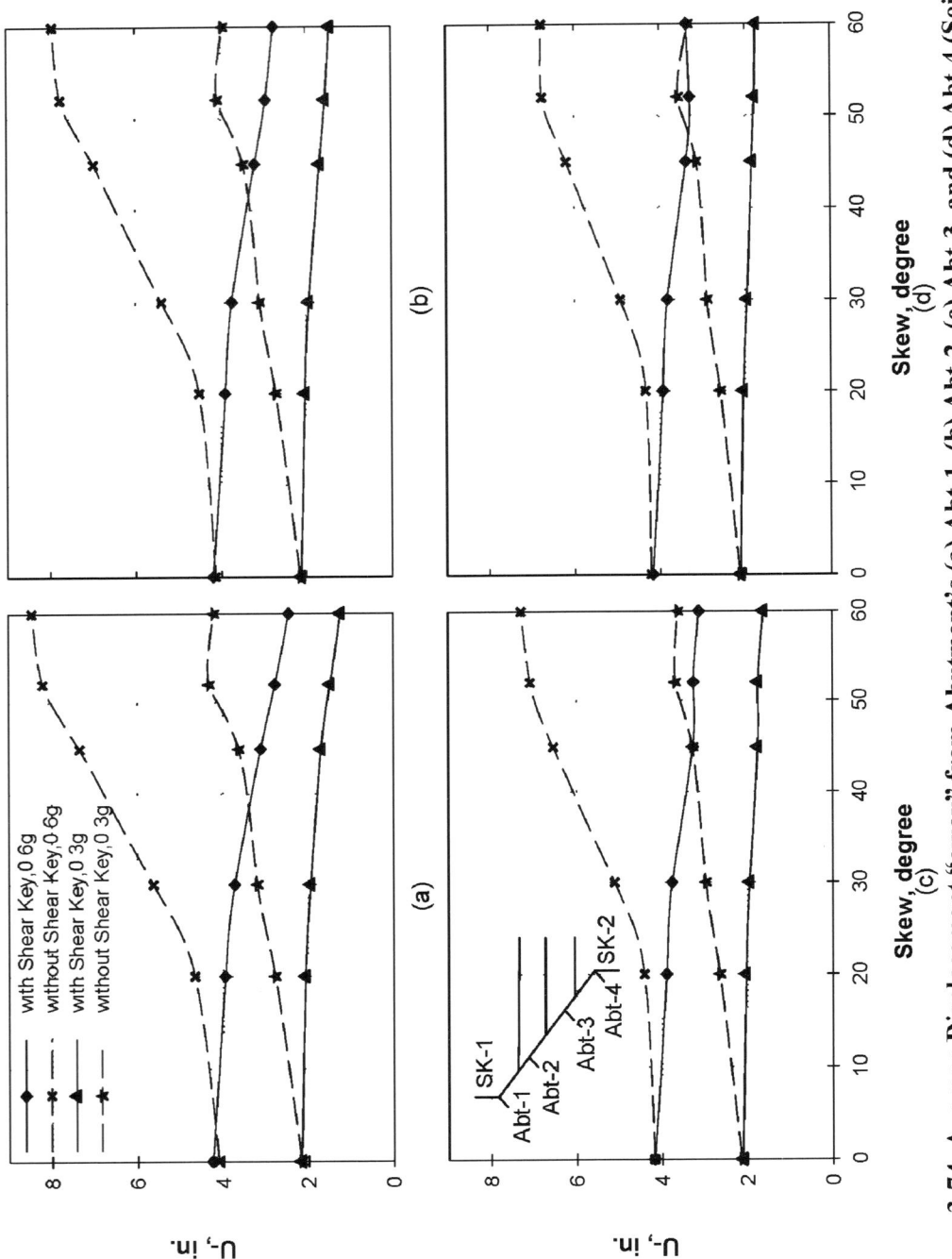

Figure 3-74. Average Displacement "away" from Abutment's (a) Abt-1, (b) Abt-2, (c) Abt-3, and (d) Abt-4 (Soil-D)

218

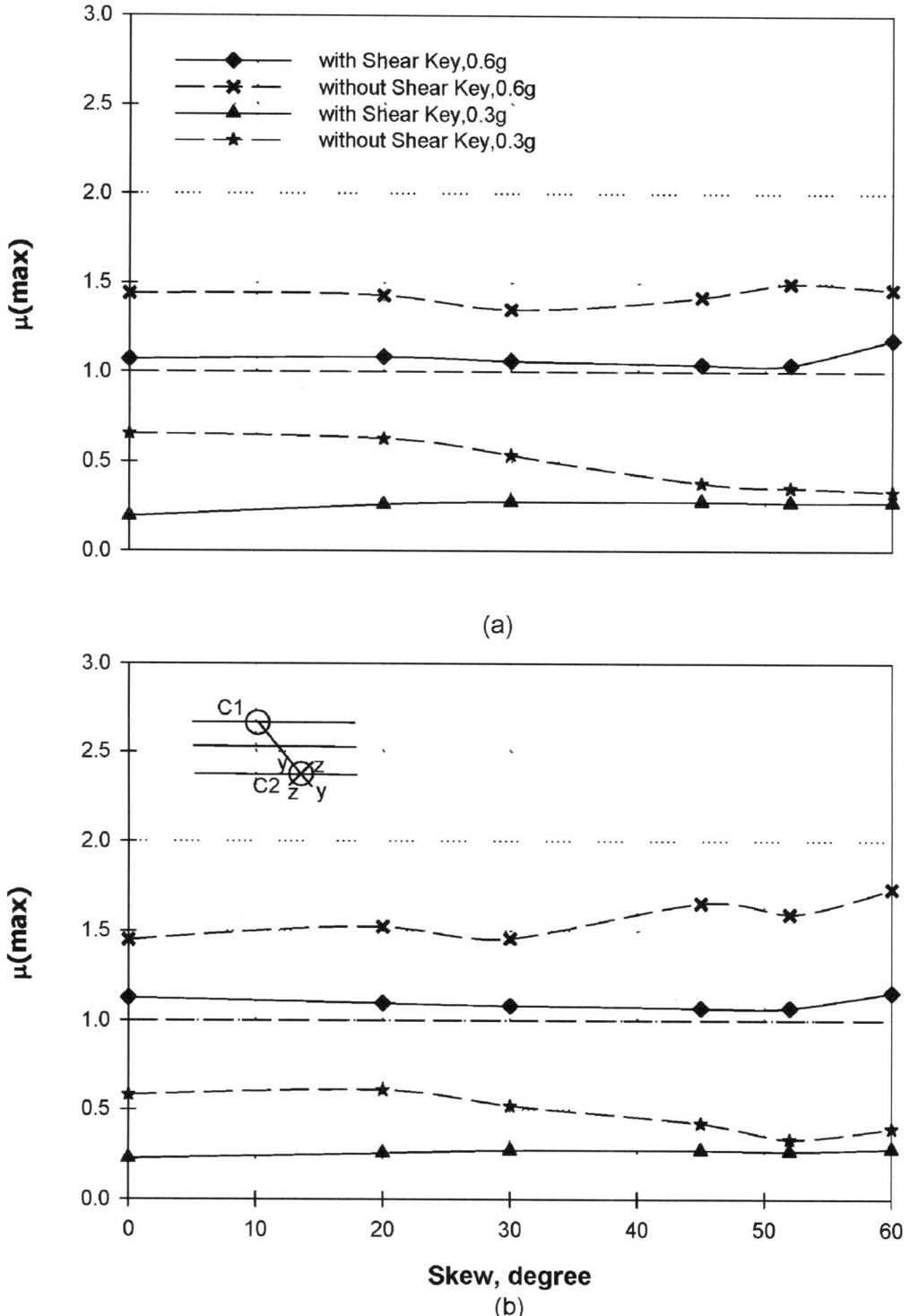

(a)

(b)

Skew, degree

Figure 3-75. Maximum Average Ductility Ratio of (a) C1 and (b) C2 (Soil-D)

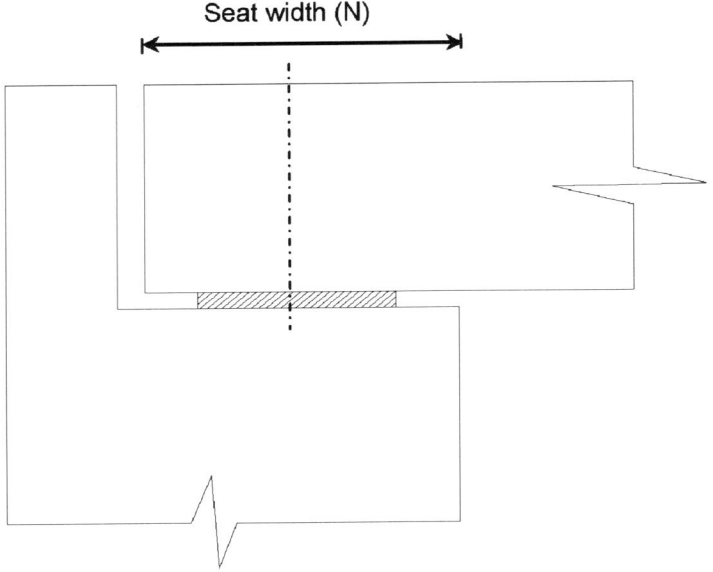

Figure 3-76. Abutment Seat width

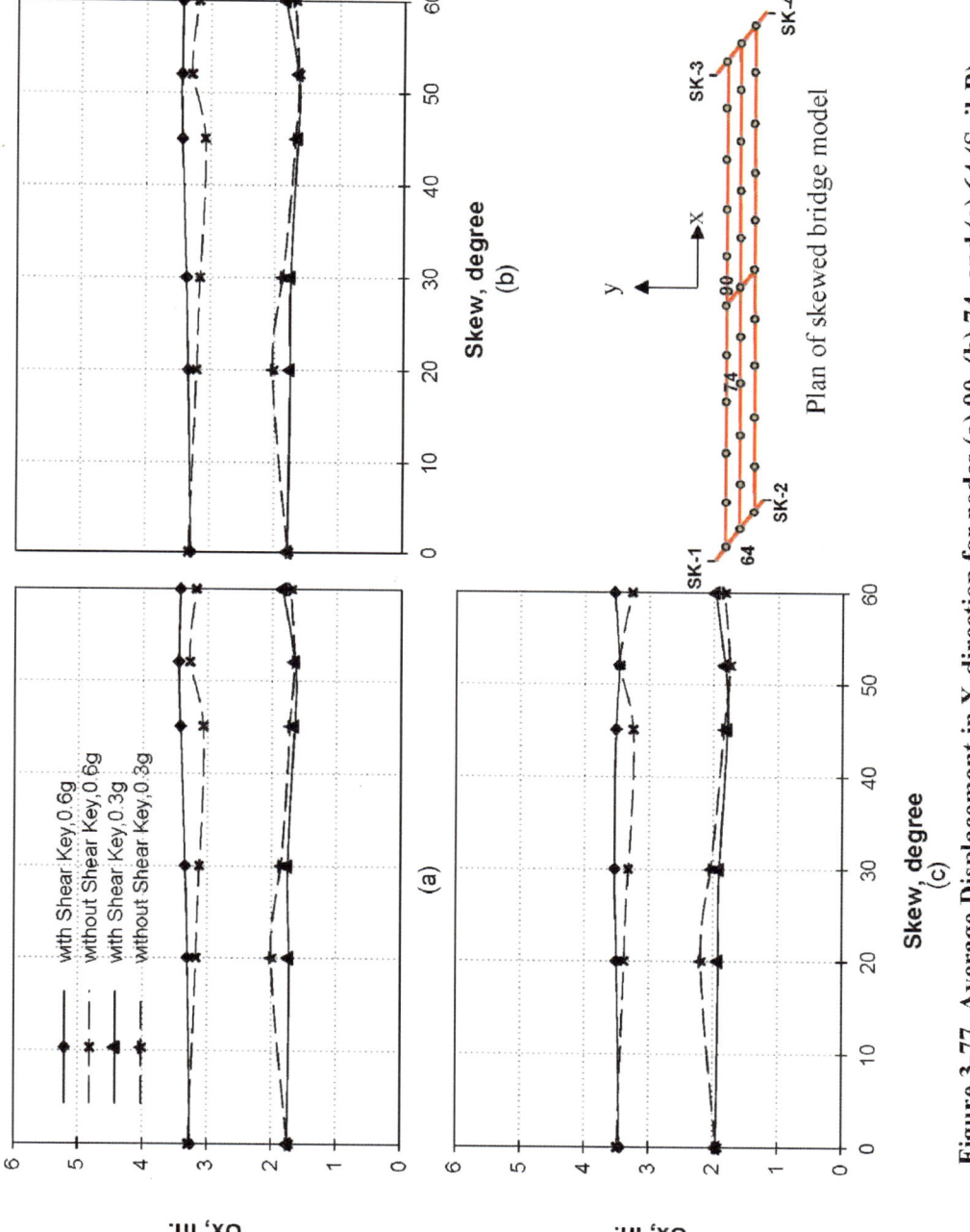

Figure 3-77. Average Displacement in X-direction for nodes (a) 90, (b) 74, and (c) 64 (Soil-B)

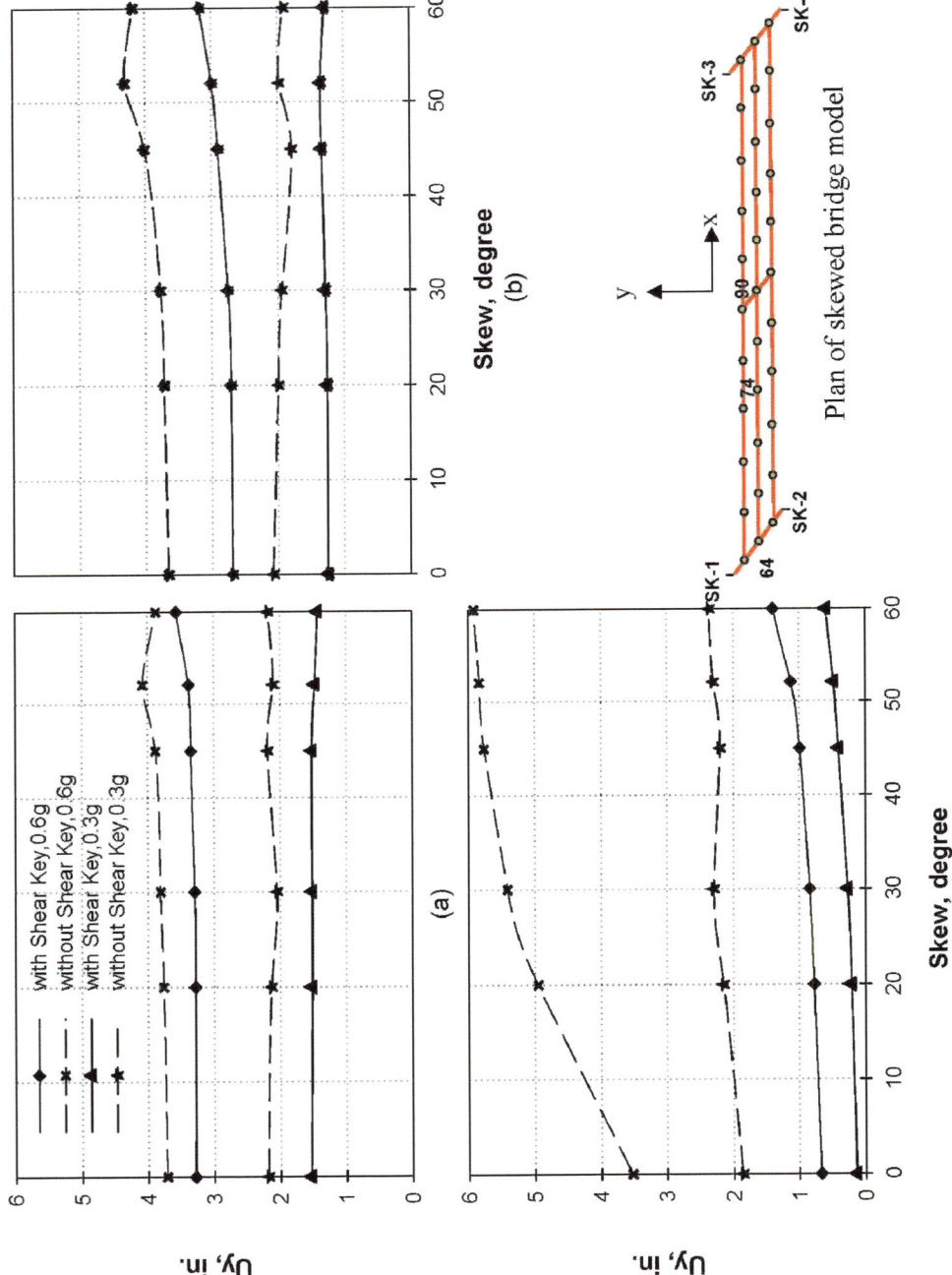

Figure 3-78. Average Displacement in Y-direction for nodes (a) 90, (b) 74, and (c) 64 (Soil-B)

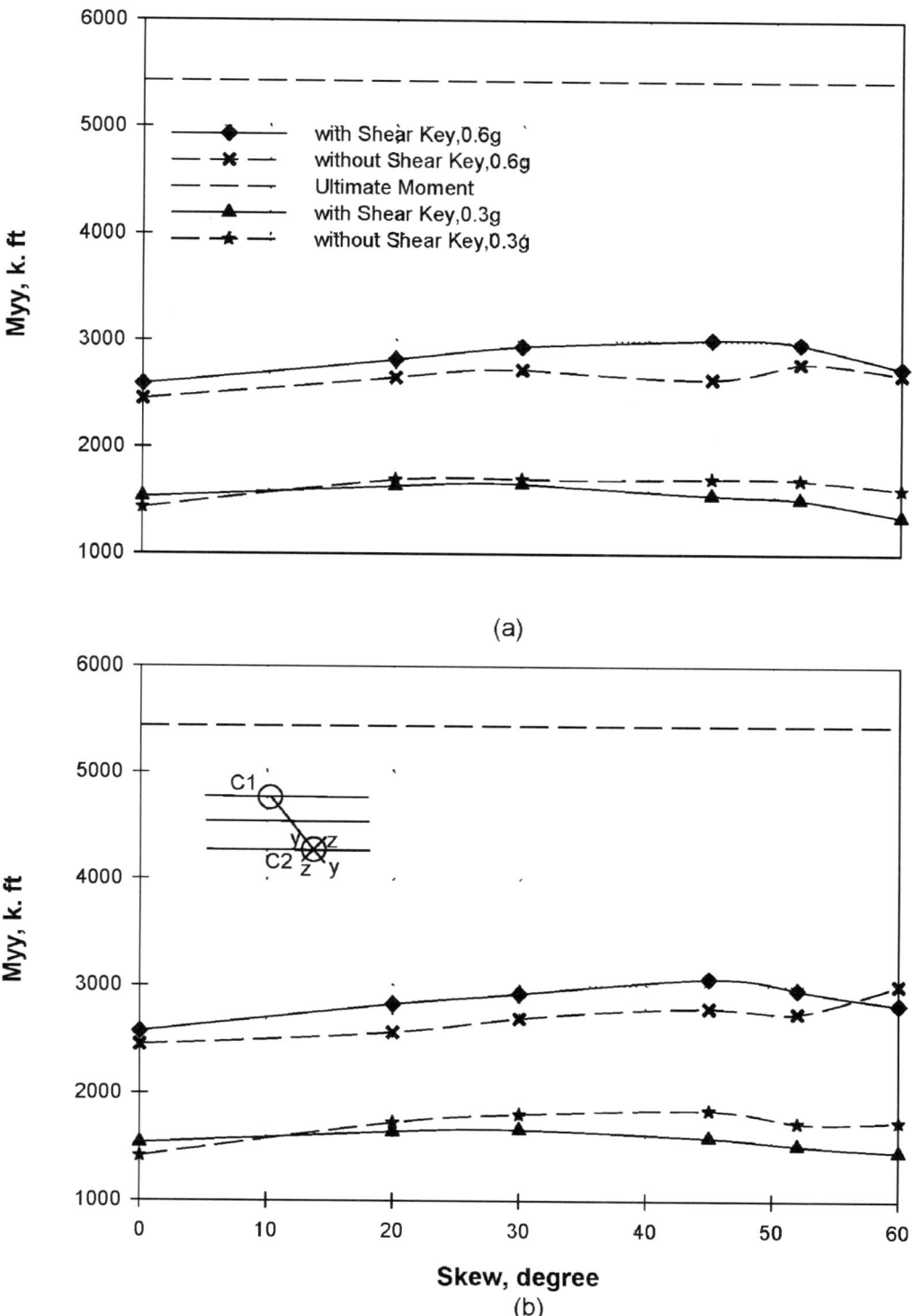

(a)

(b)

Figure 3-79. Average Moment in Y-Directions of (a) C1, and (b) C2 (Soil-B)

224

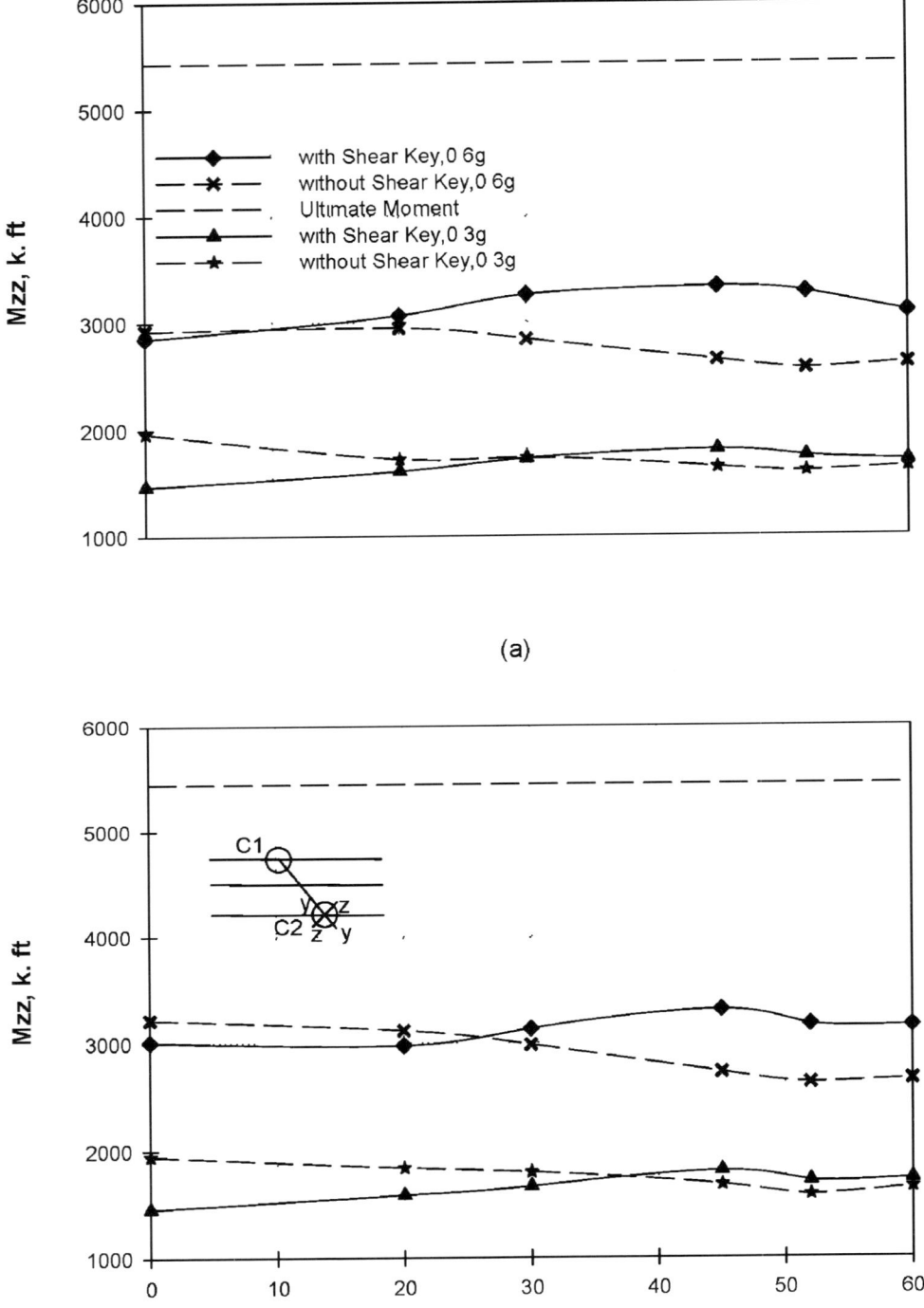

(a)

(b)

Figure 3-80. Average Moment in Z-Directions of (a) C1, and (b) C2 (Soil-B)

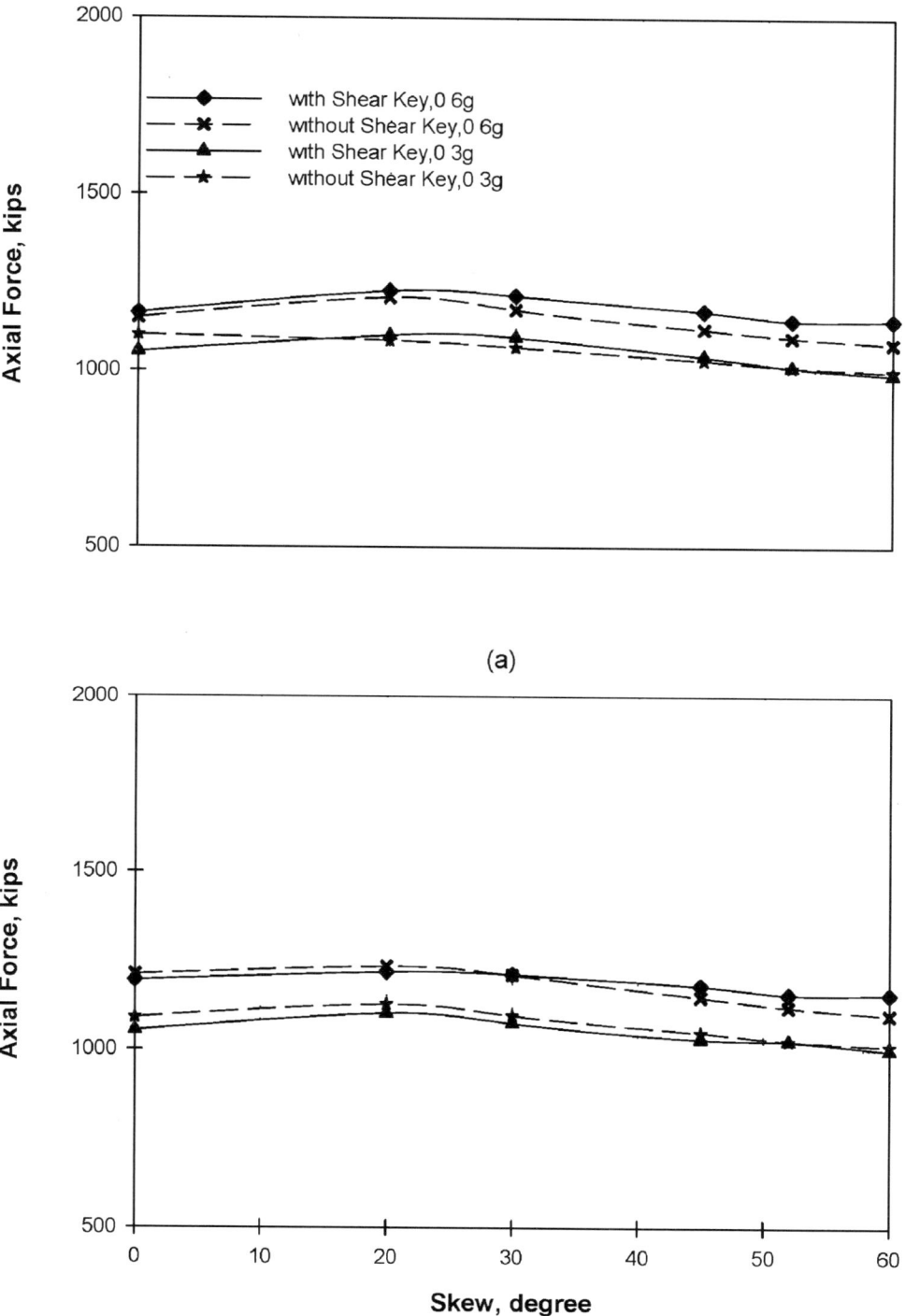

(a)

(b)

Figure 3-81. Average Axial Force in (a) C1, and (b) C2 (Soil-B)

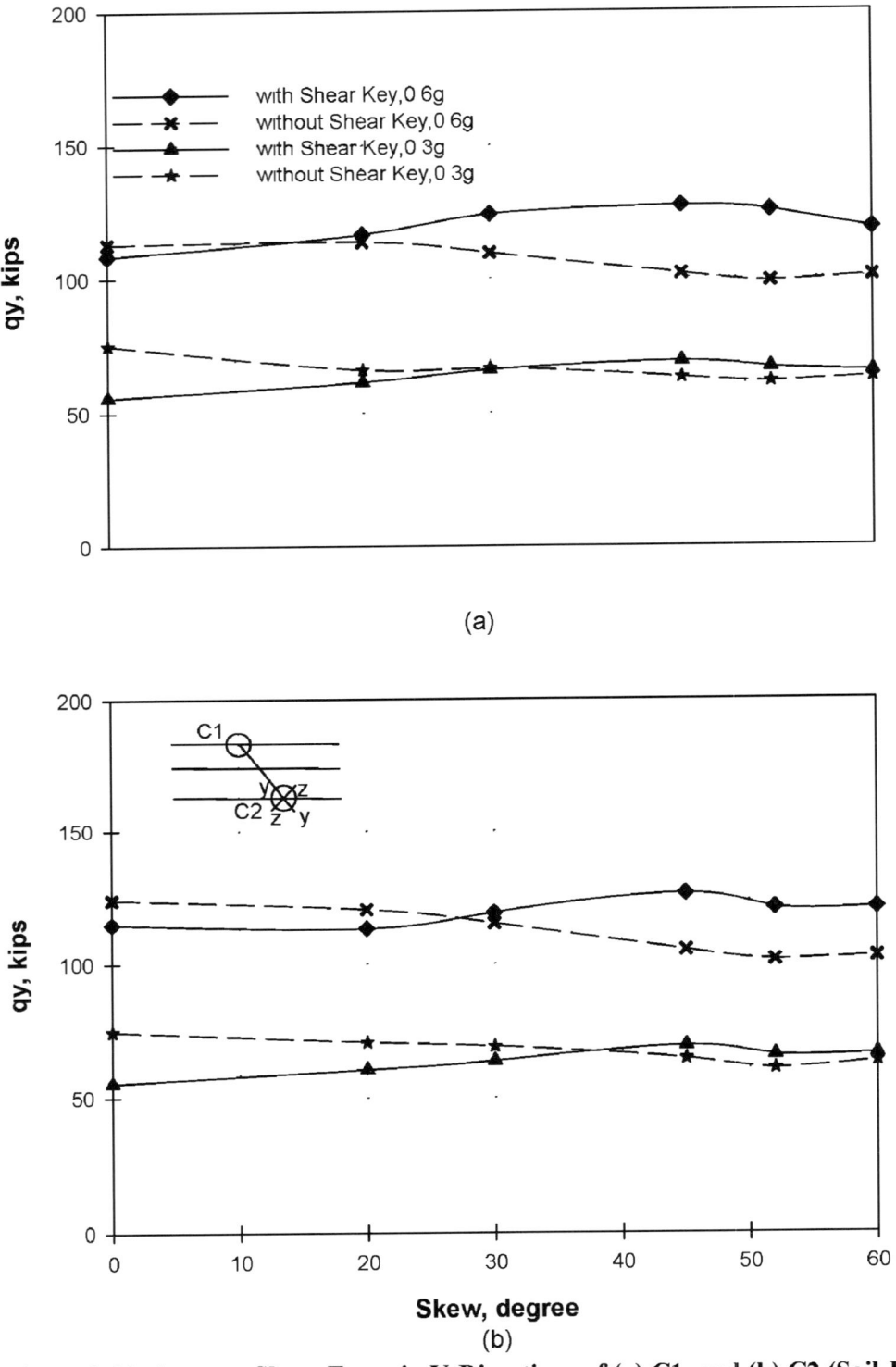

(a)

(b)

Figure 3-82. Average Shear Force in Y-Directions of (a) C1, and (b) C2 (Soil-B)

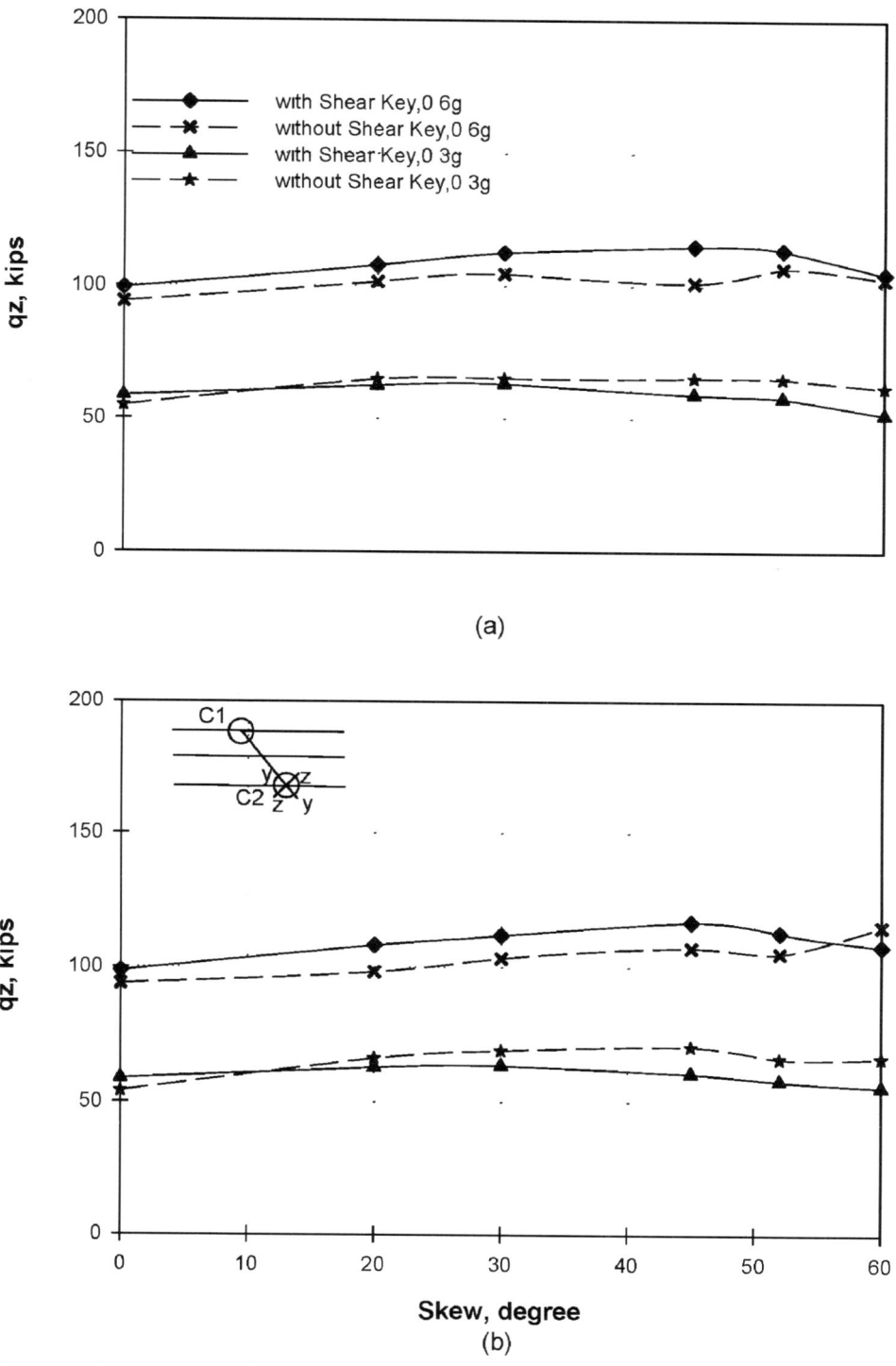

Figure 3-83. Average Shear Force in Z-Directions of (a) C1, and (b) C2 (Soil-B)

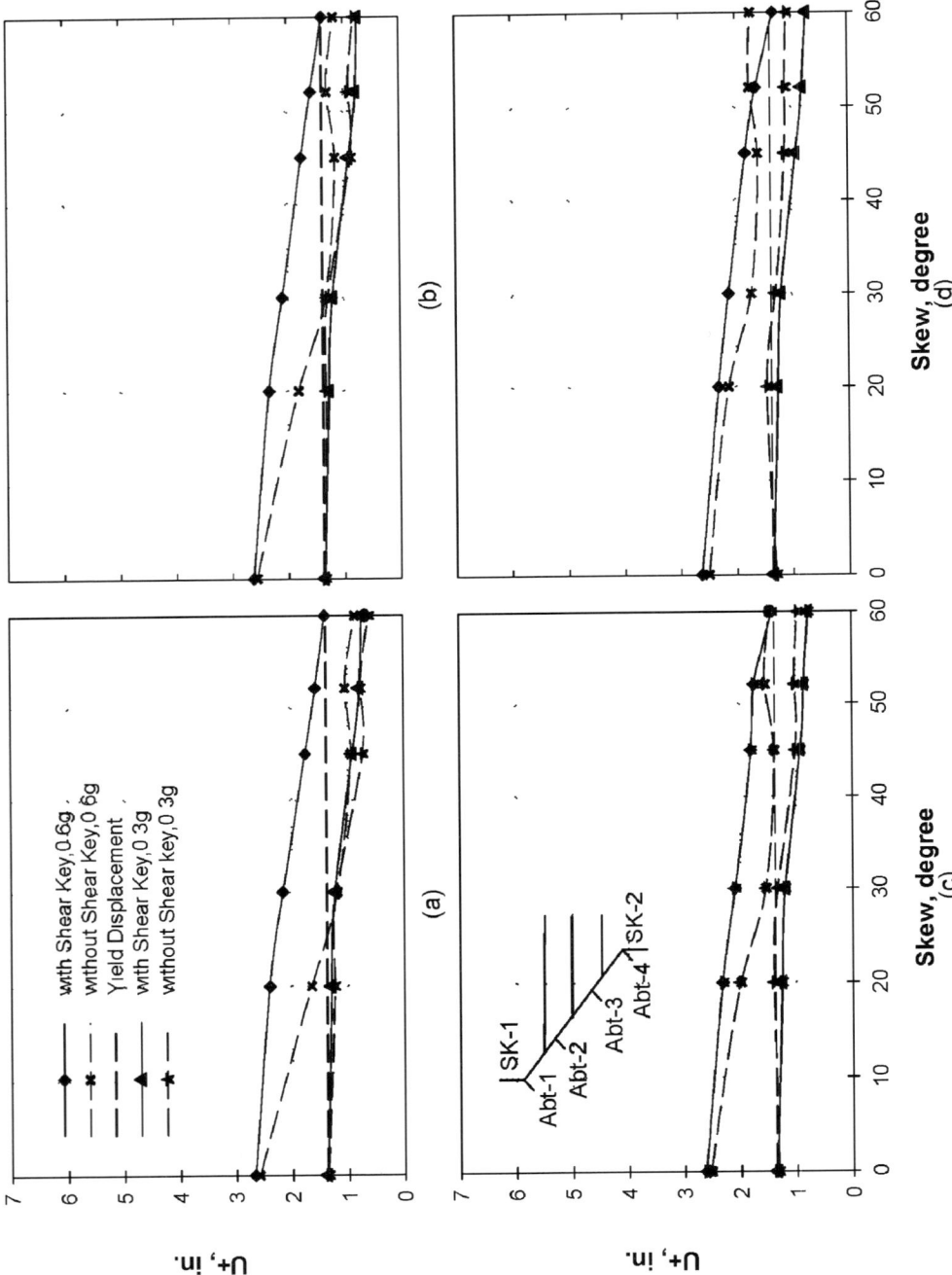

Figure 3-84. Average Displacement "into" Abutment's (a) Abt-1, (b) Abt-2, (c) Abt-3, and (d) Abt-4 (Soil-B)

228

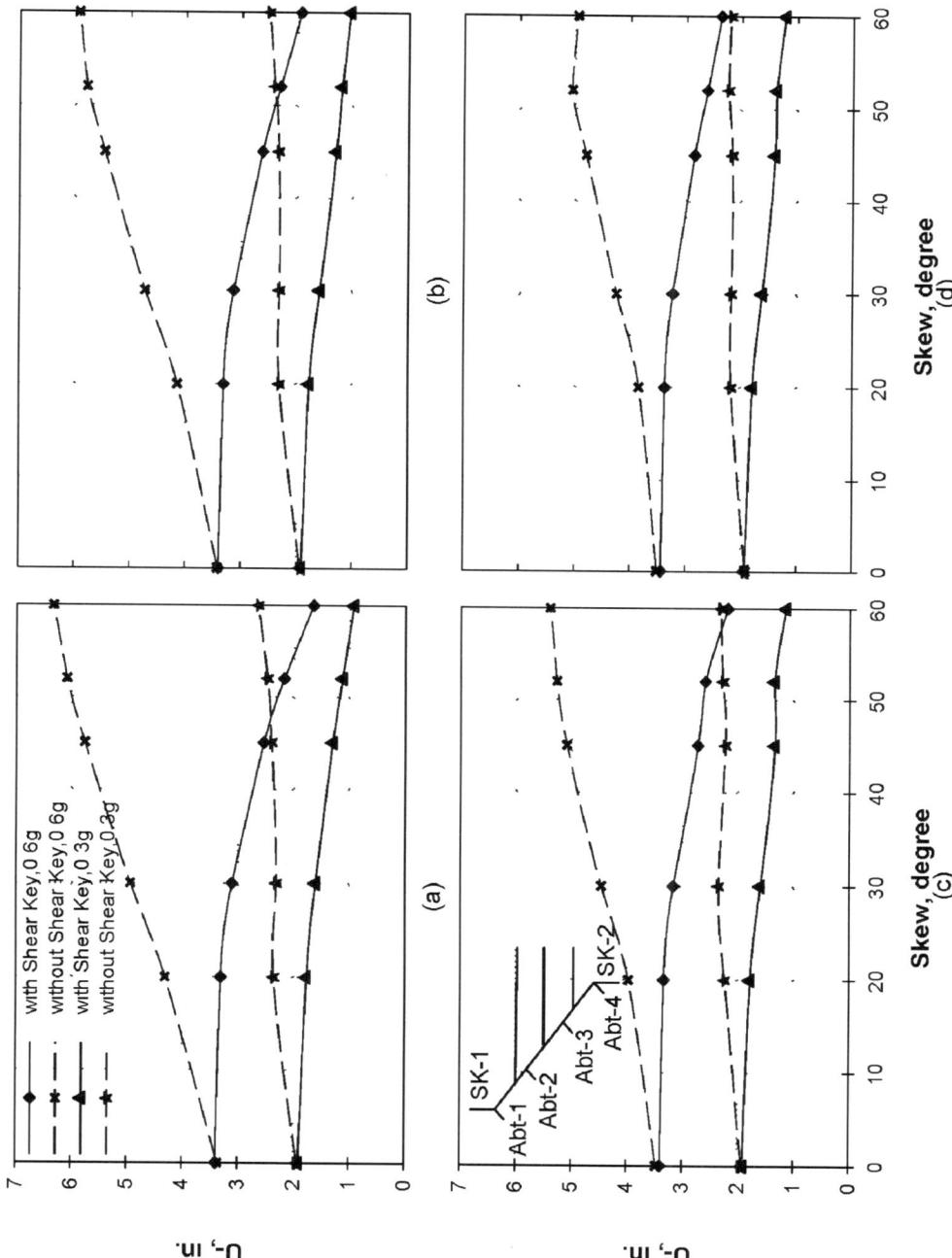

Figure 3-85. Average Displacement "away" from Abutment's (a) Abt-1, (b) Abt-2, (c) Abt-3, and (d) Abt-4 (Soil-B)

229

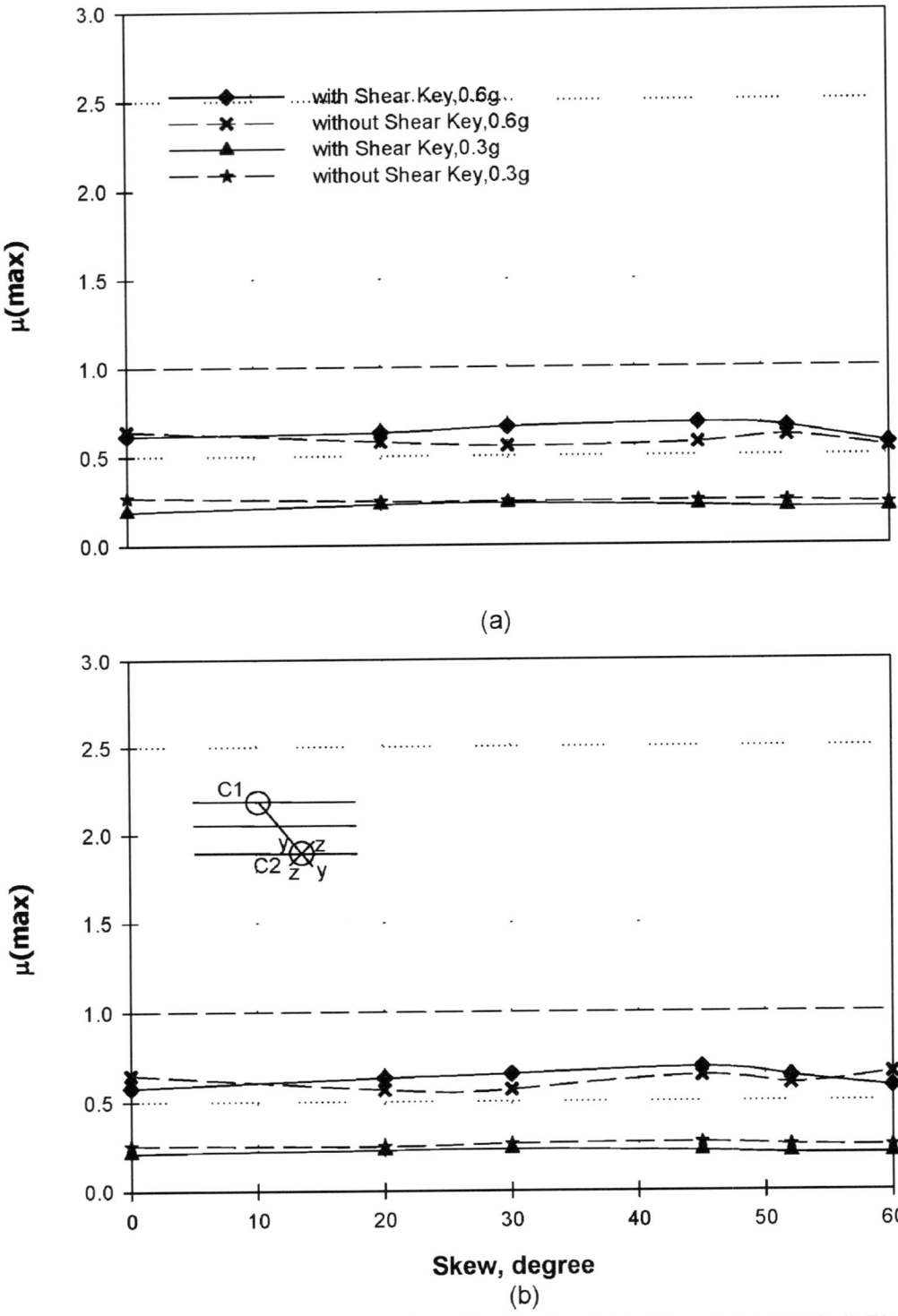

(a)

(b)

Figure 3-86. Maximum Average Ductility Ratio of (a) C1 and (b) C2 (Soil-B)

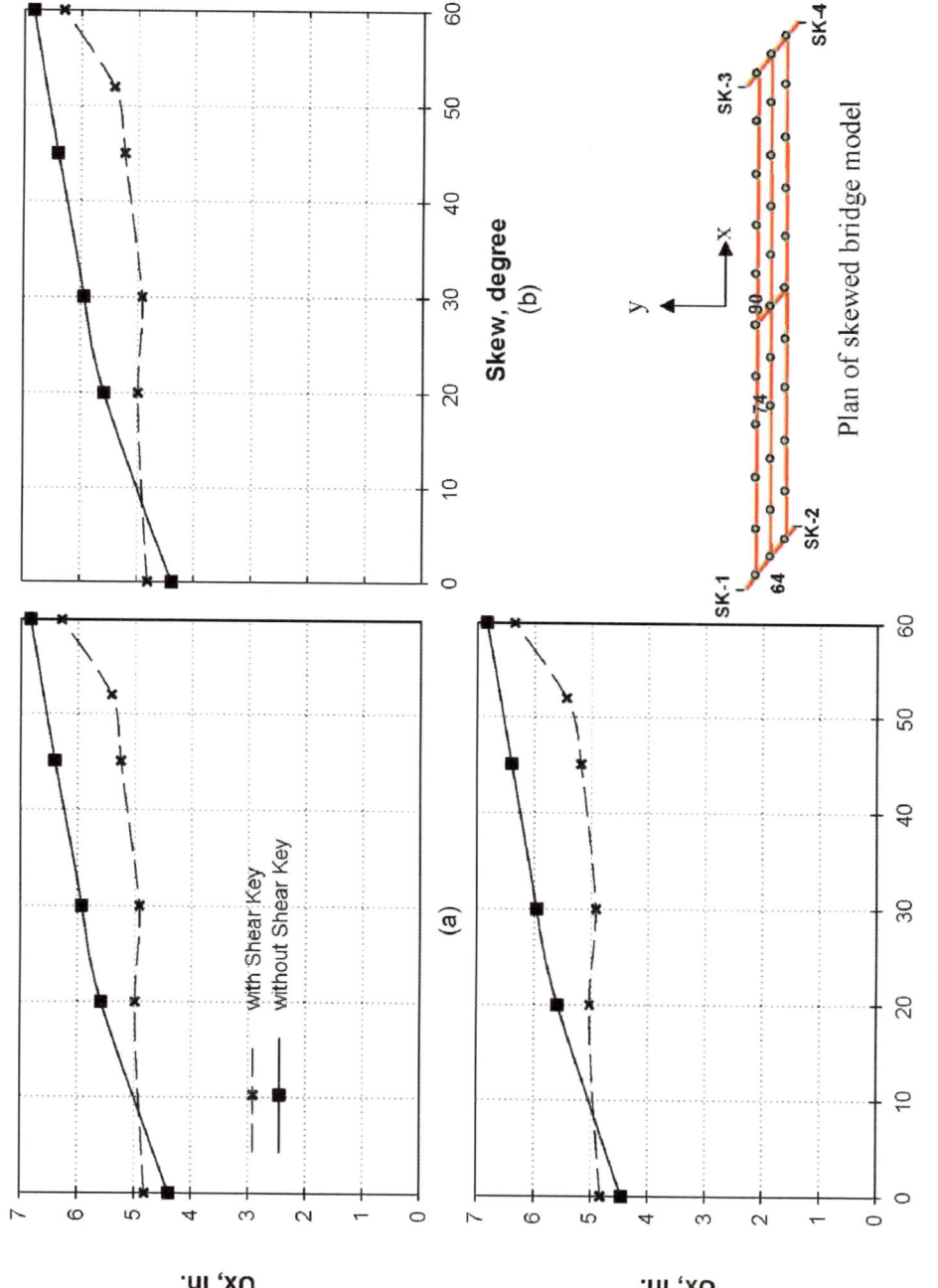

Figure 3-87. Average Displacement in X-direction for nodes (a) 90, (b) 74, and (c) 64 (0.6g-Soil-D-W/L=0.54)

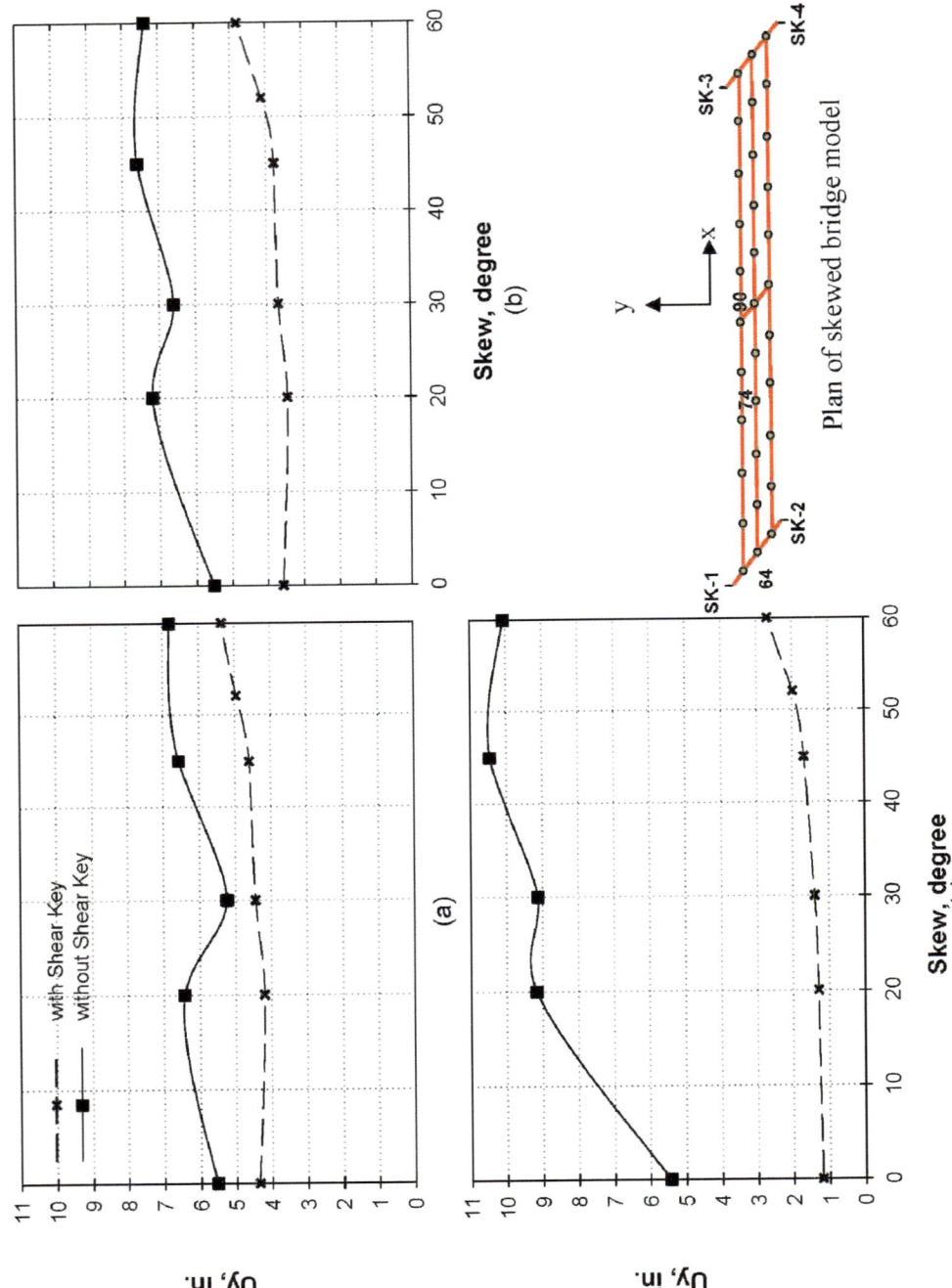

Figure 3-88. Average Displacement in Y-direction for nodes (a) 90, (b) 74, and (c) 64 (0.6g-Soil-D-W/L=0.54)

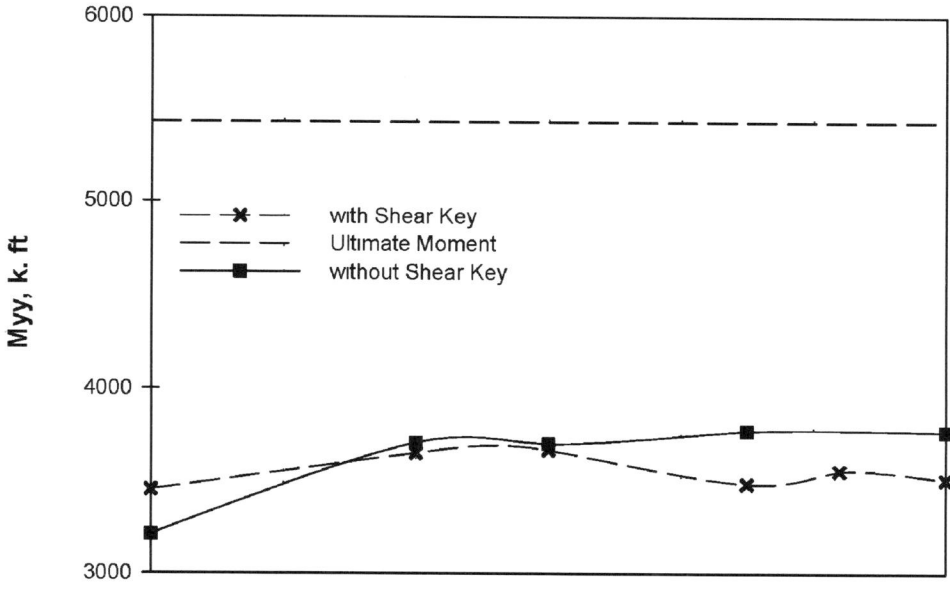

(a)

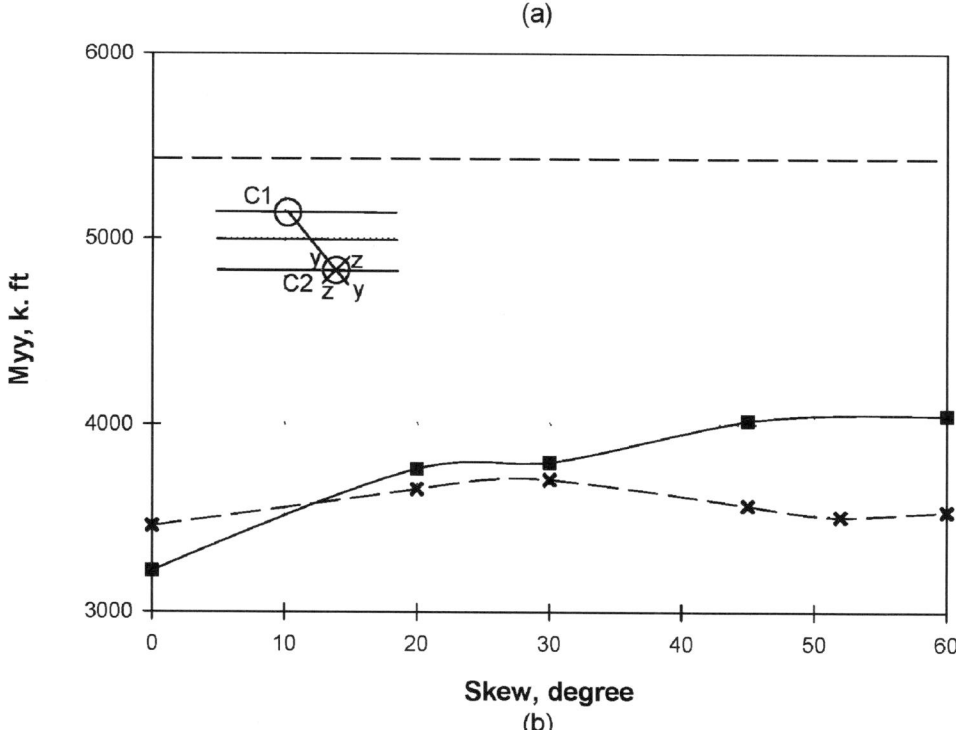

(b)

Figure 3-89. Average Moment in Y-Directions of (a) C1, and (b) C2 (0.6g-Soil-D-W/L=0.54)

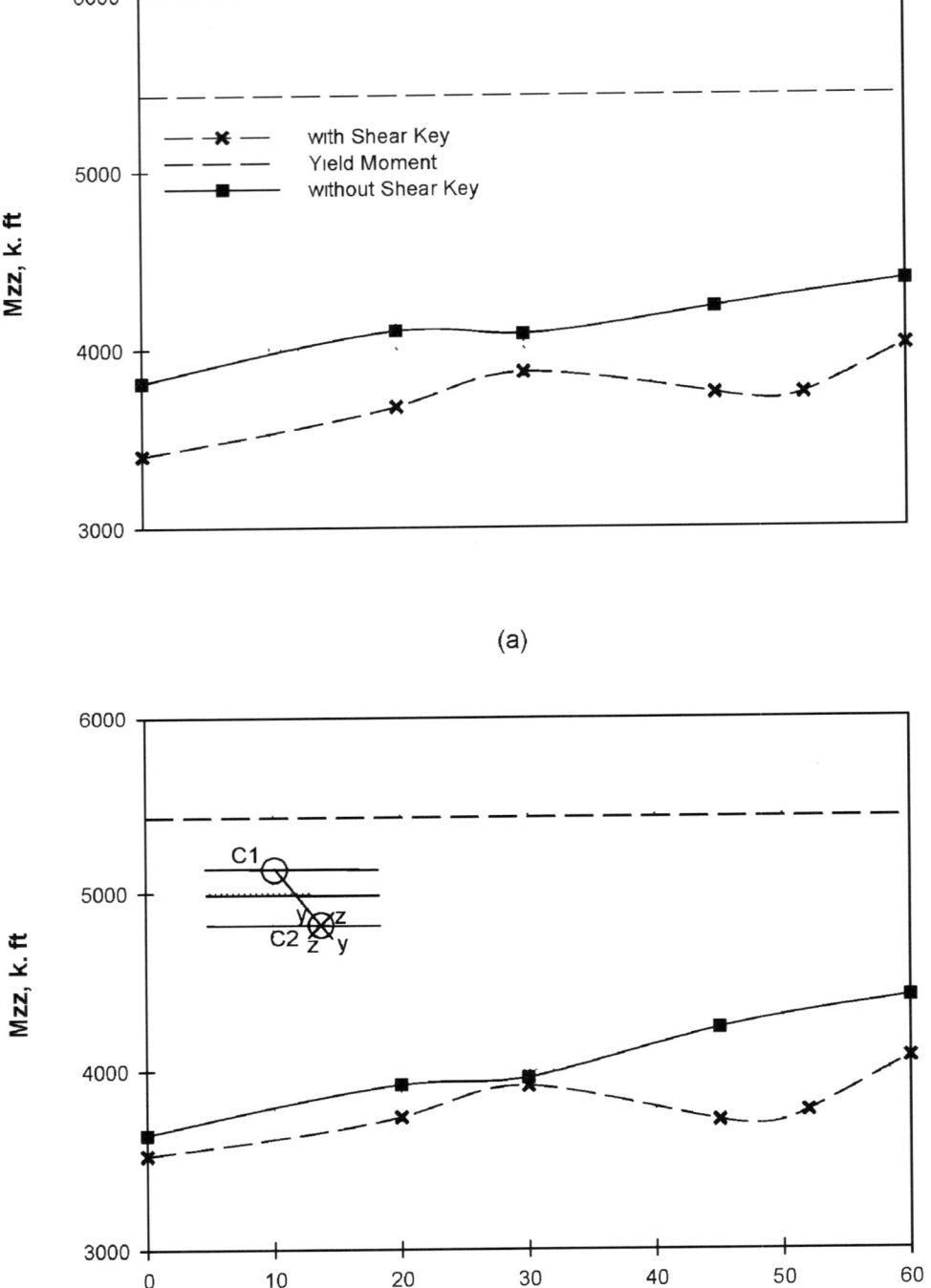

(a)

(b)

Figure 3-90. Average Moment in Z-Directions of (a) C1, and (b) C2 (0.6g-Soil-D-W/L=0.54)

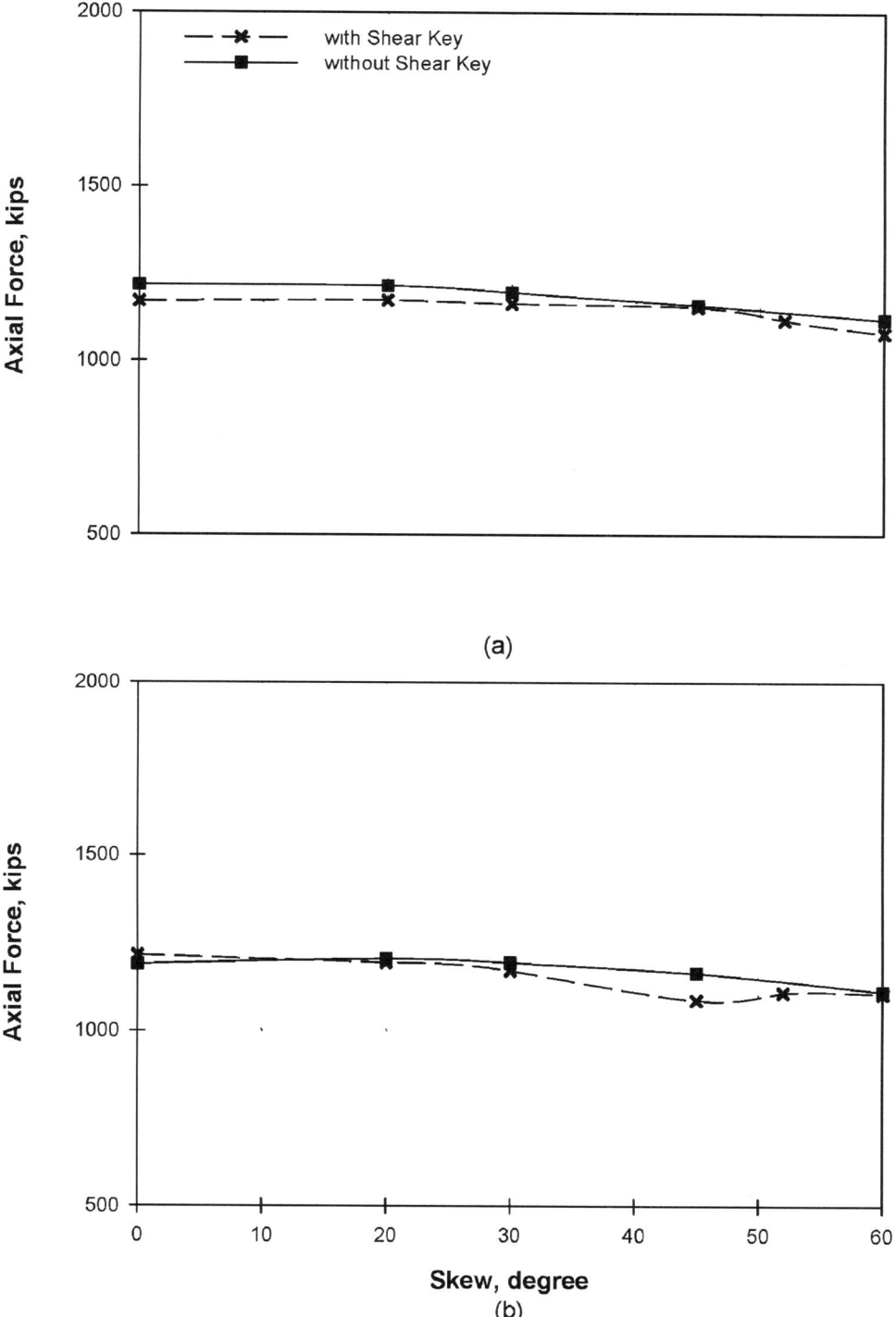

(a)

(b)

Figure 3-91. Average Axial Force in (a) C1, and (b) C2 (0.6g-Soil-D-W/L=0.54)

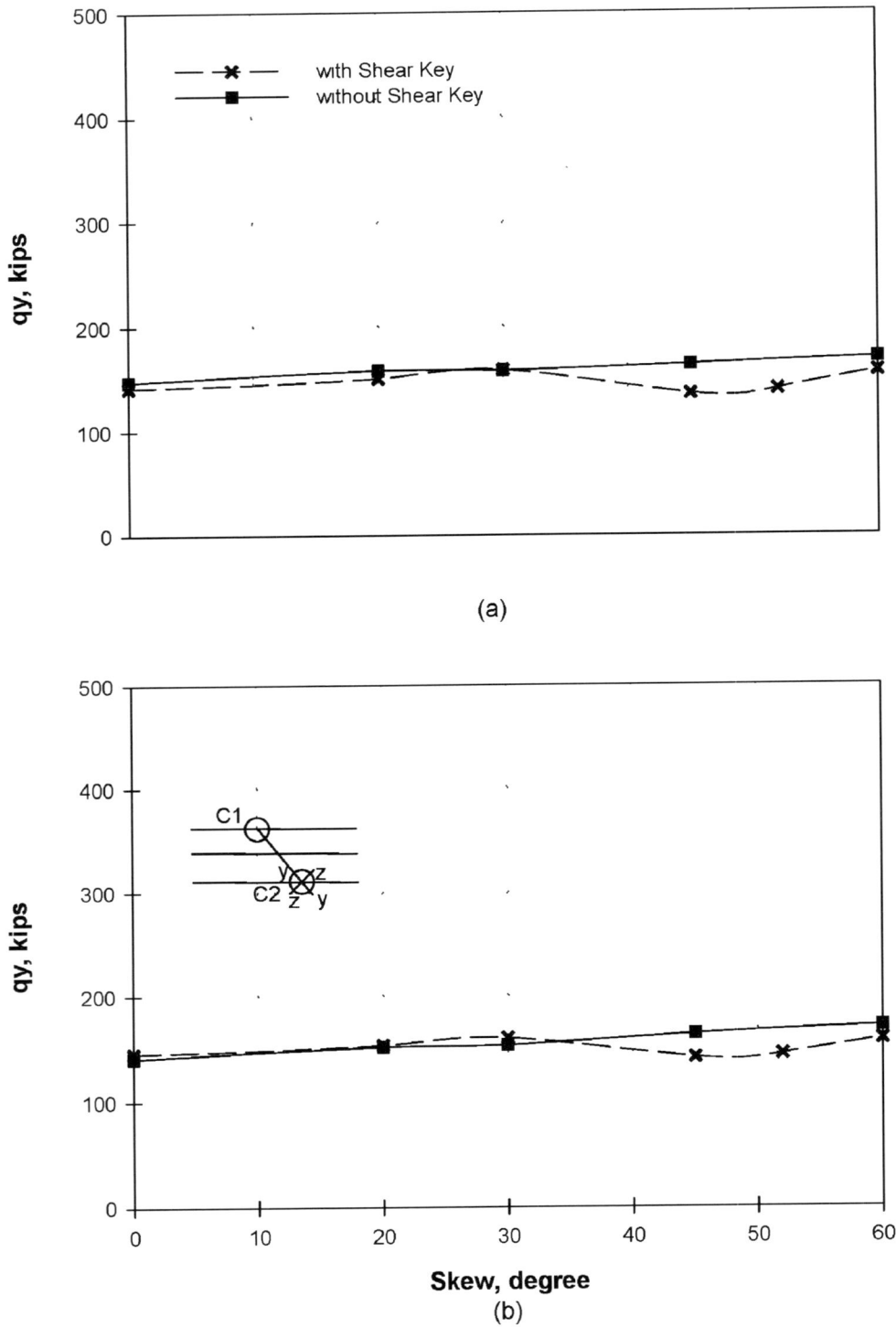

(a)

(b)

Figure 3-92. Average Shear Force in Y-Directions of (a) C1, and (b) C2 (0.6g-Soil-D-W/L=0.54)

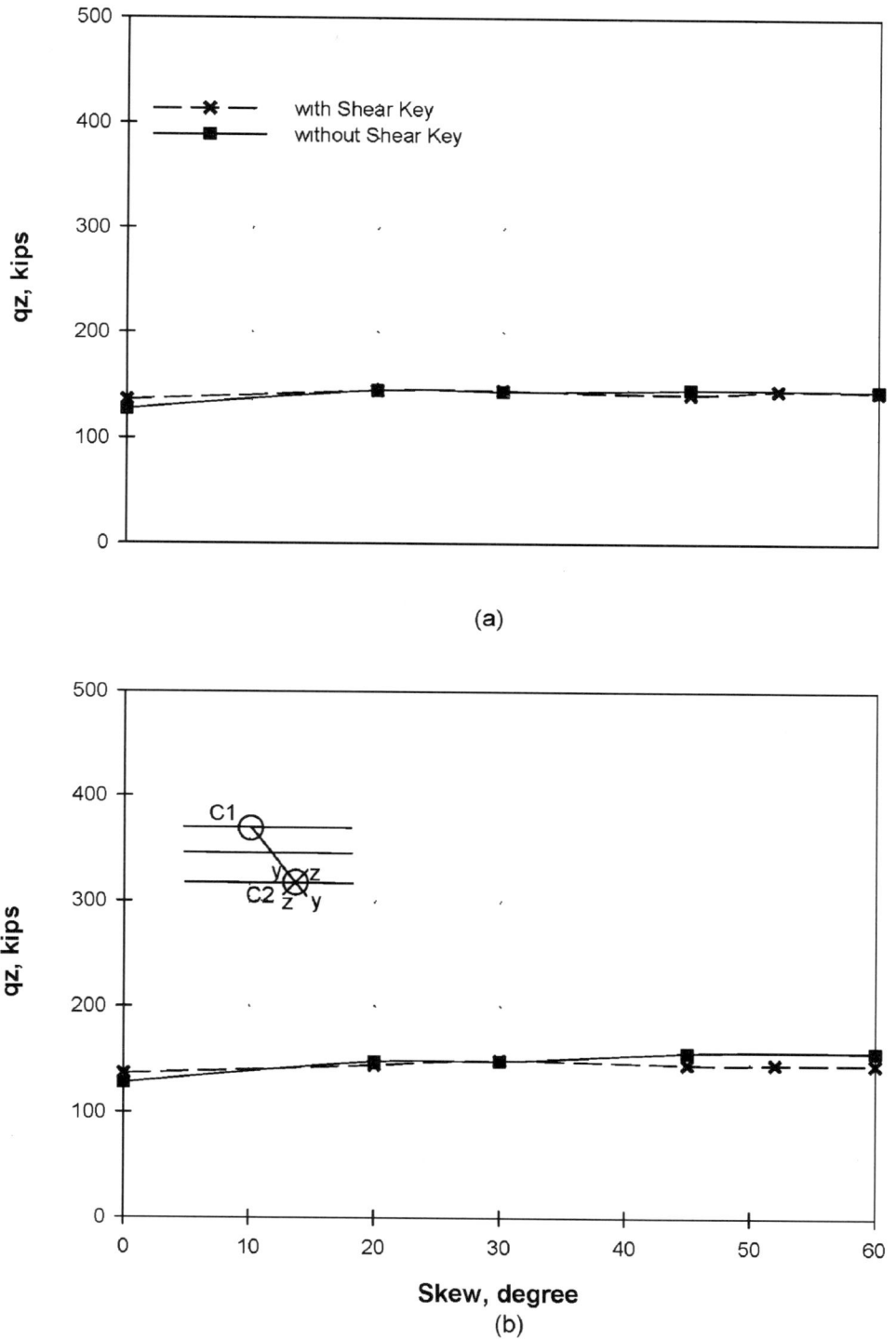

(a)

(b)

Figure 3-93. Average Shear Force in Z-Directions of (a) C1, and (b) C2 (0.6g-Soil-D-W/L=0.54)

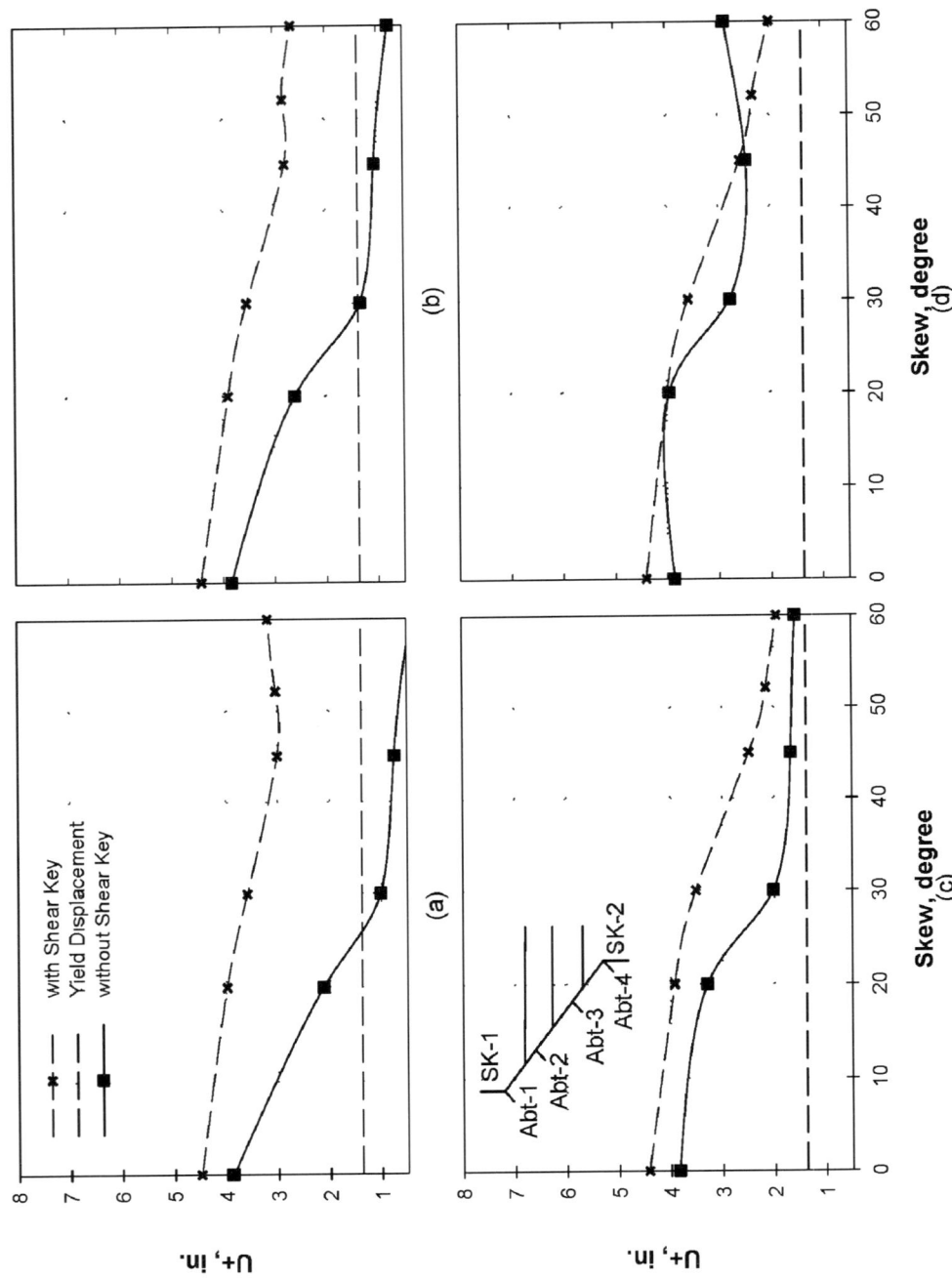

Figure 3-94. Average Displacement "into" Abutment's (a) Abt-1, (b) Abt-2, (c) Abt-3, and (d) Abt-4 (0.6g-Soil-D-W/L=0.54)

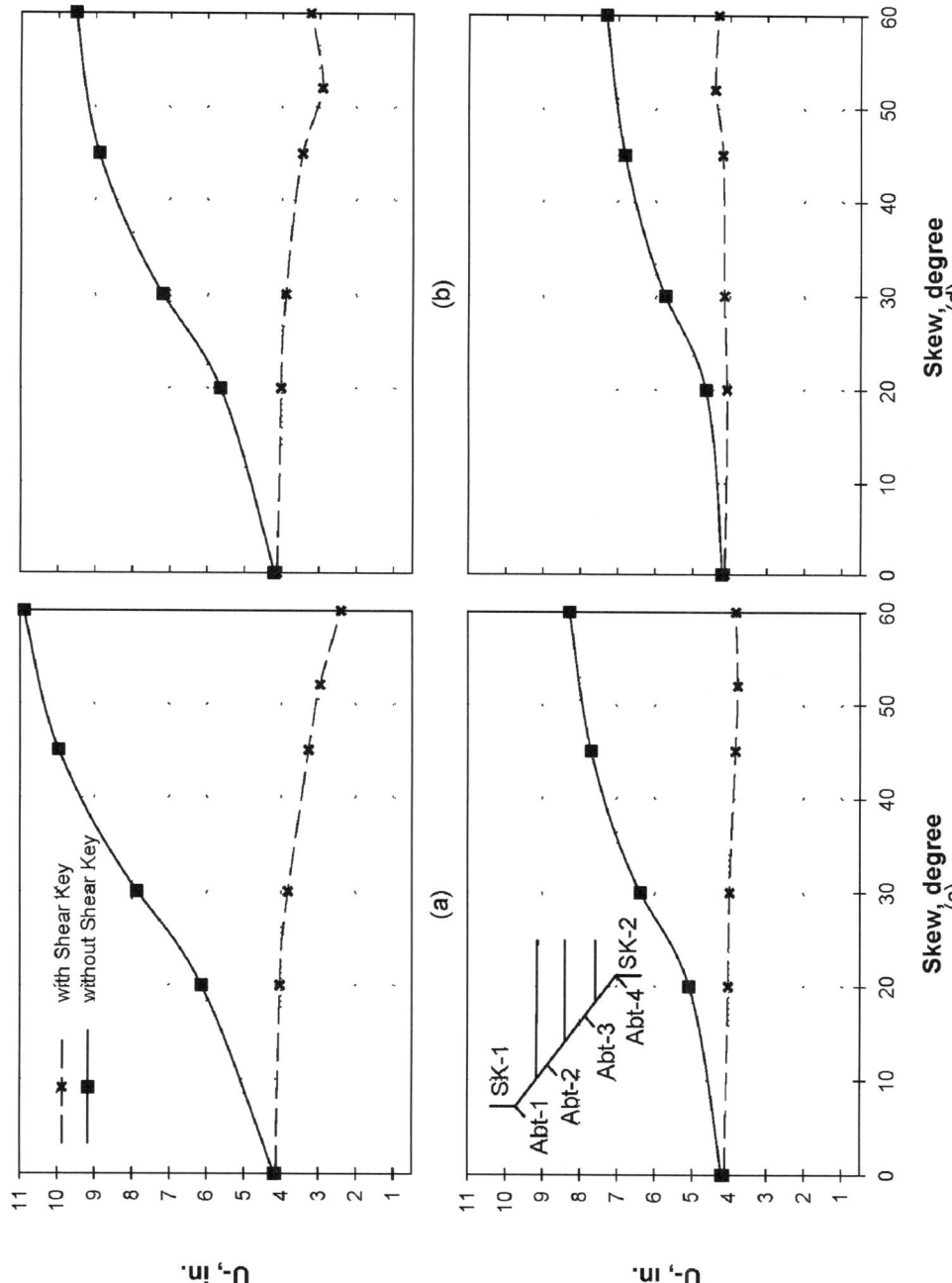

Figure 3-95. Average Displacement "away" Abutment's (a) Abt-1, (b) Abt-2, (c) Abt-3, and (d) Abt-4 (0.6g-Soil-D-W/L=0.54)

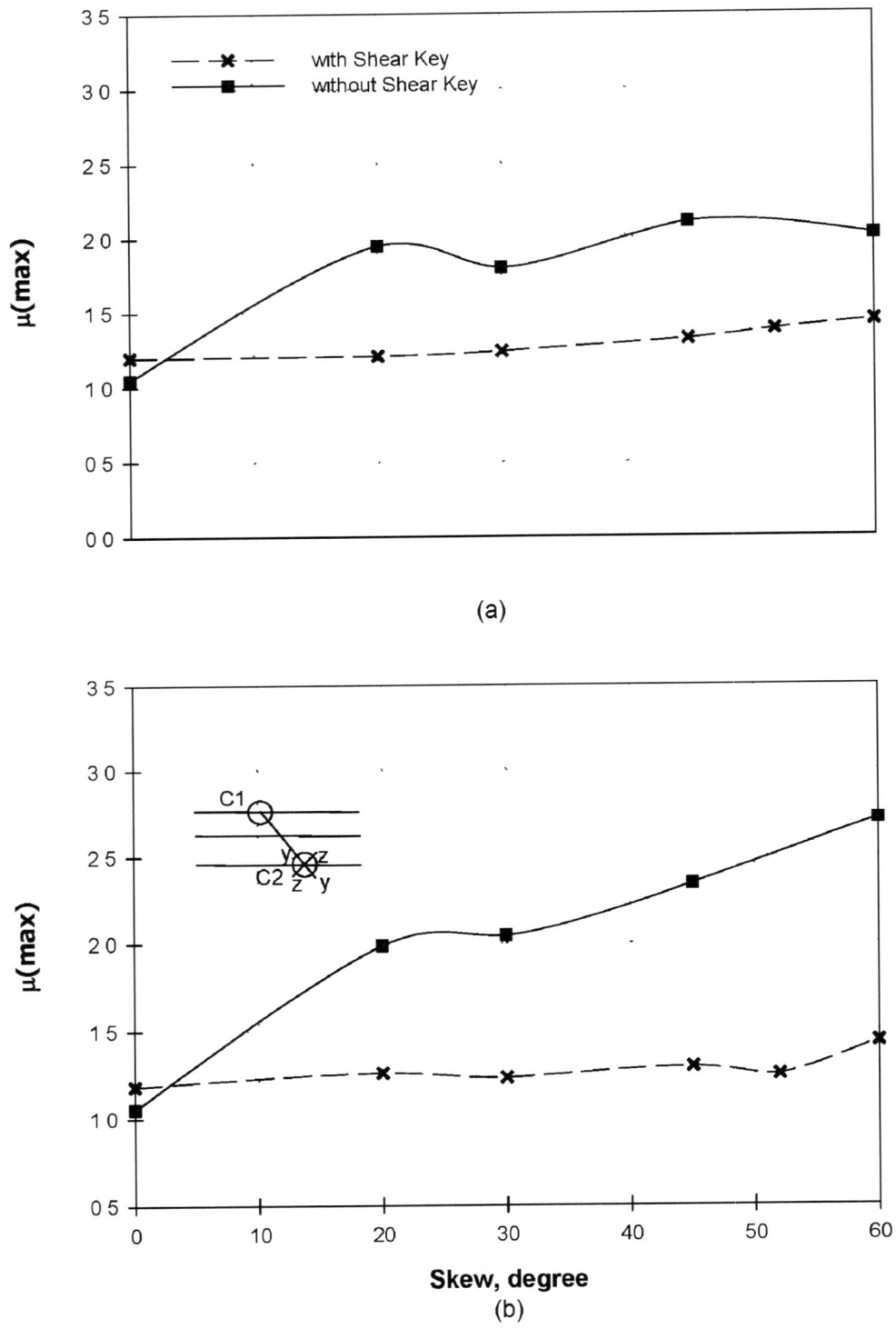

(a)

(b)

Figure 3-96. Maximum Average Ductility Ratio of (a) C1 and (b) C2 (0.6g-Soil-D-W/L=0.54)

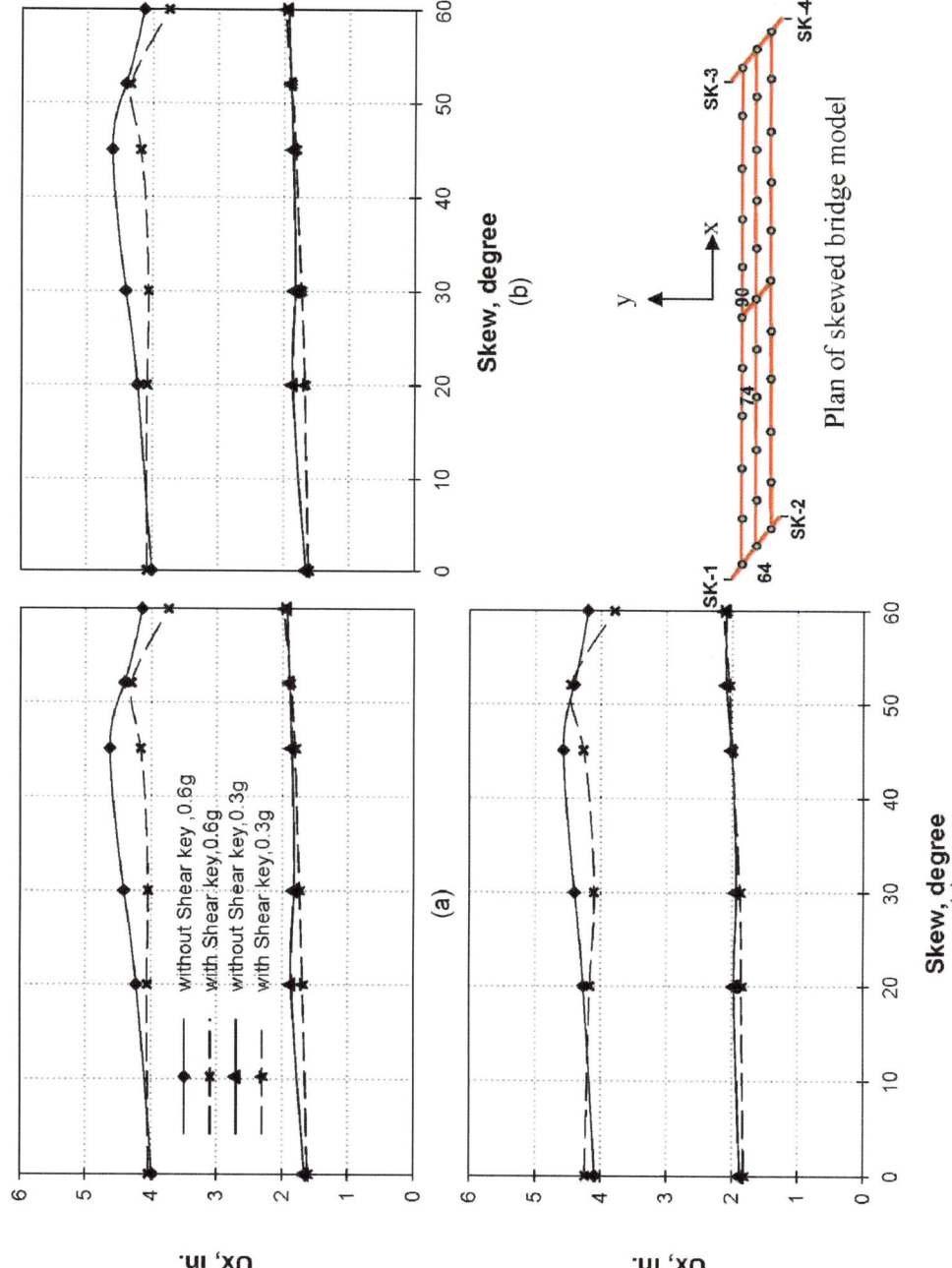

Figure 3-97. Average Displacement in X-direction for nodes (a) 90, (b) 74, and (c) 64 (Soil-D-F)

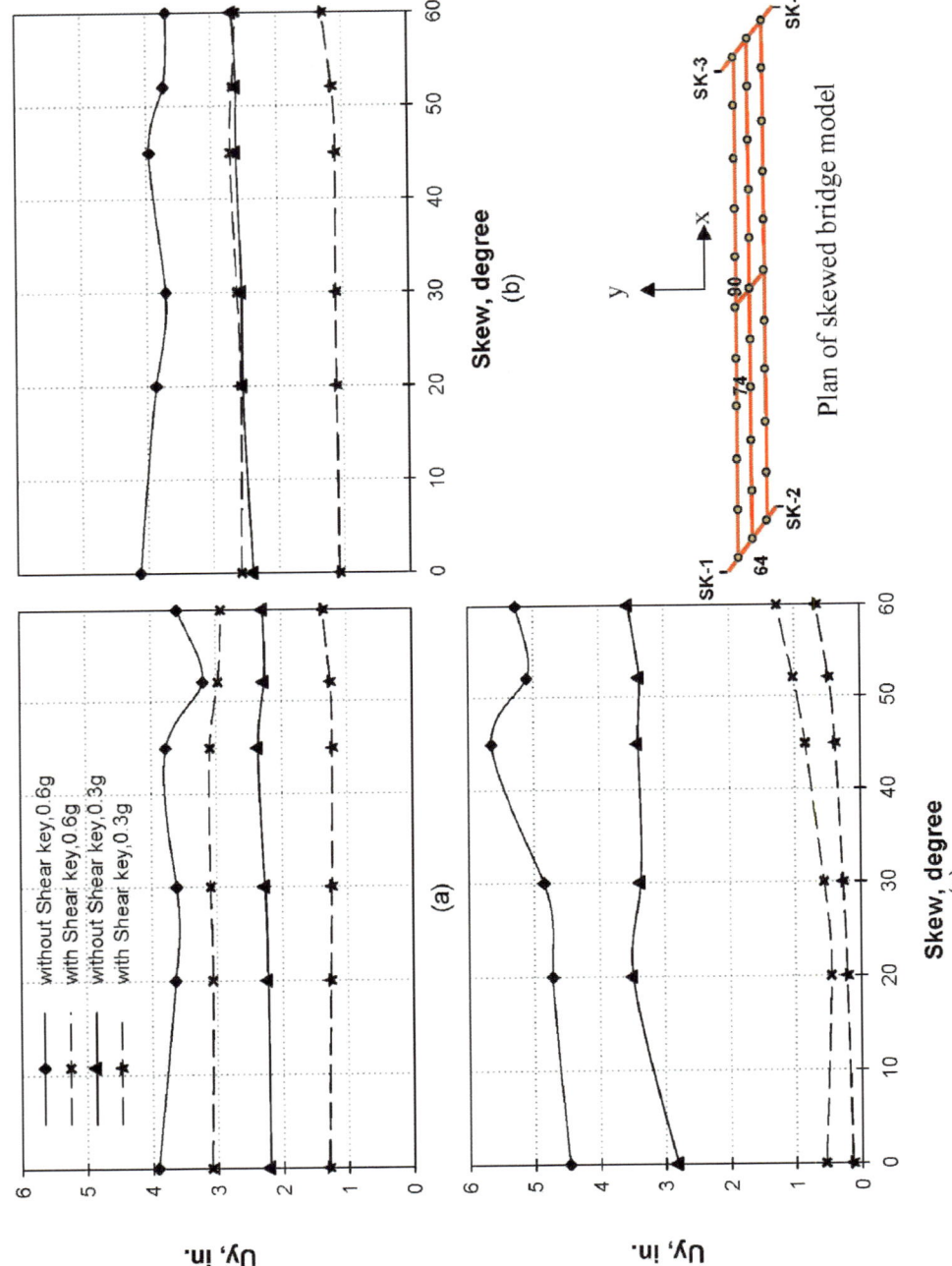

Figure 3-98. Average Displacement in Y-direction for nodes (a) 90, (b) 74, and (c) 64 (Soil-D-F)

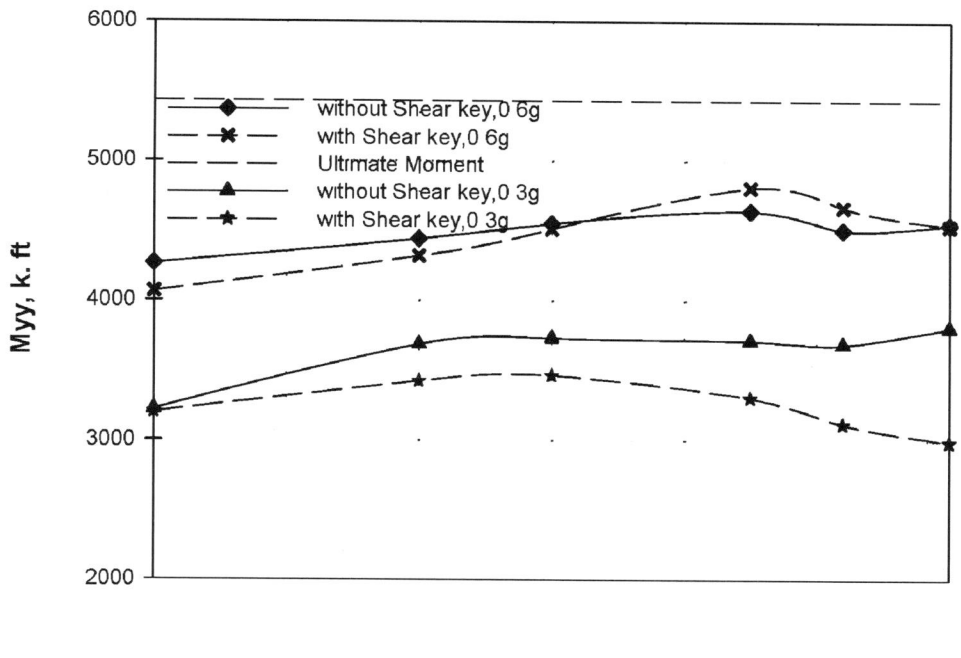

(a)

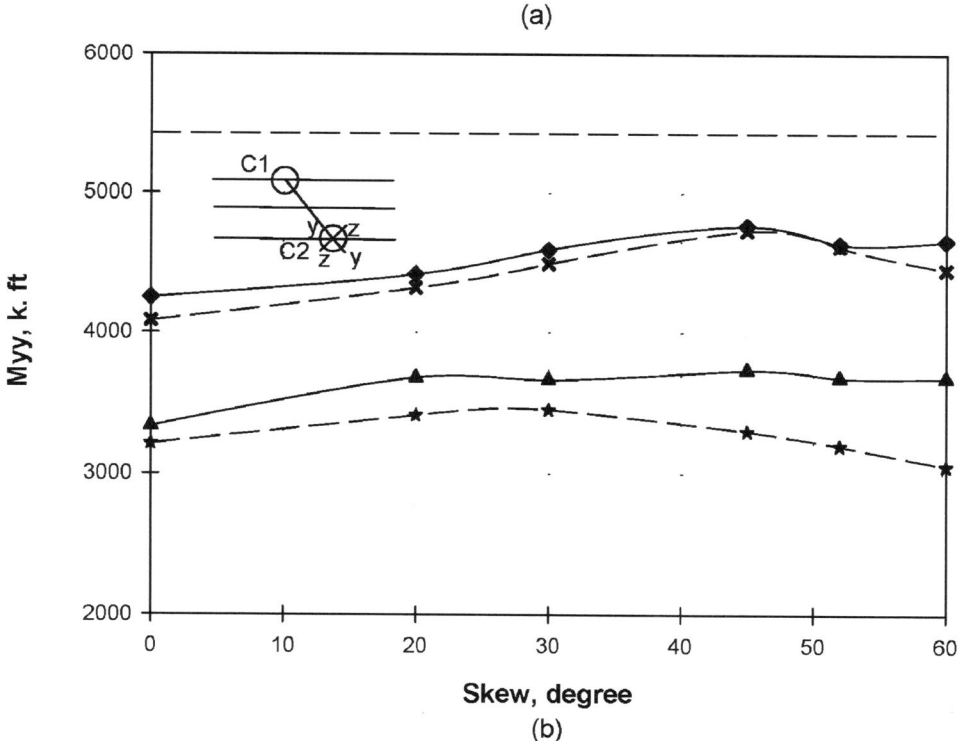

Skew, degree

(b)

Figure 3-99. Average Moment in Y-Directions of (a) C1, and (b) C2 (Soil-D-F)

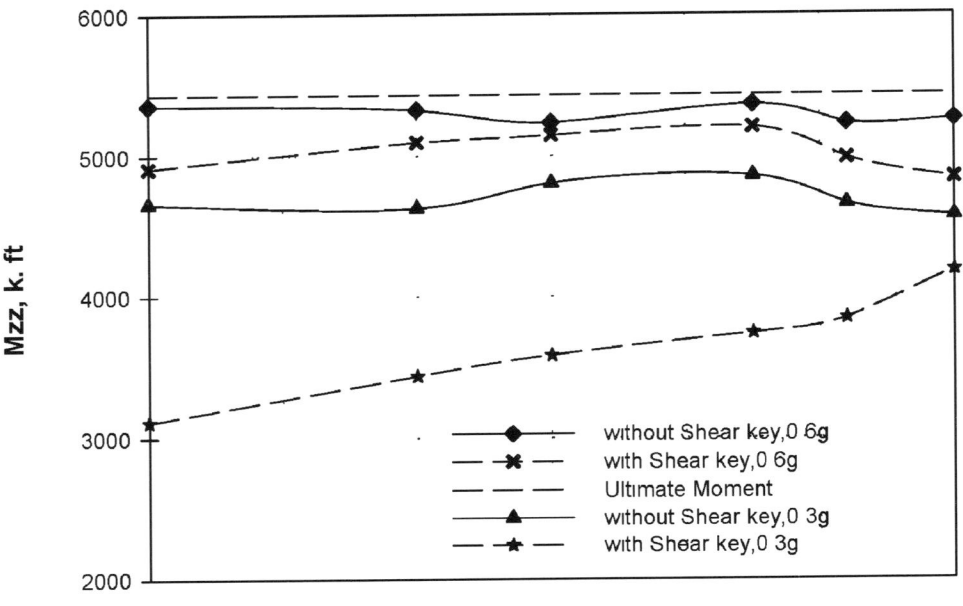

(a)

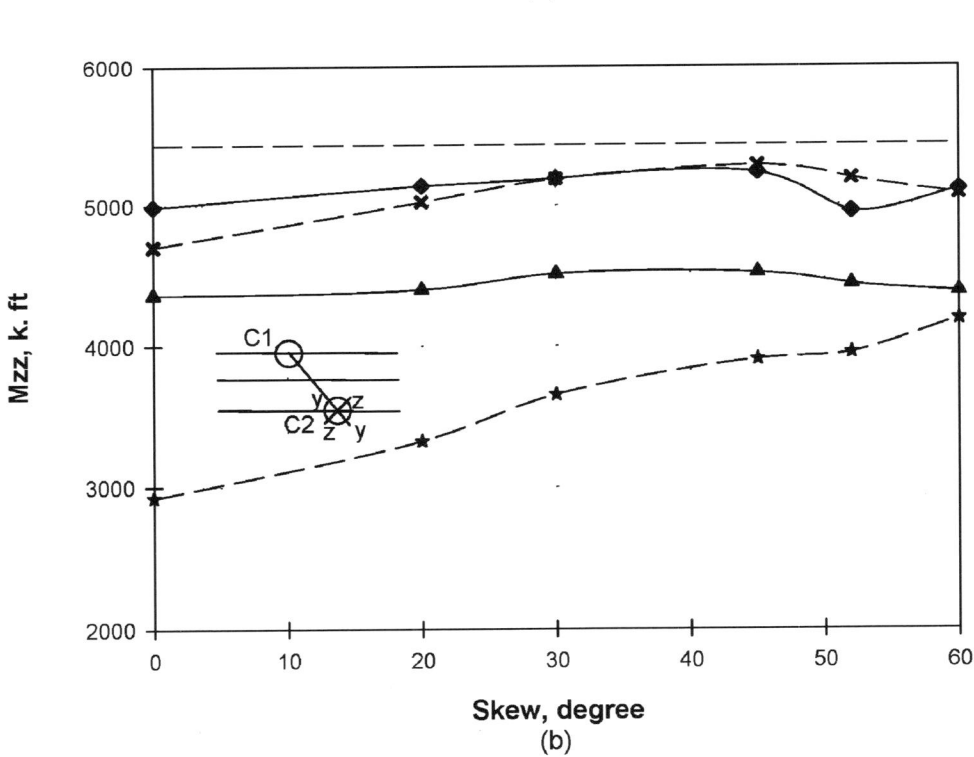

Skew, degree

(b)

Figure 3-100. Average Moment in Z-Directions of (a) C1, and (b) C2 (Soil-D-F)

245

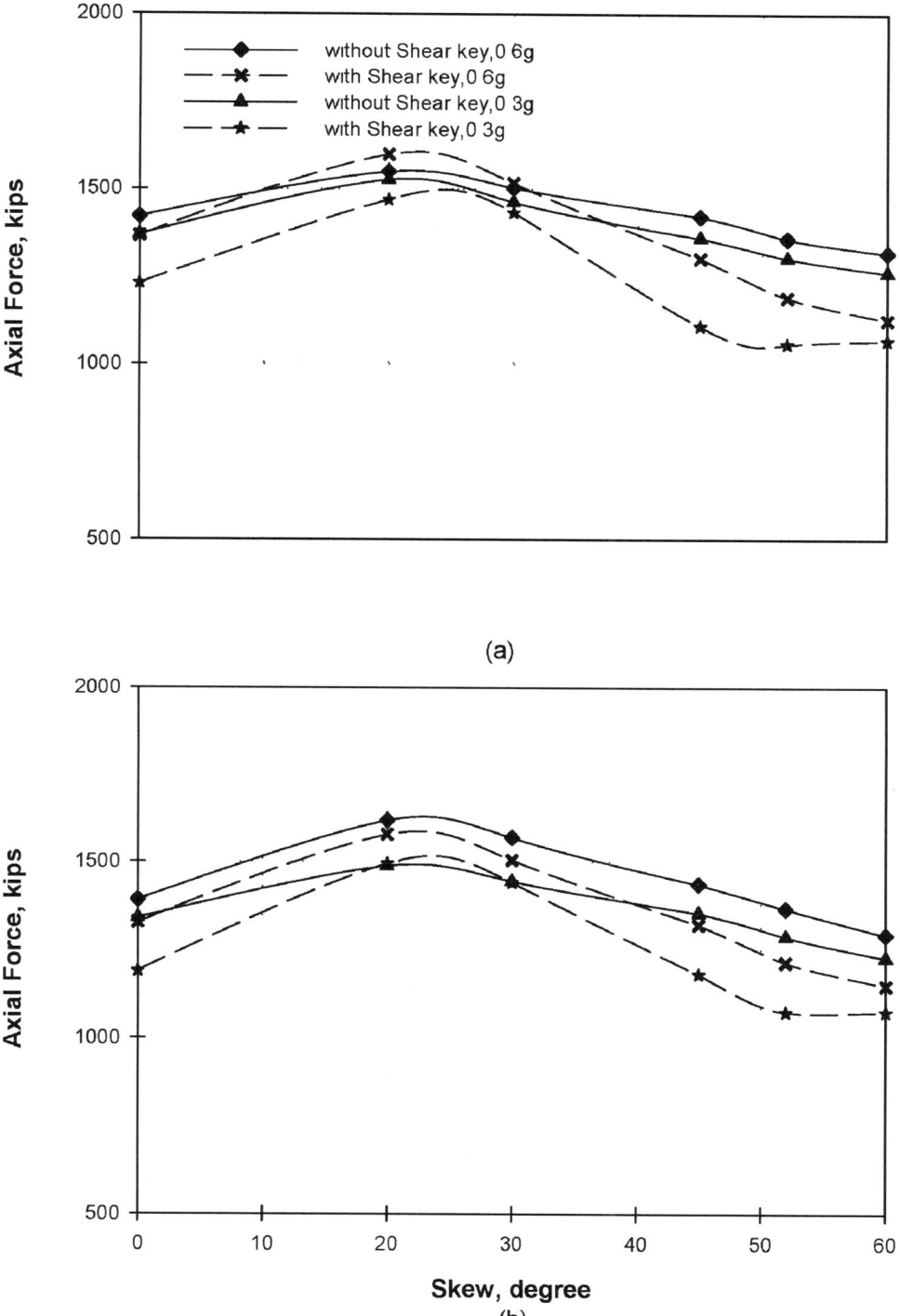

(a)

(b)

Figure 3-101. Average Axial Force in (a) C1, and (b) C2 (Soil-D-F)

246

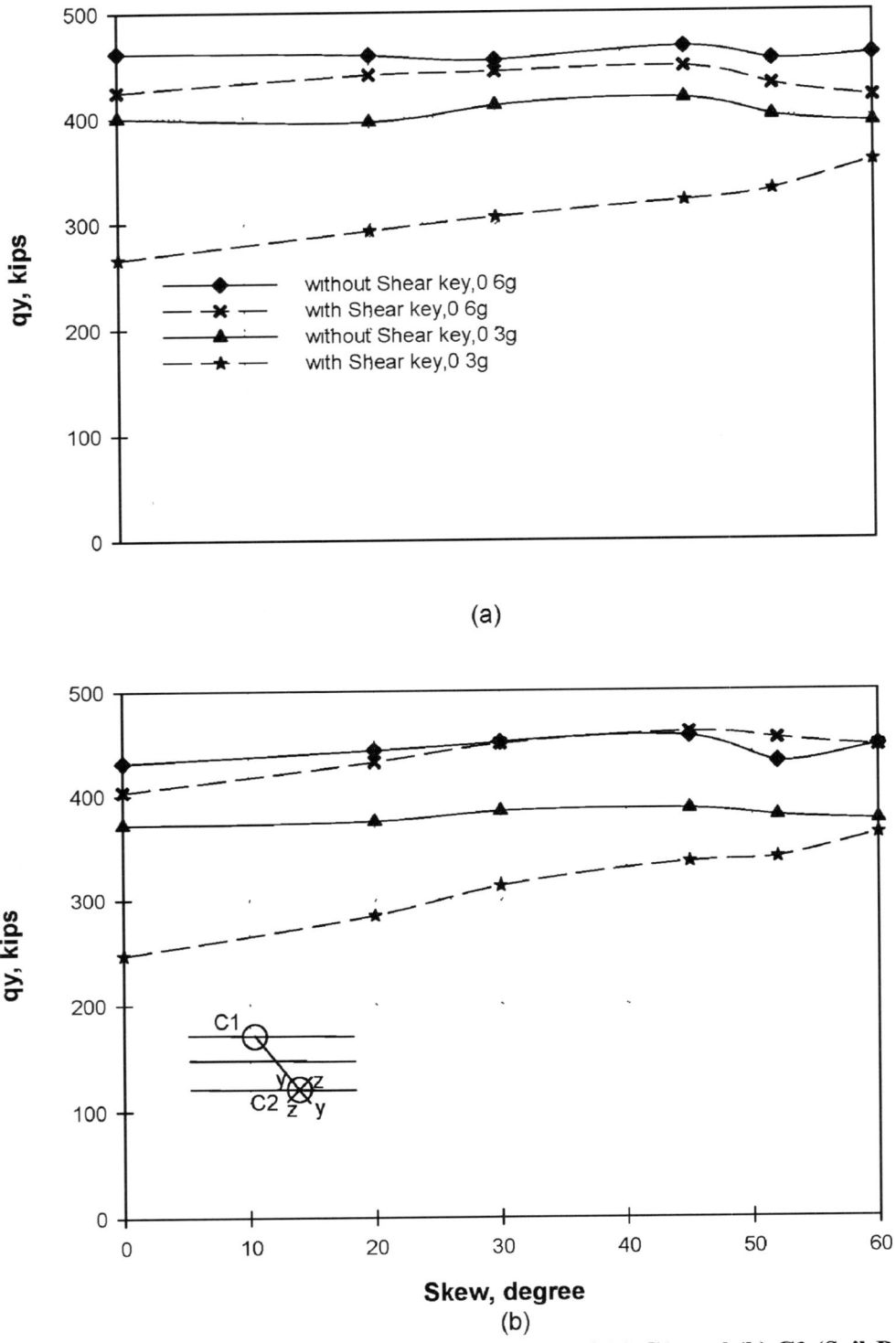

(a)

(b)

Figure 3-102. Average Shear Force in Y-Directions of (a) C1, and (b) C2 (Soil-D-F)

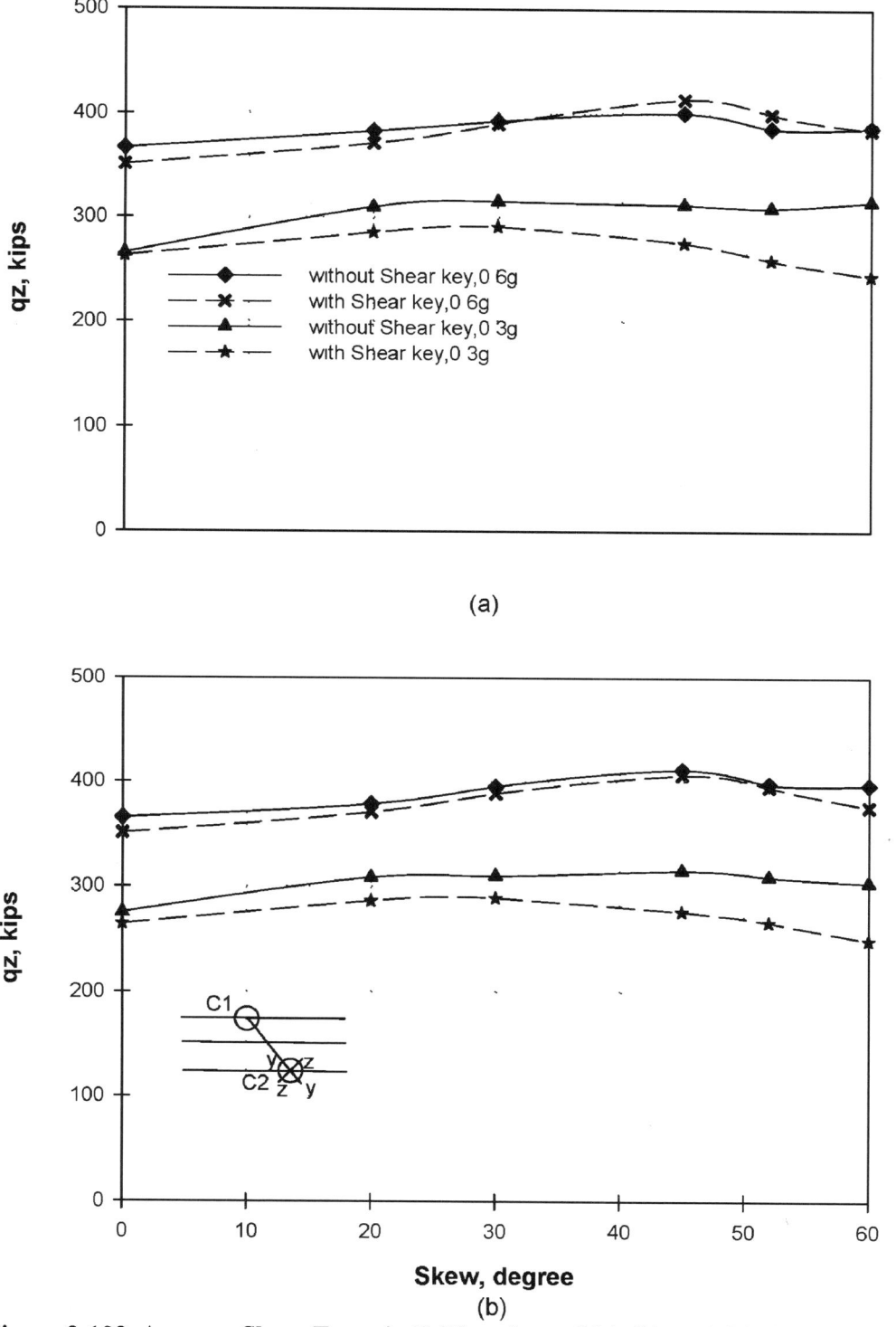

(a)

(b)

Figure 3-103. Average Shear Force in Z-Directions of (a) C1, and (b) C2 (Soil-D-F)

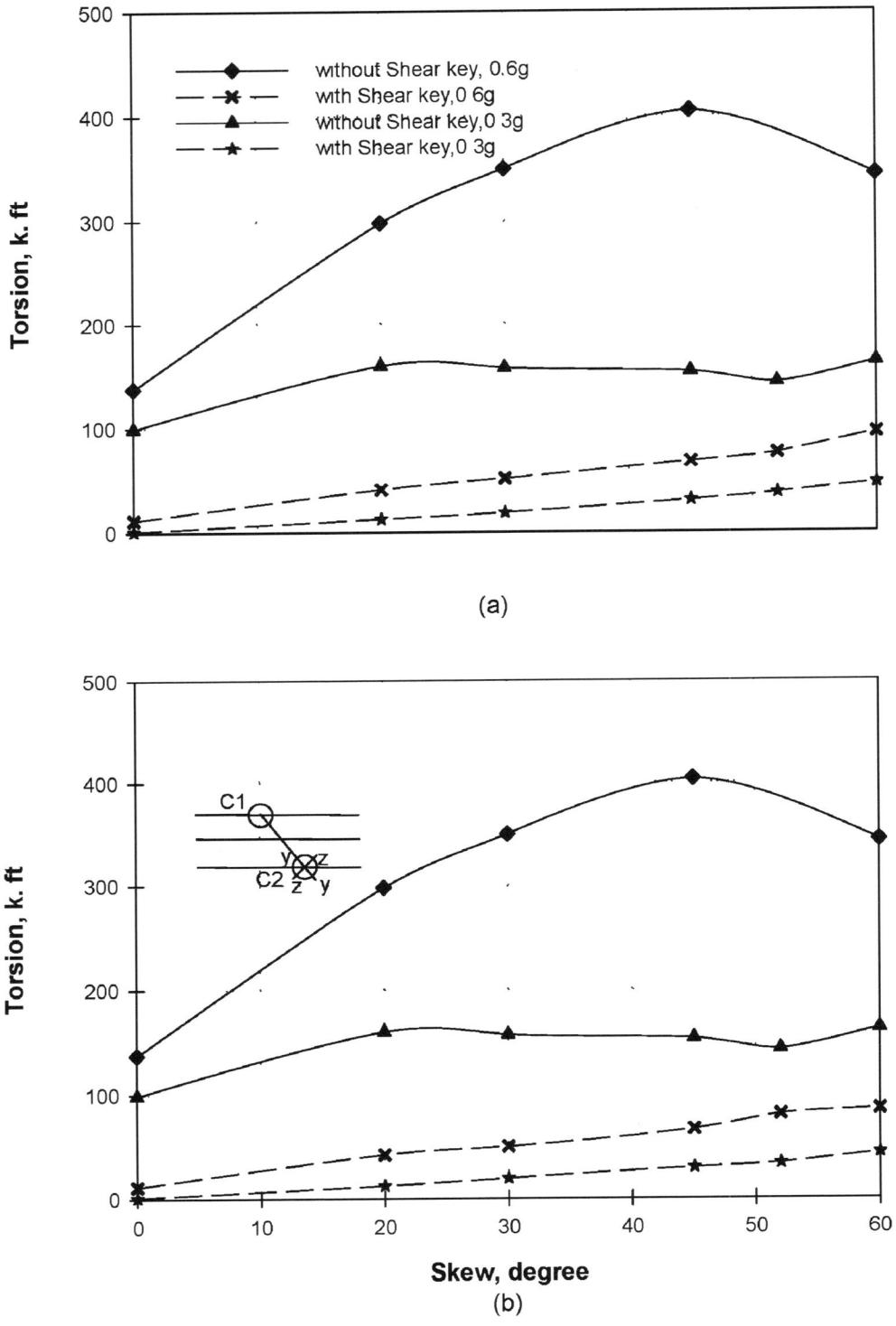

(a)

(b)

Figure 3-104. Average Torsion of (a) C1, and (b) C2 (Soil-D-F)

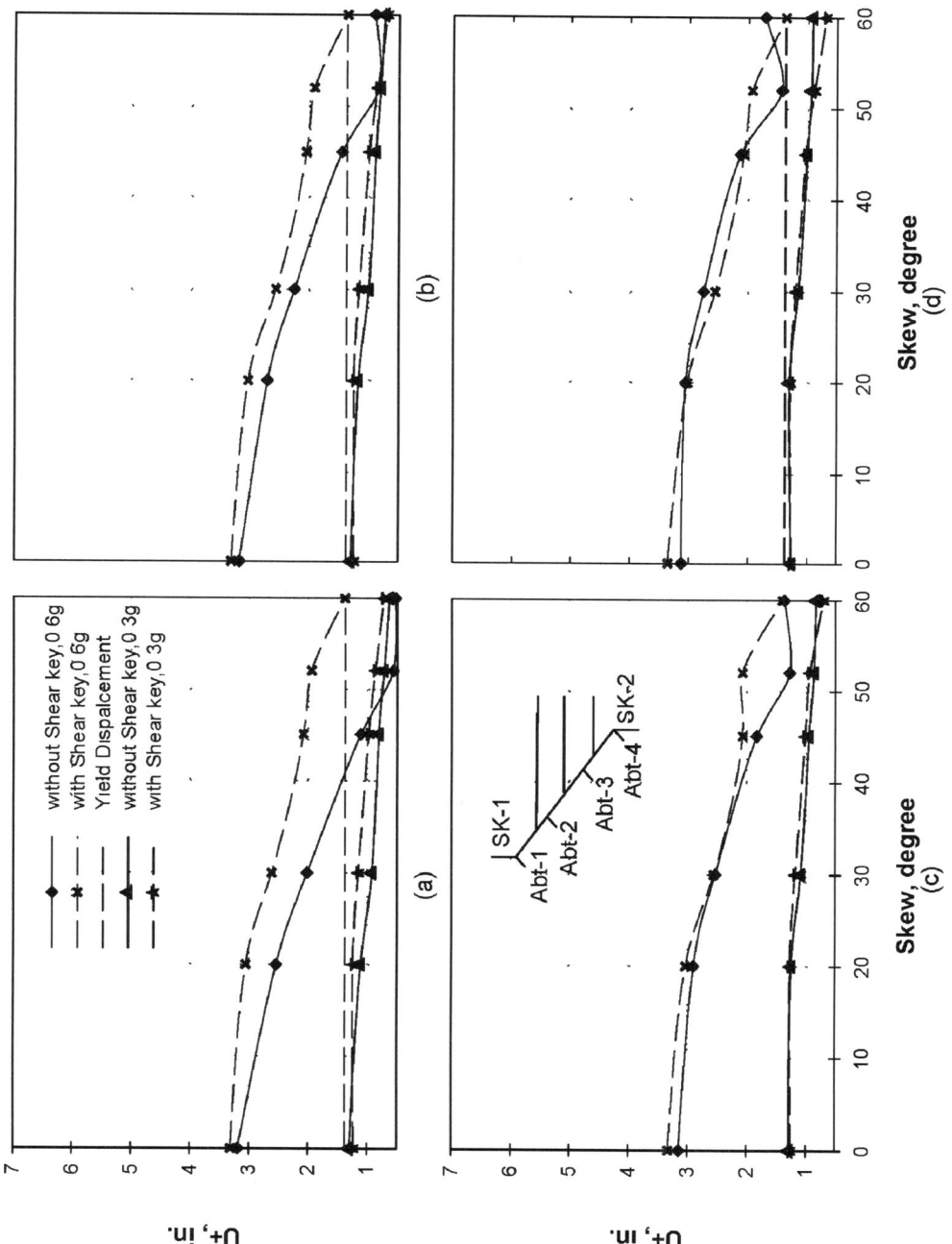

Figure 3-105. Average Displacement "into" Abutment's (a) Abt-1, (b) Abt-2, (c) Abt-3, and (d) Abt-4 (Soil-D-F)

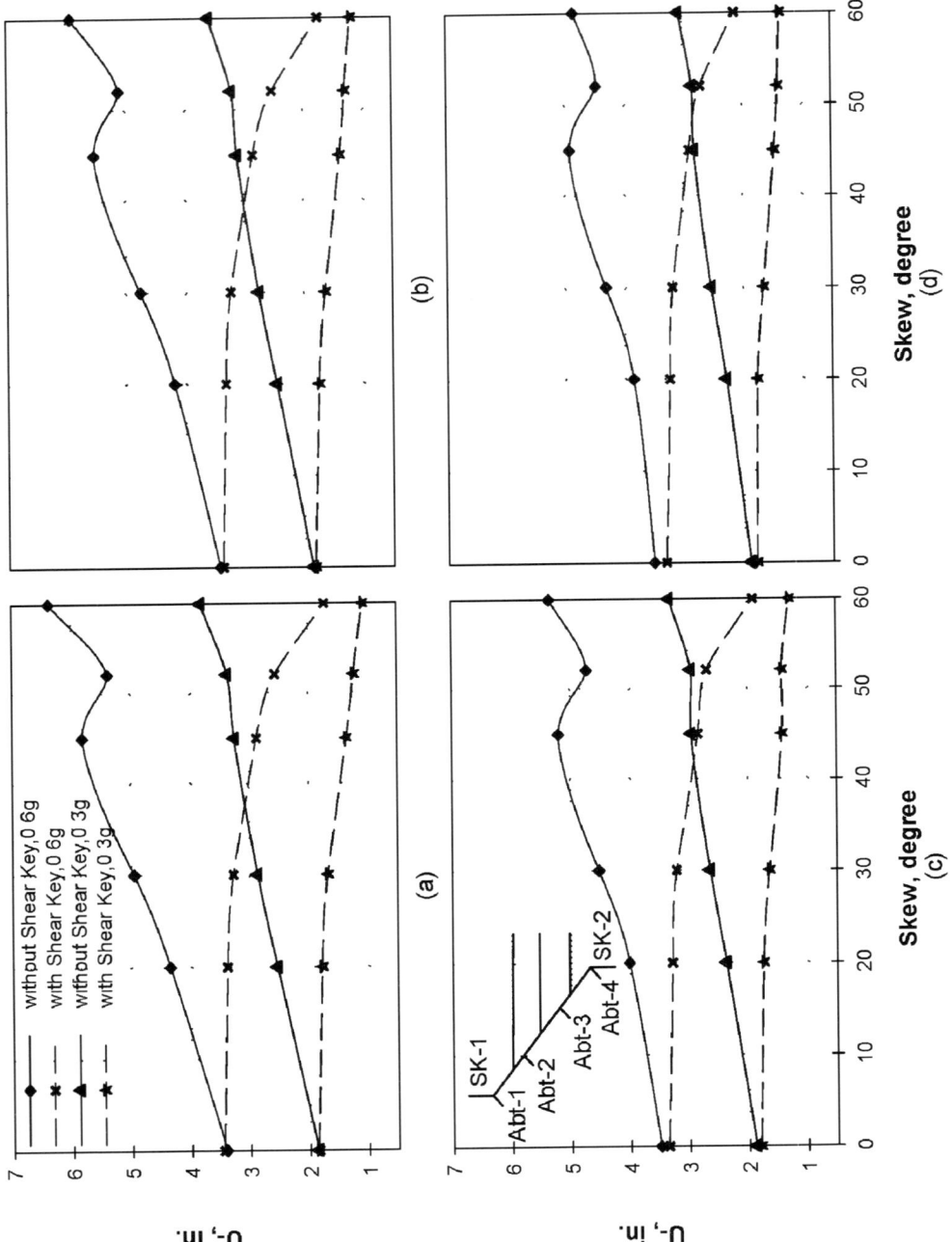

Figure 3-106. Average Displacement "away" from Abutment's (a) Abt-1, (b) Abt-2, (c) Abt-3, and (d) Abt-4 (Soil-D-F)

250

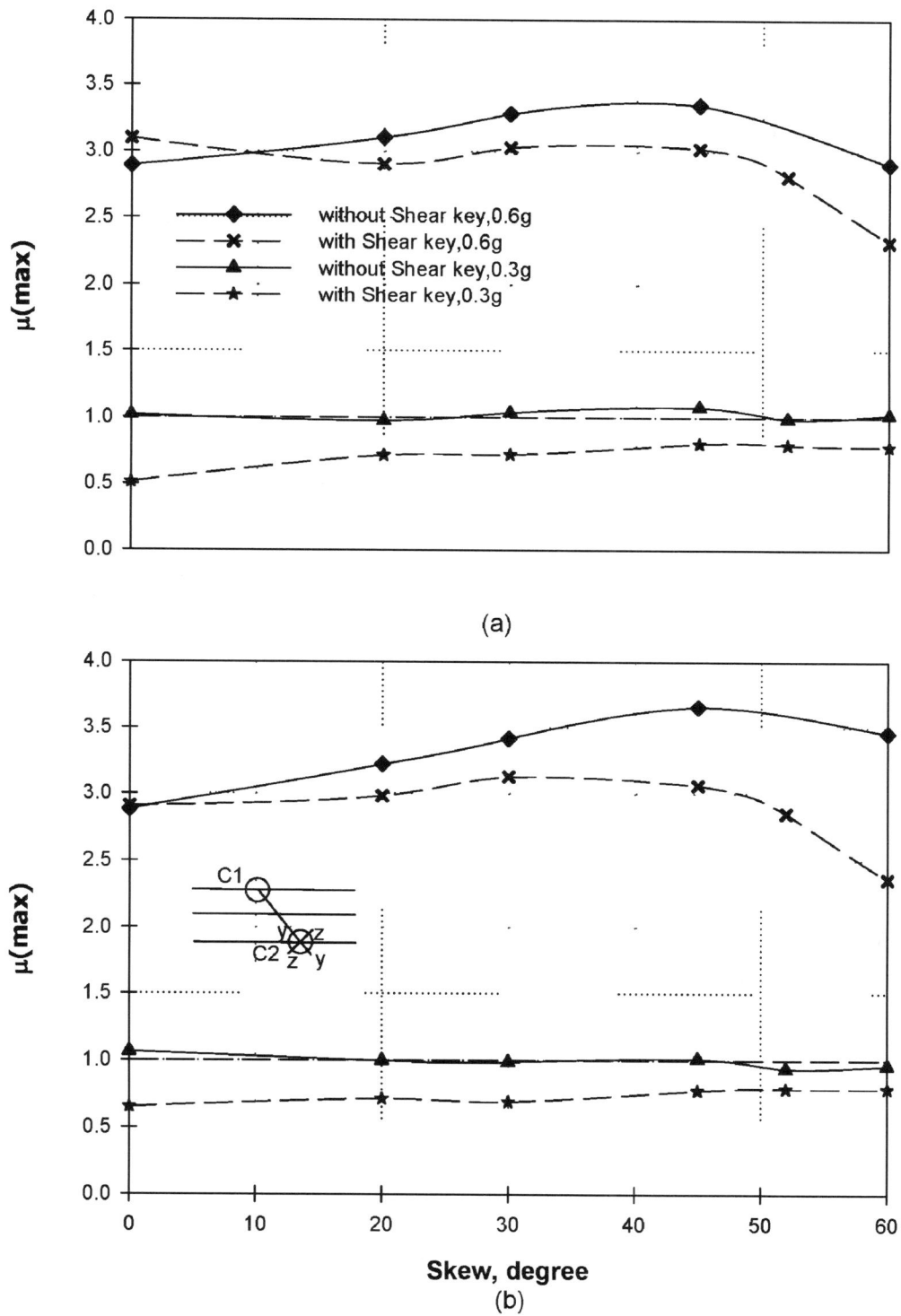

Figure 3-107. Maximum Average Ductility Ratio of (a) C1 and (b) C2 (Soil-D-F)

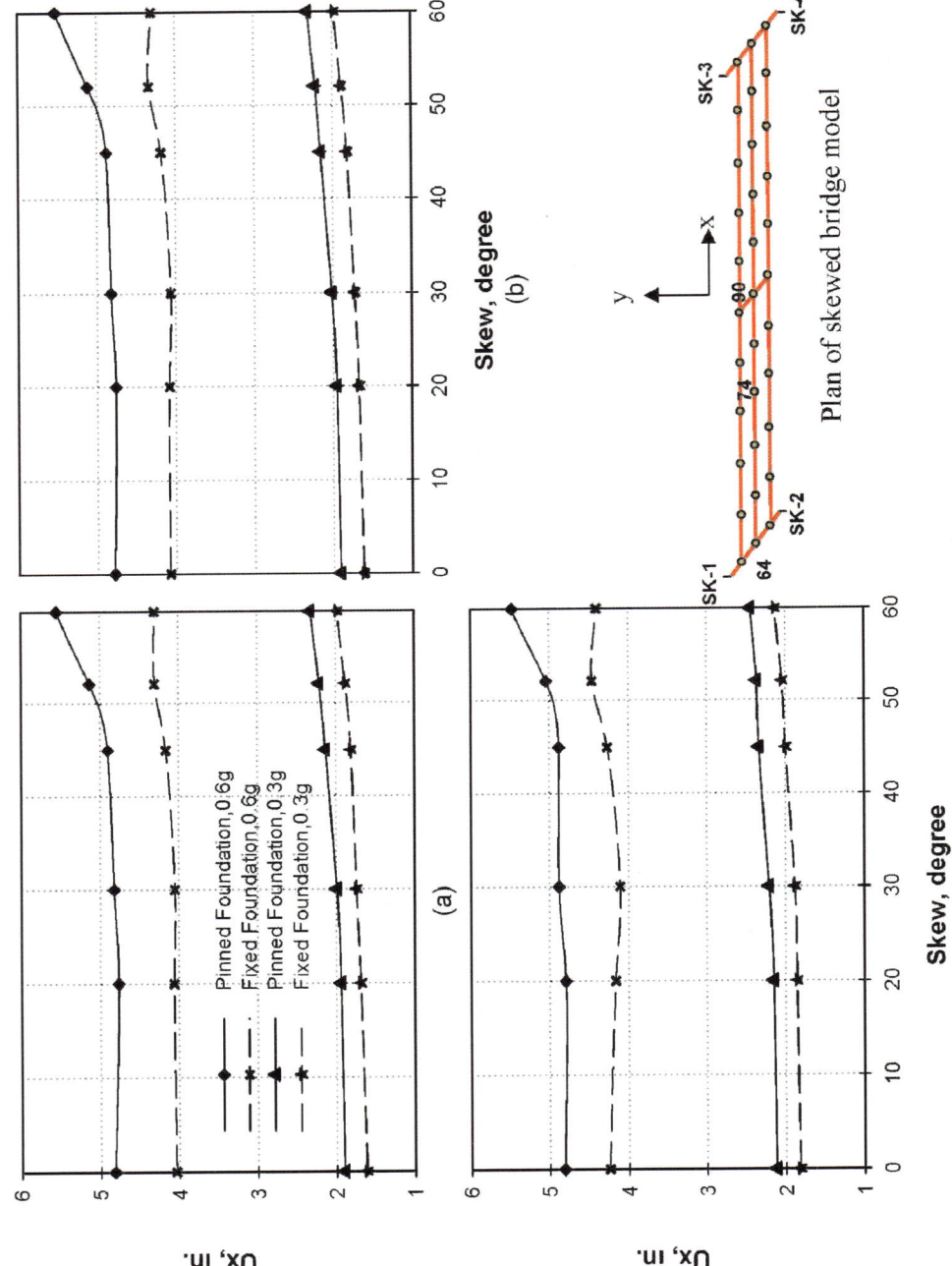

Figure 3-108. Average Displacement in X-direction for nodes (a) 90, (b) 74, and (c) 64 (Soil-D)

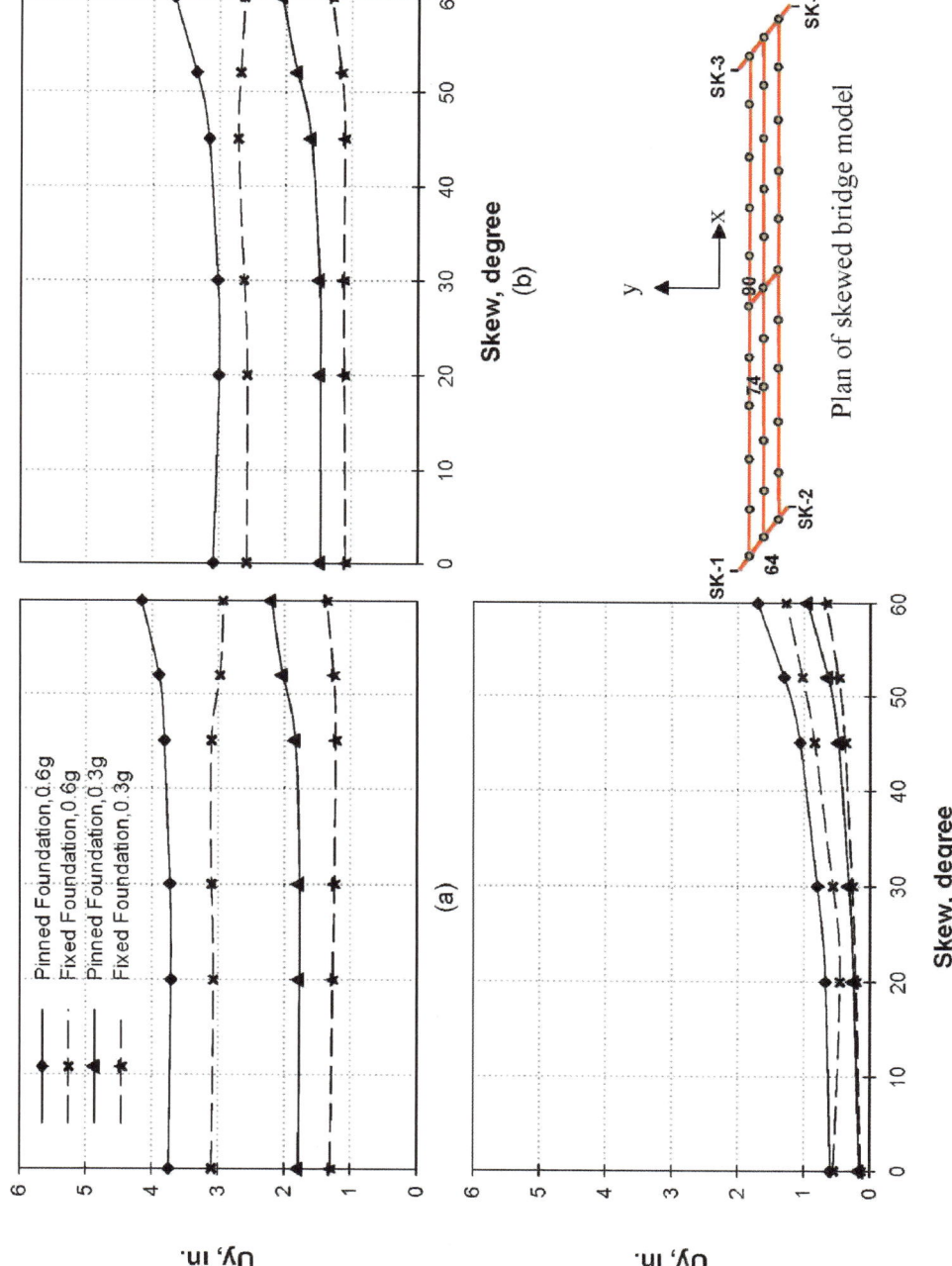

Figure 3-109. Average Displacement in Y-direction for nodes (a) 90, (b) 74, and (c) 64 (Soil-D)

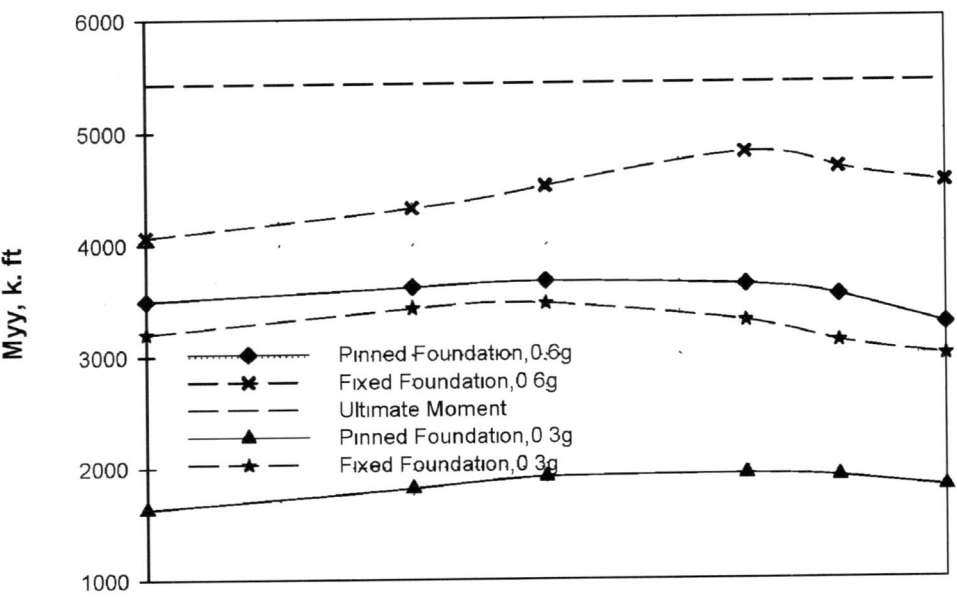

(a)

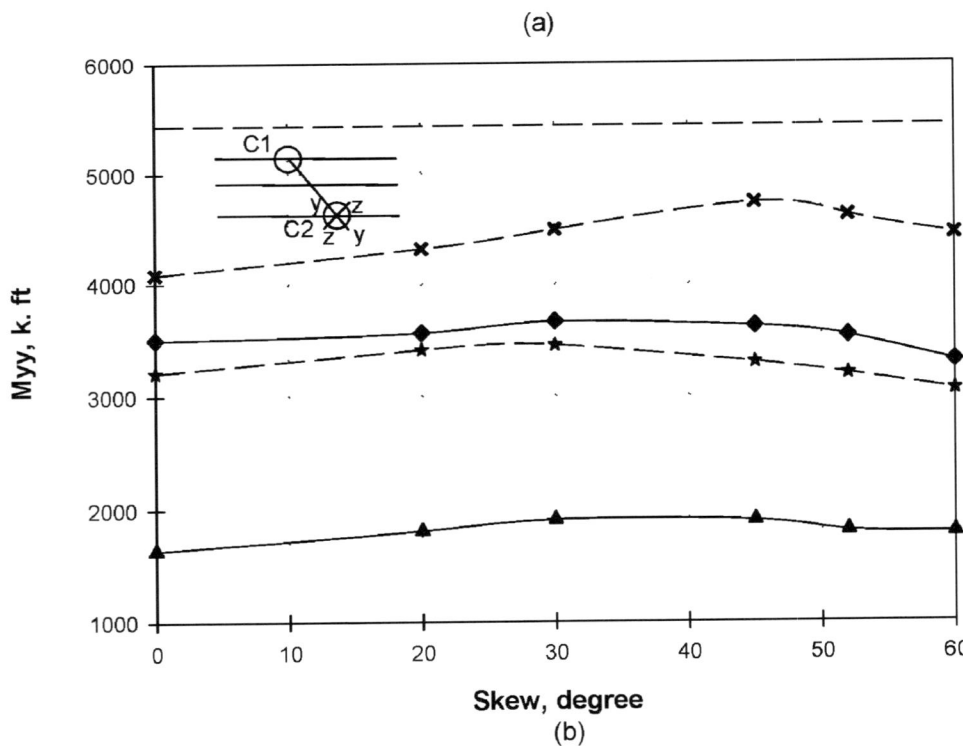

Skew, degree

(b)

Figure 3-110. Average Moment in Y-Directions of (a) C1, and (b) C2 (Soil-D)

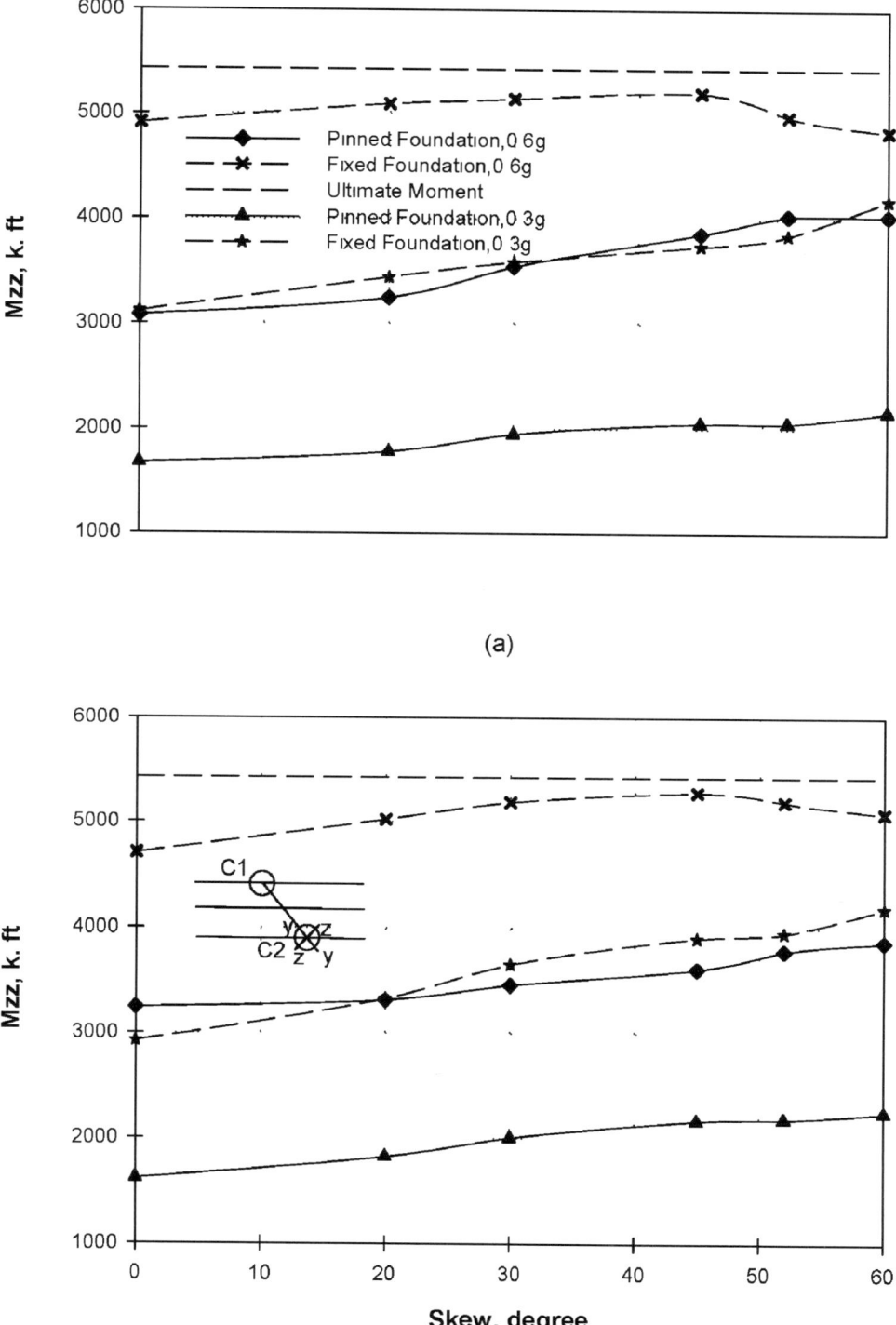

(a)

(b)

Figure 3-111. Average Moment in Z-Directions of (a) C1, and (b) C2 (Soil-D)

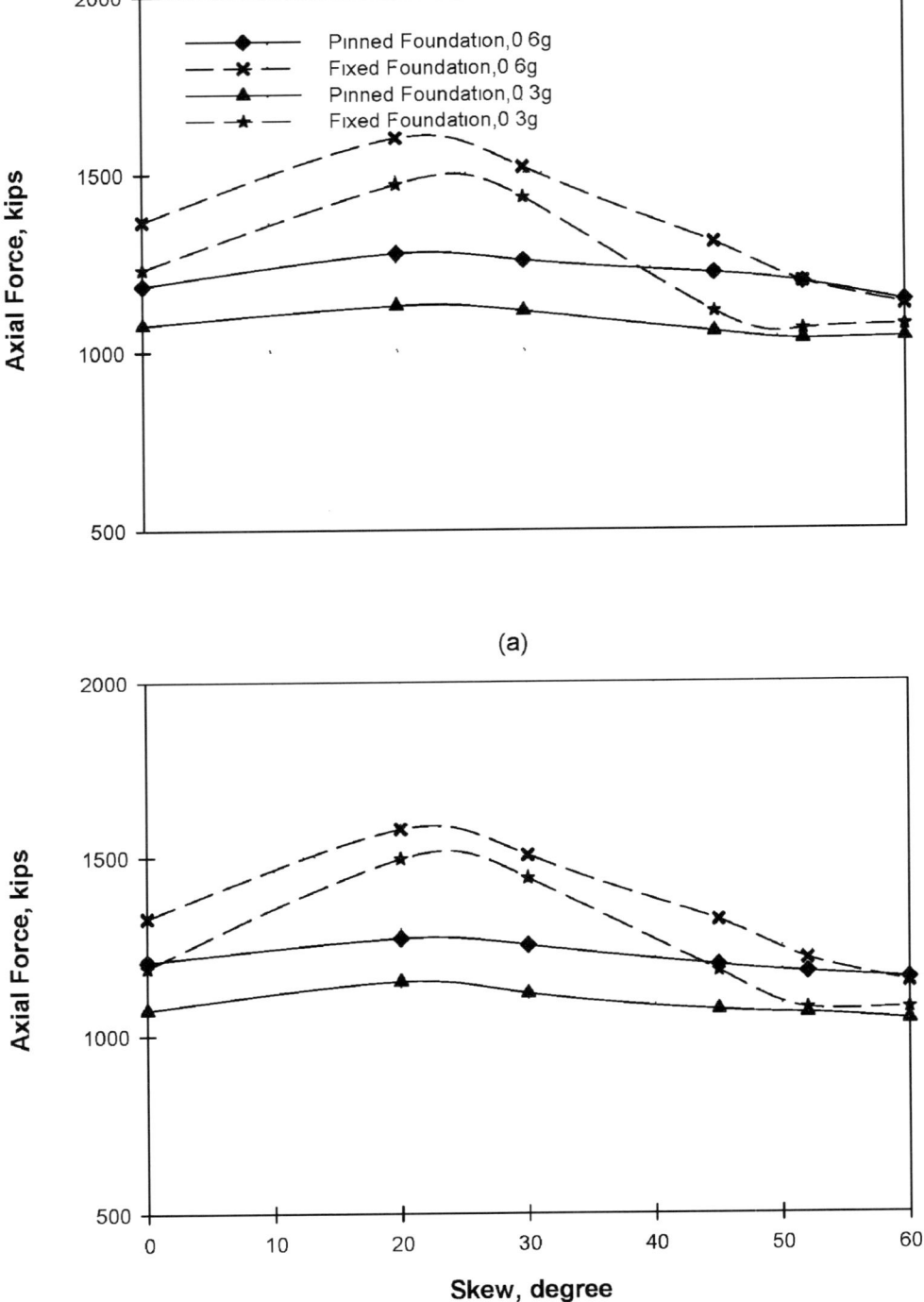

(a)

(b)

Figure 3-112. Average Axial Force in (a) C1, and (b) C2 (Soil-D)

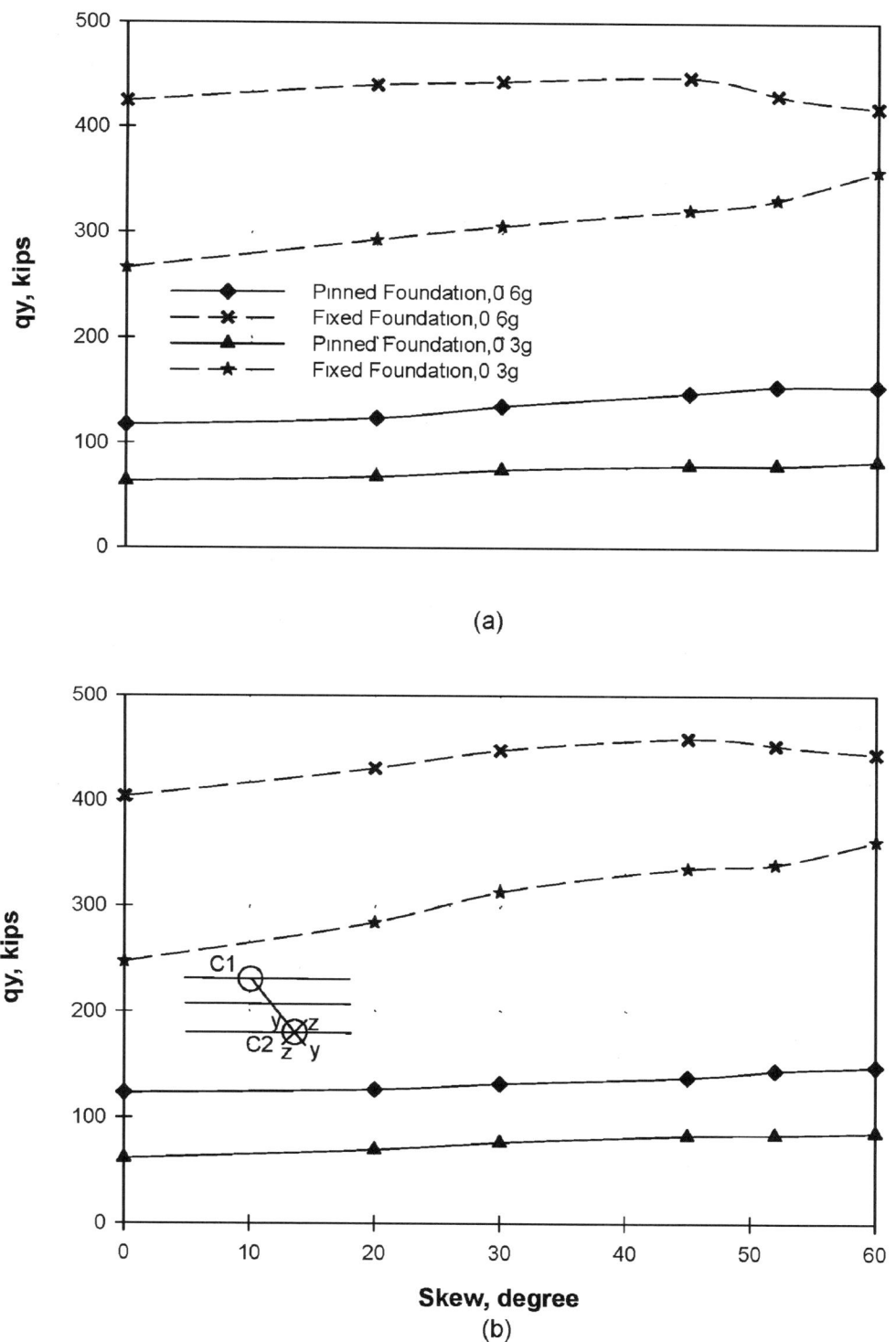

(a)

(b)

Figure 3-113. Average Shear Force in Y-Directions of (a) C1, and (b) C2 (Soil-D)

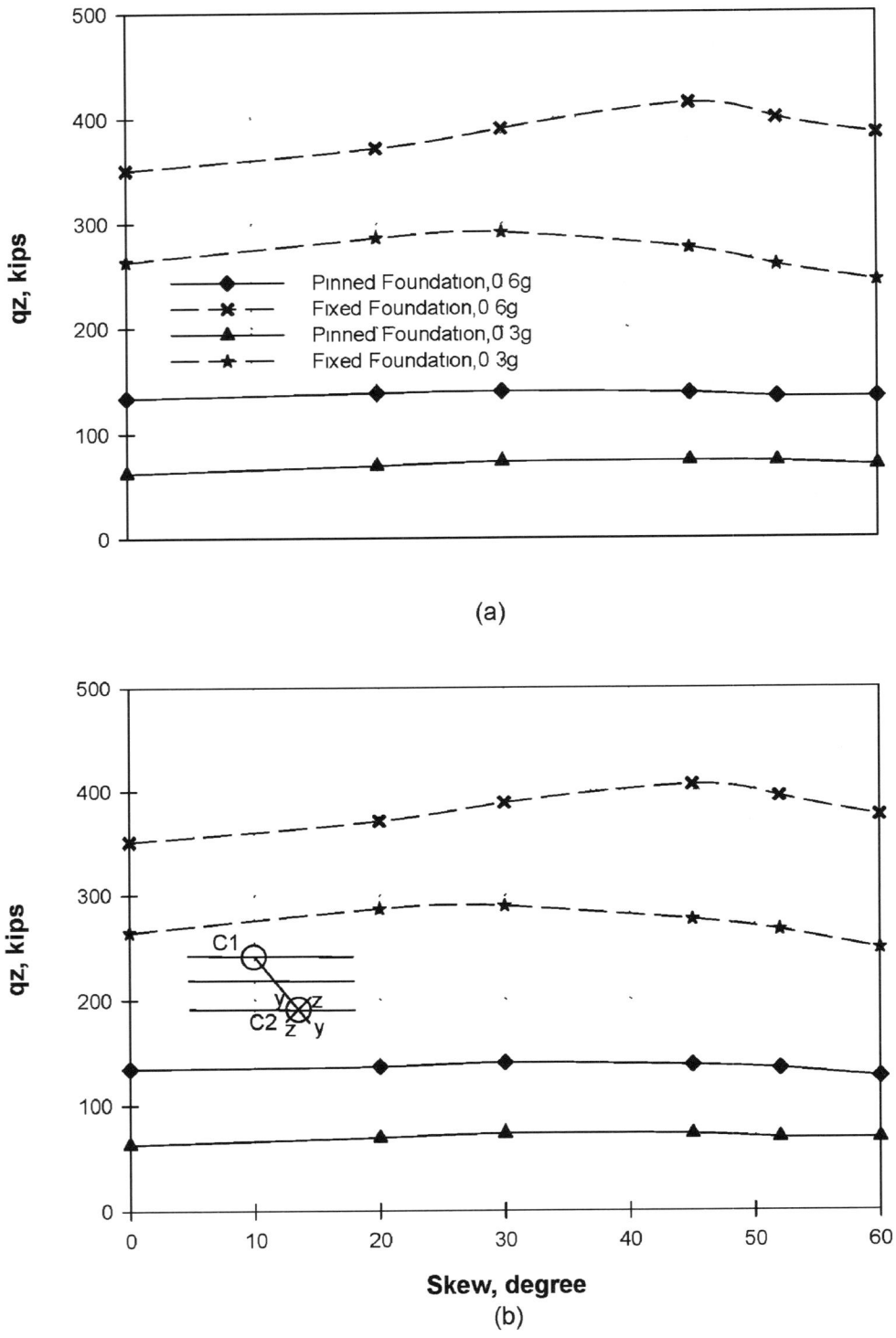

(a)

(b)

Figure 3-114. Average Shear Force in Z-Directions of (a) C1, and (b) C2 (Soil-D)

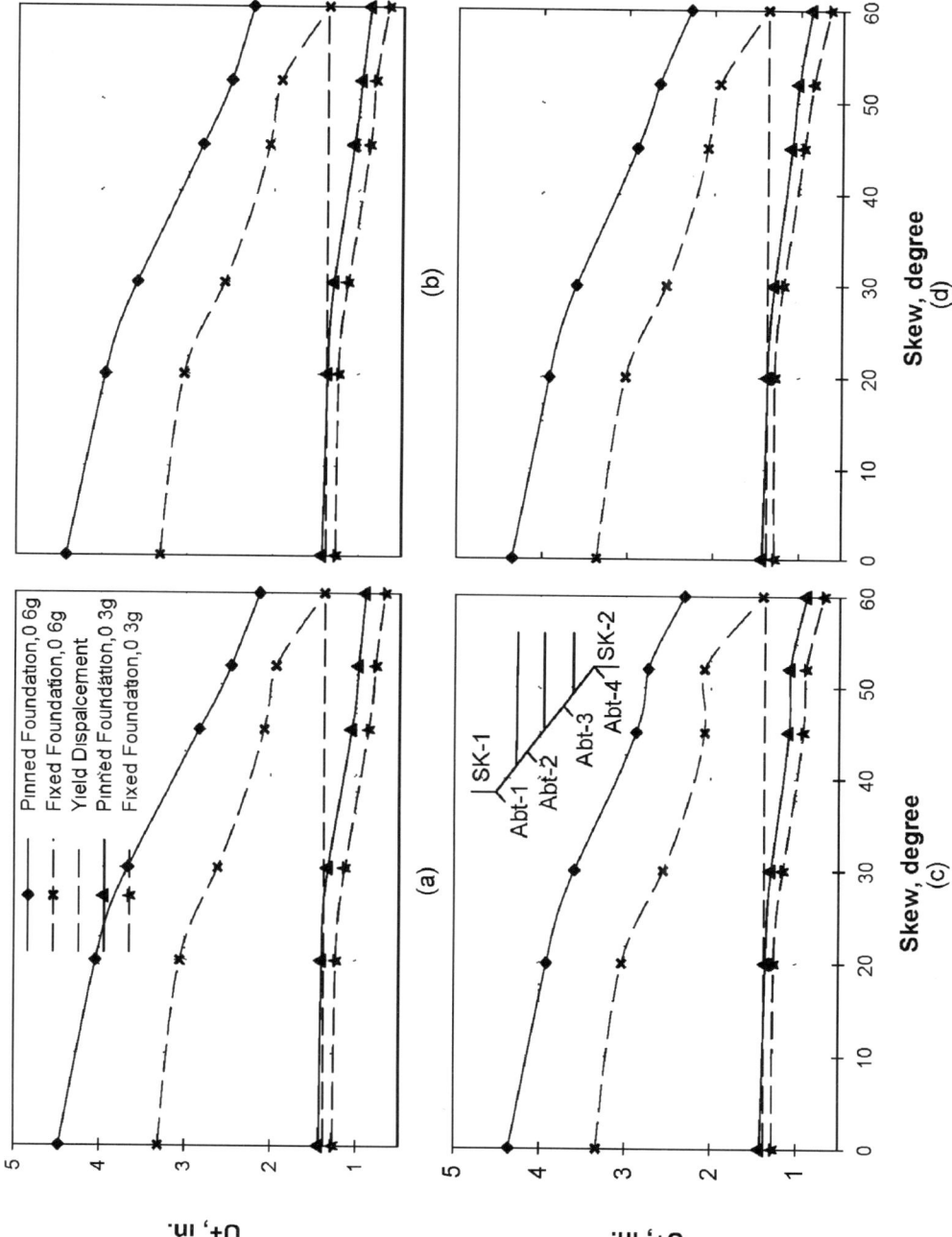

Figure 3-115. Average Displacement "into" Abutment's (a) Abt-1, (b) Abt-2, (c) Abt-3, and (d) Abt-4 (Soil-D)

259

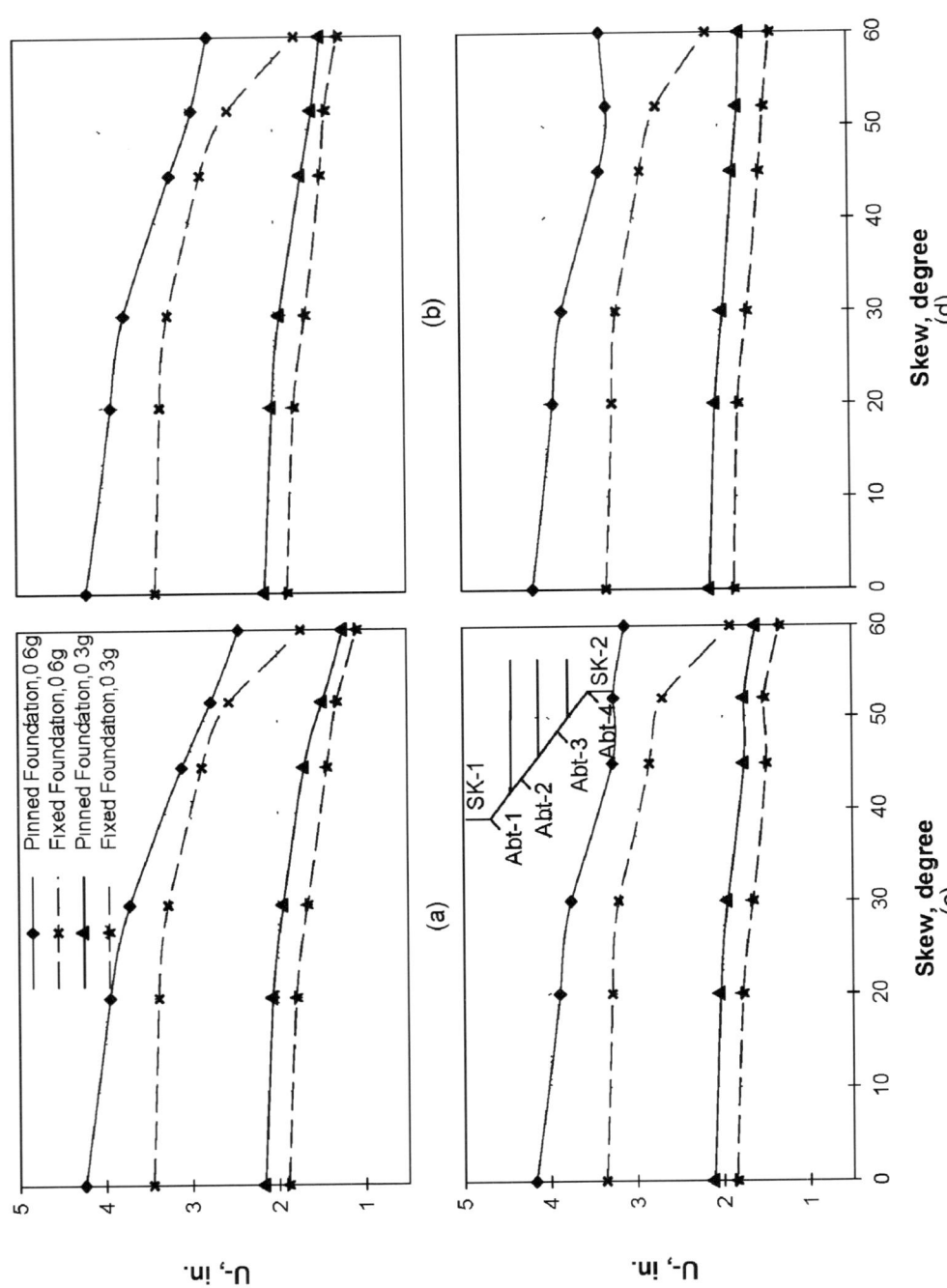

Figure 3-116. Average Displacement "away" from Abutment's (a) Abt-1, (b) Abt-2, (c) Abt-3, and (d) Abt-4 (Soil-D)

260

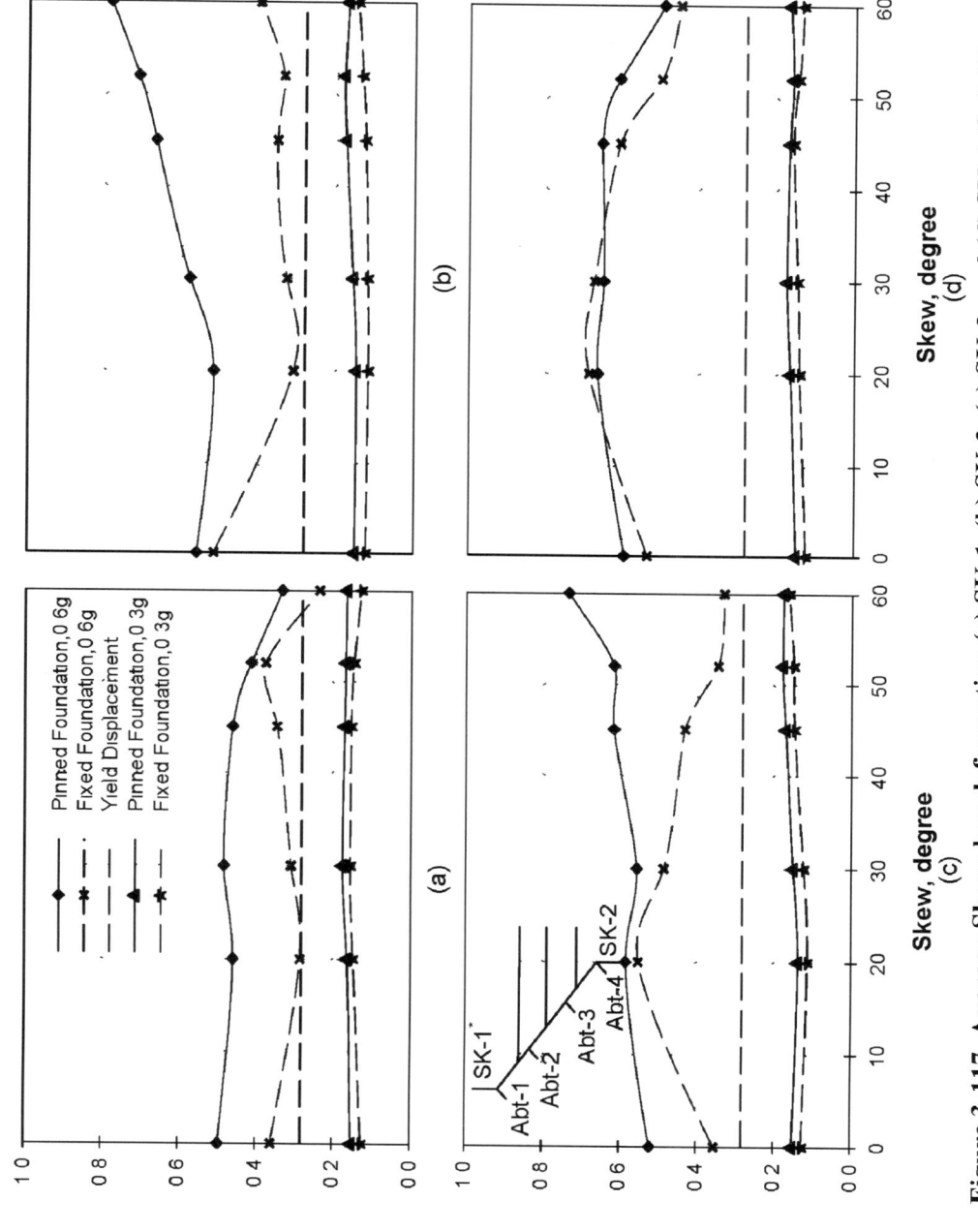

Figure 3-117. Average Shear-key deformation (a) SK-1, (b) SK-2, (c) SK-3, and (d) SK-4 (Soil-D)

*SK-3 and SK-4 are at the other abutment

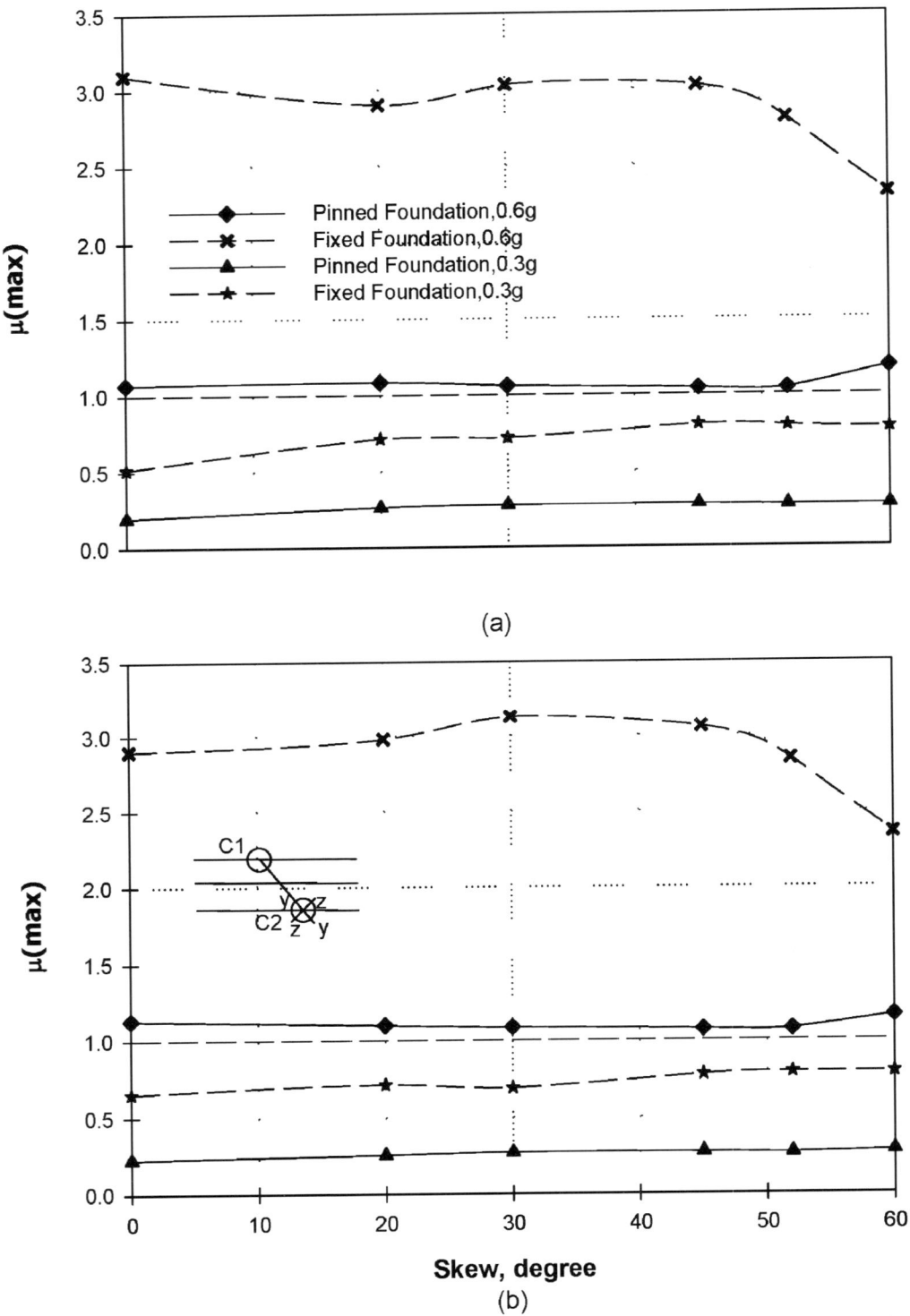

(a)

(b)

Figure 3-118. Maximum Average Ductility Ratio of (a) C1 and (b) C2 (Soil-D)

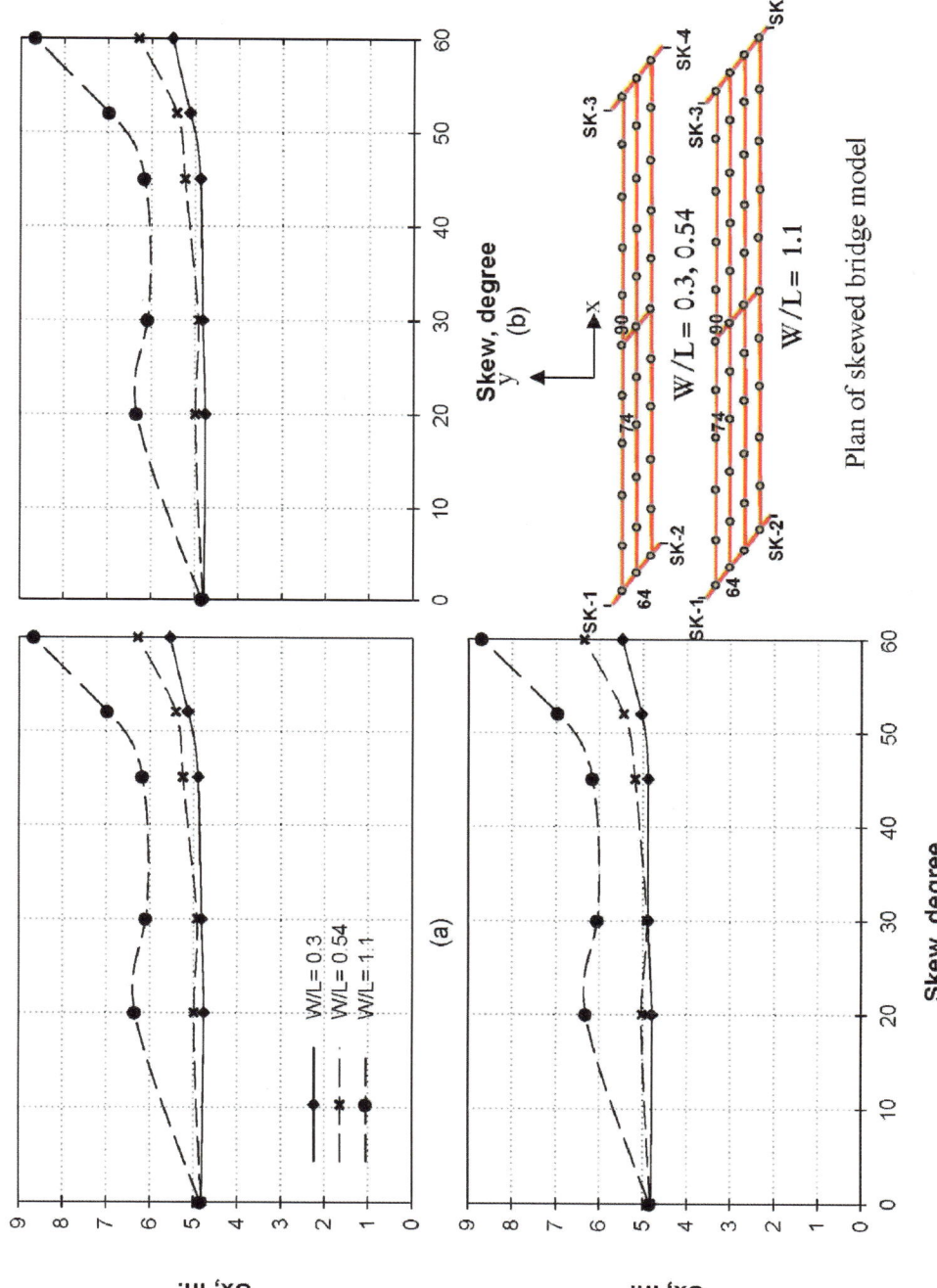

Figure 3-119. Average Displacement in X-direction for nodes (a) 90, (b) 74, and (c) 64 (Soil-D-0.6g)

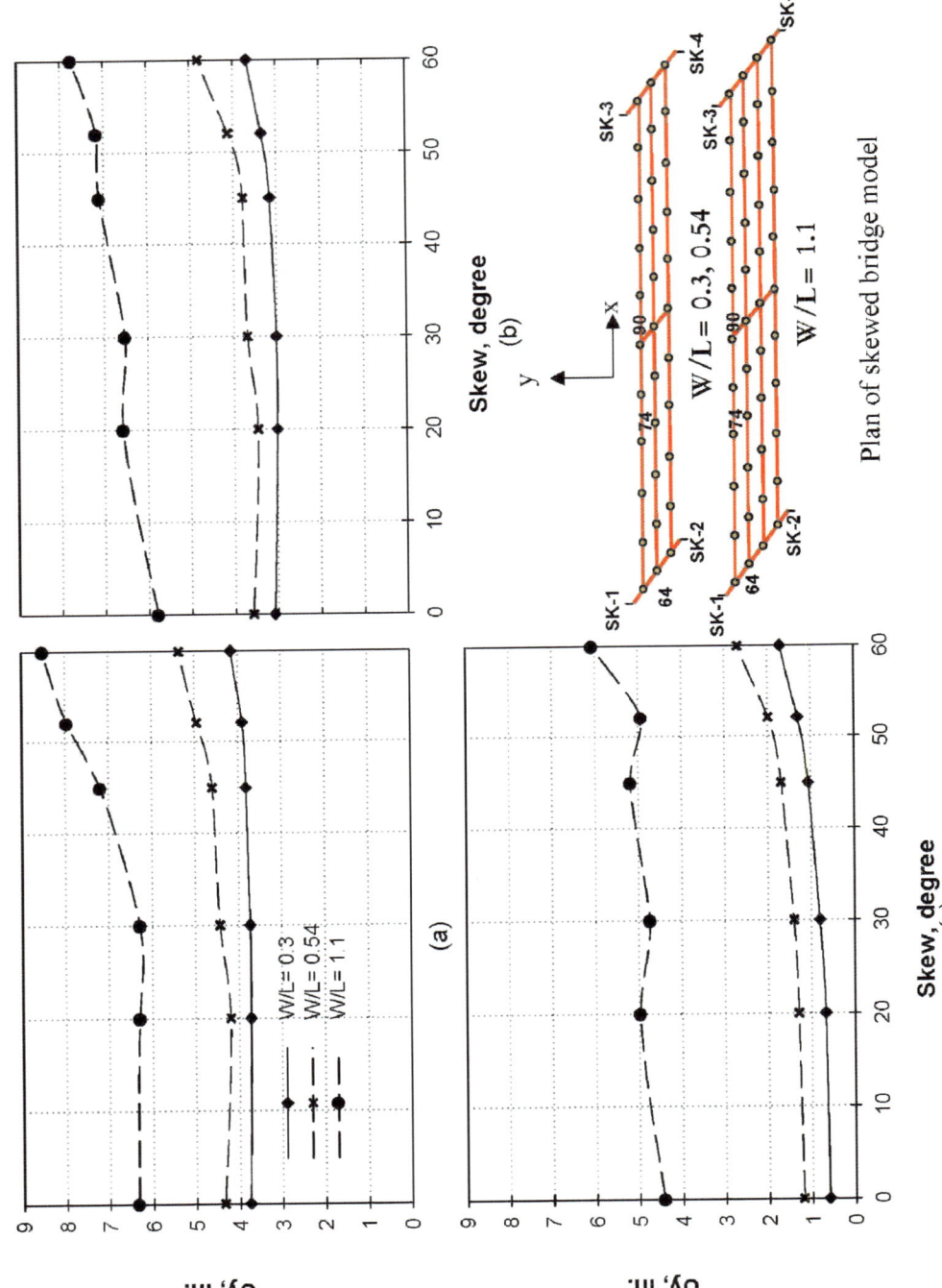

Figure 3-120. Average Displacement in Y-direction for nodes (a) 90, (b) 74, and (c) 64 (Soil-D-0.6g)

265

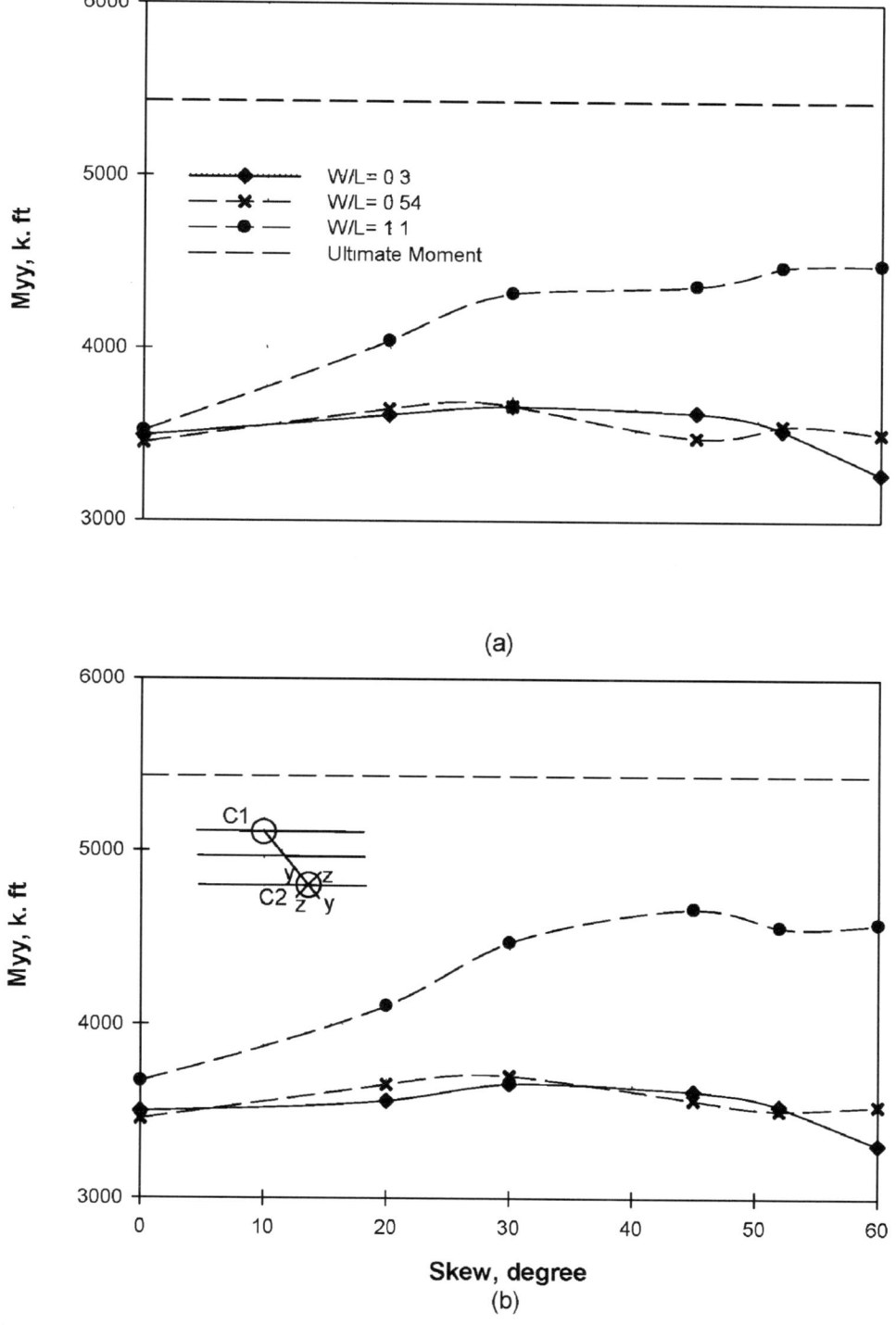

(a)

(b)

Figure 3-121. Average Moment in Y-Directions of (a) C1, and (b) C2 (Soil-D-0.6g)

266

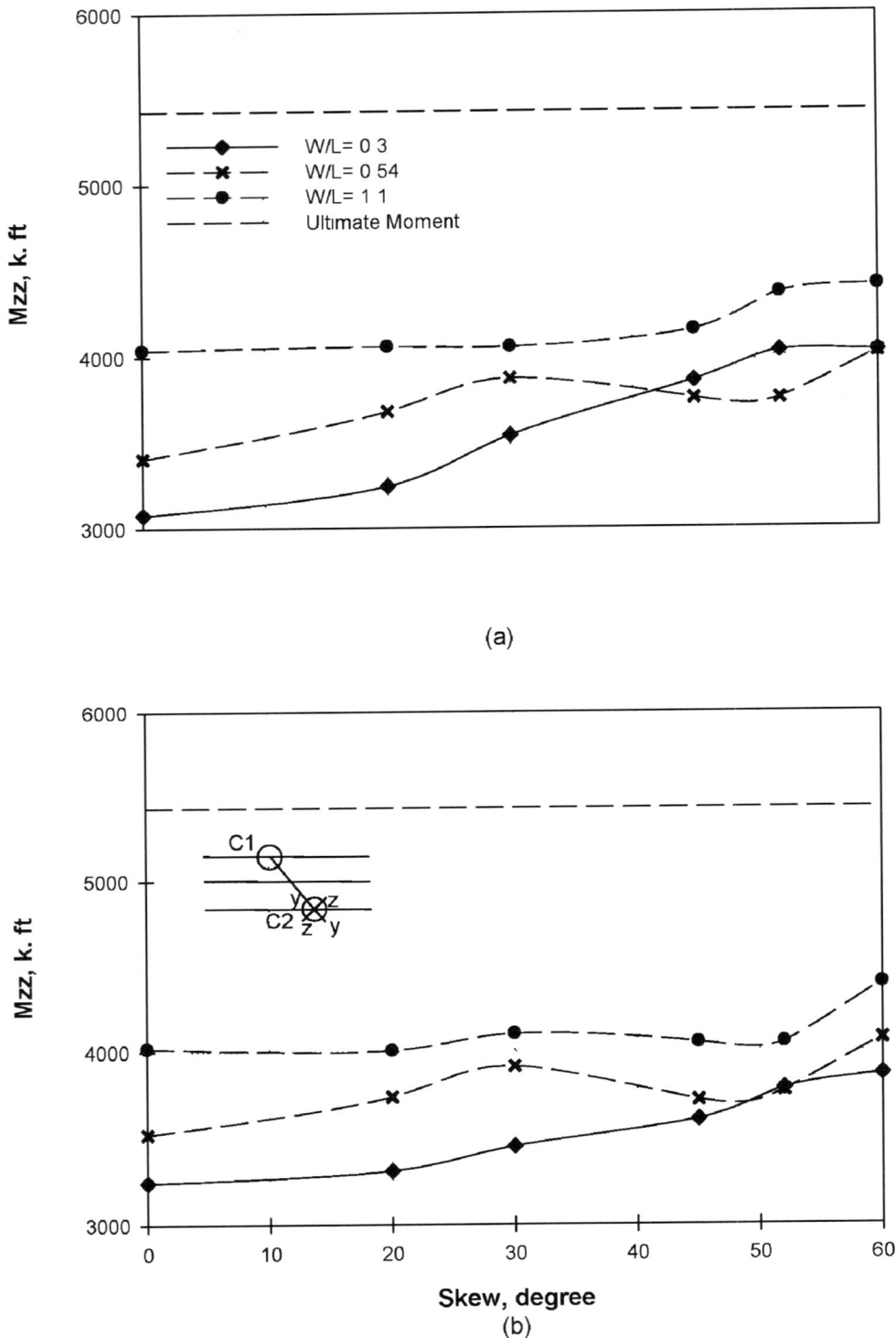

(a)

(b)

Figure 3-122. Average Moment in Z-Directions of (a) C1, and (b) C2 (Soil-D-0.6g)

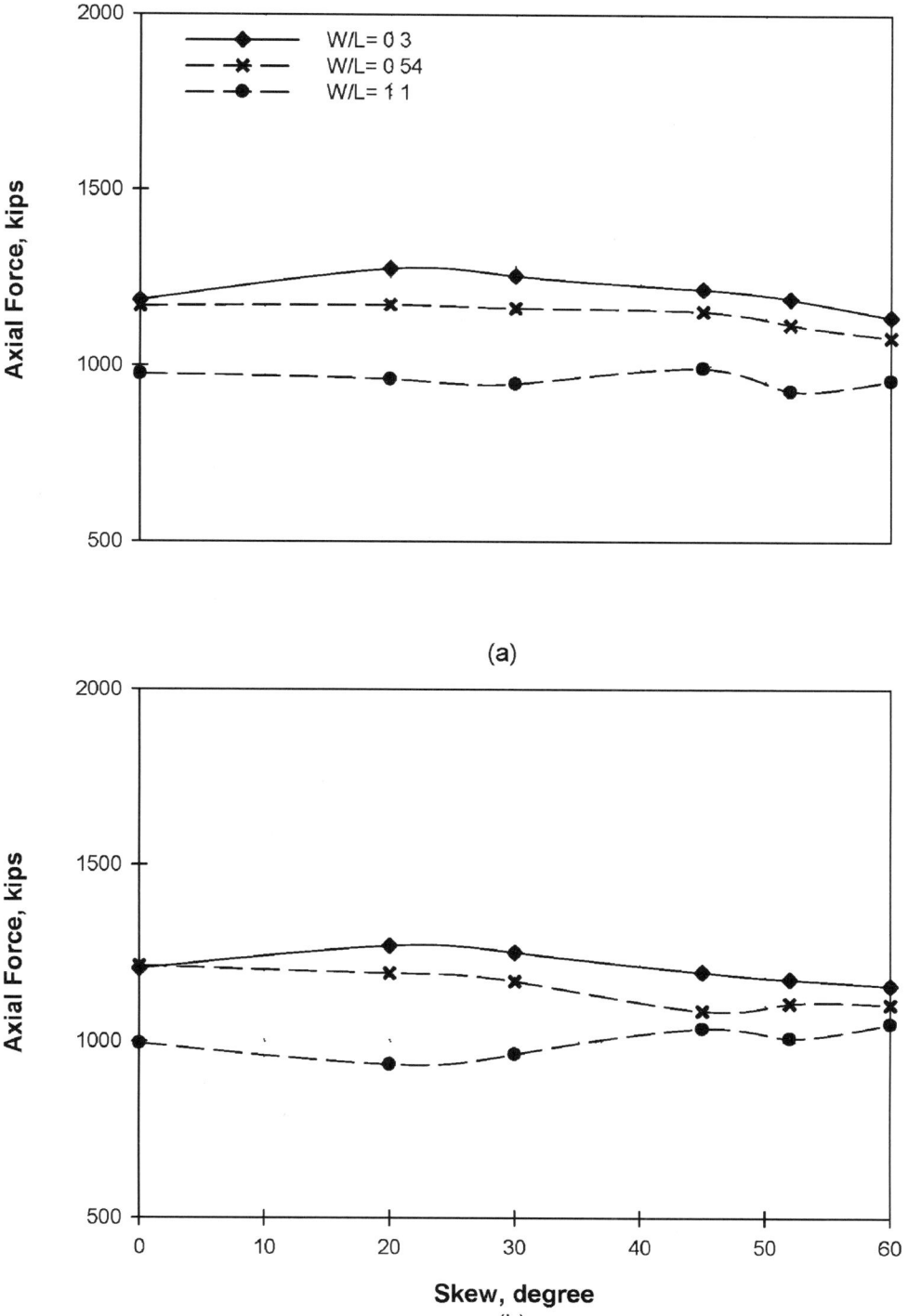

(a)

(b)

Figure 3-123. Average Axial Force in (a) C1, and (b) C2 (Soil-D-0.6g)

268

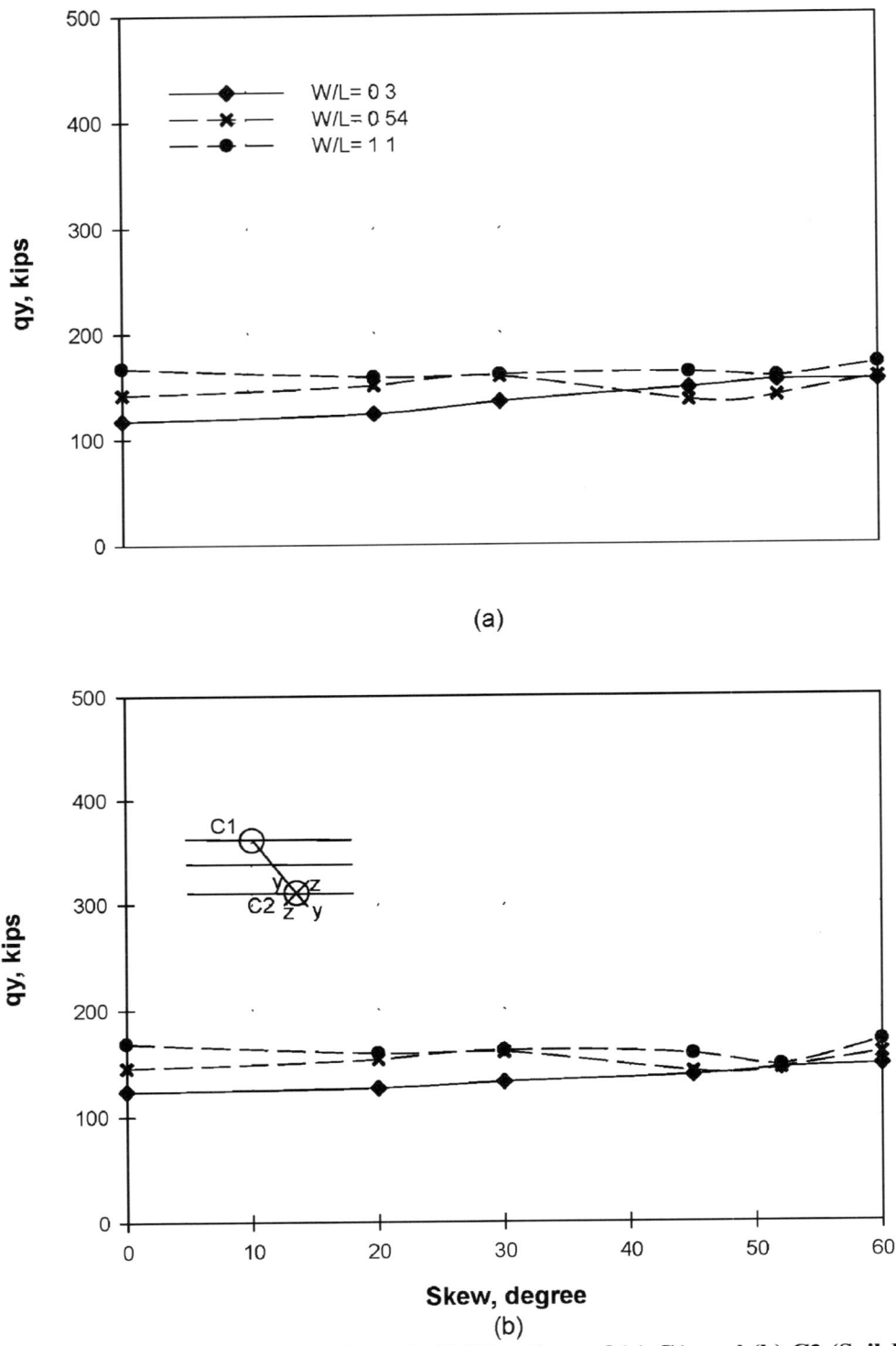

(a)

(b)

Skew, degree

Figure 3-124. Average Shear Force in Y-Directions of (a) C1, and (b) C2 (Soil-D-0.6g)

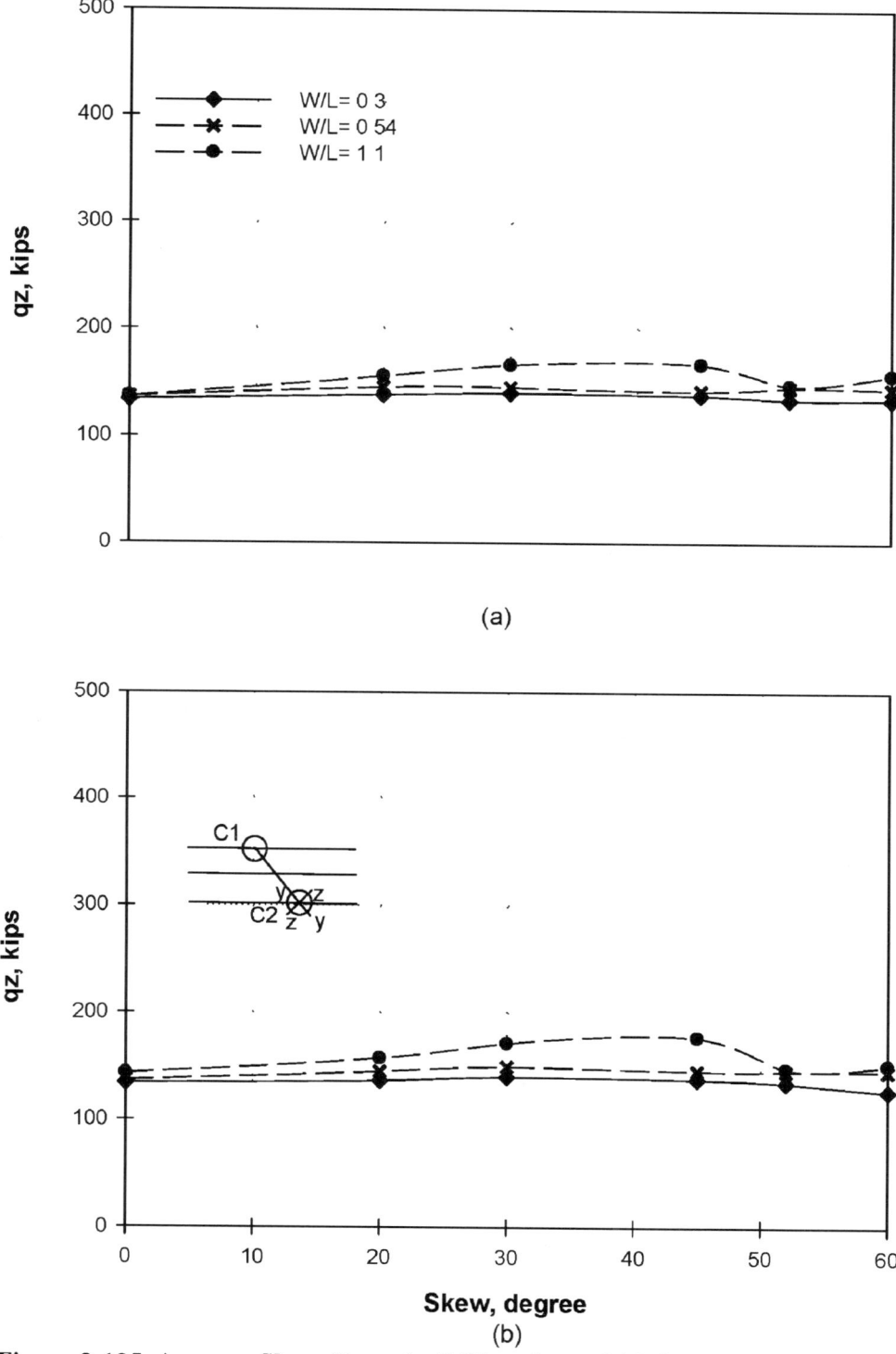

(a)

(b)

Figure 3-125. Average Shear Force in Z-Directions of (a) C1, and (b) C2 (Soil-D-0.6g)

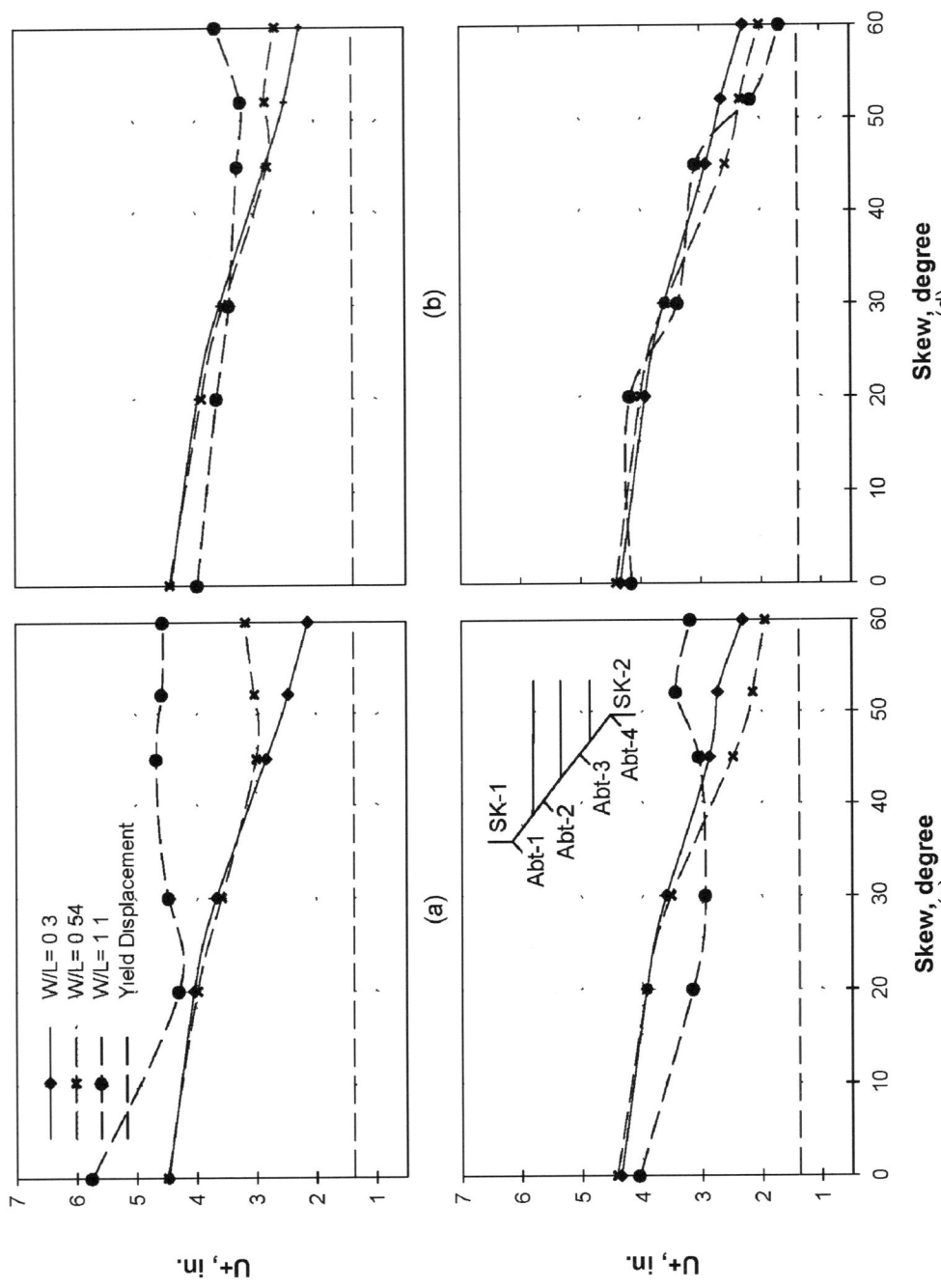

Figure 3-126. Average Displacement "into" Abutment's (a) Abt-1, (b) Abt-2, (c) Abt-3, and (d) Abt-4 (Soil-D-0.6g)

270

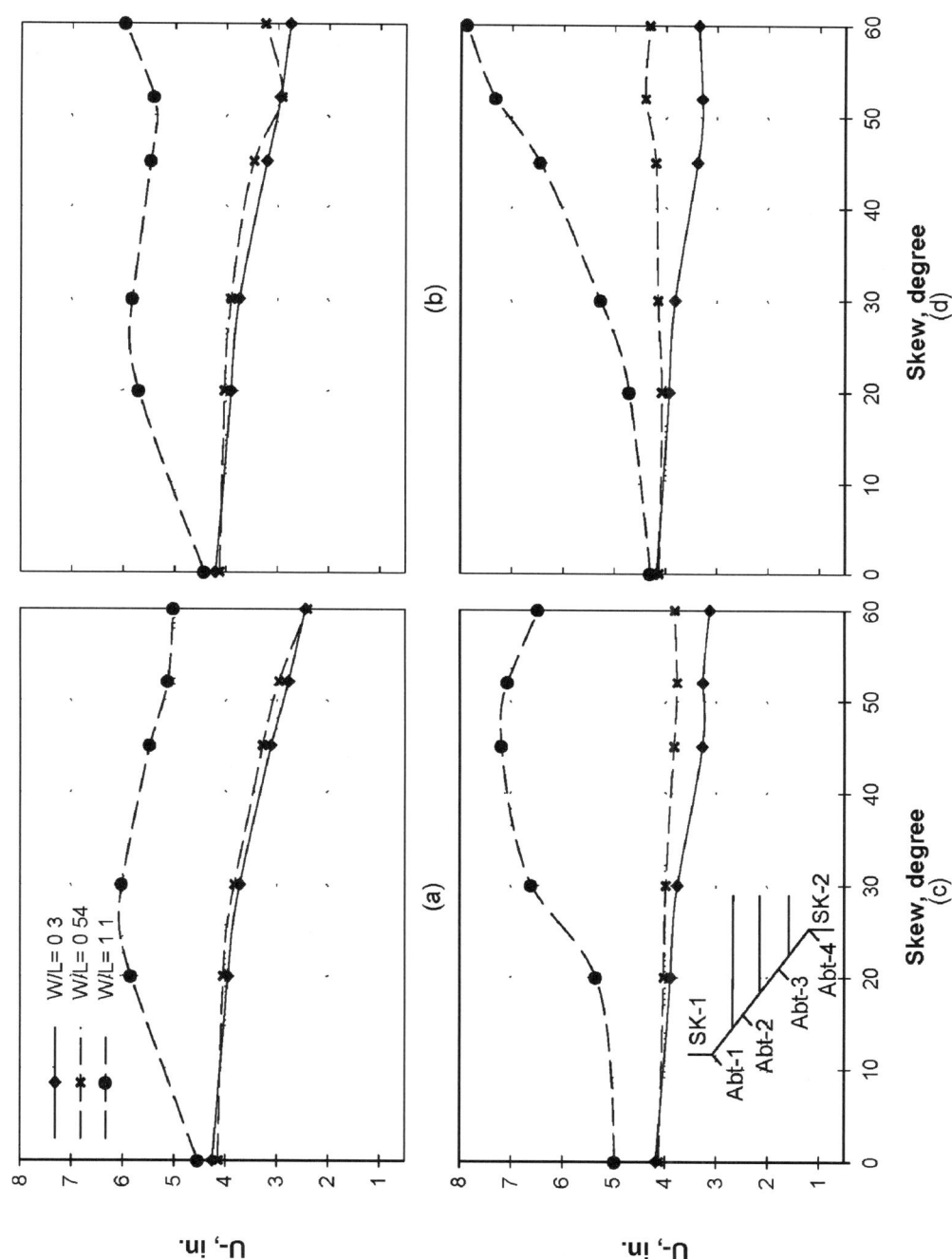

Figure 3-127. Average Displacement "away" from Abutment's (a) Abt-1, (b) Abt-2, (c) Abt-3, and (d) Abt-4 (Soil-D-0.6g)

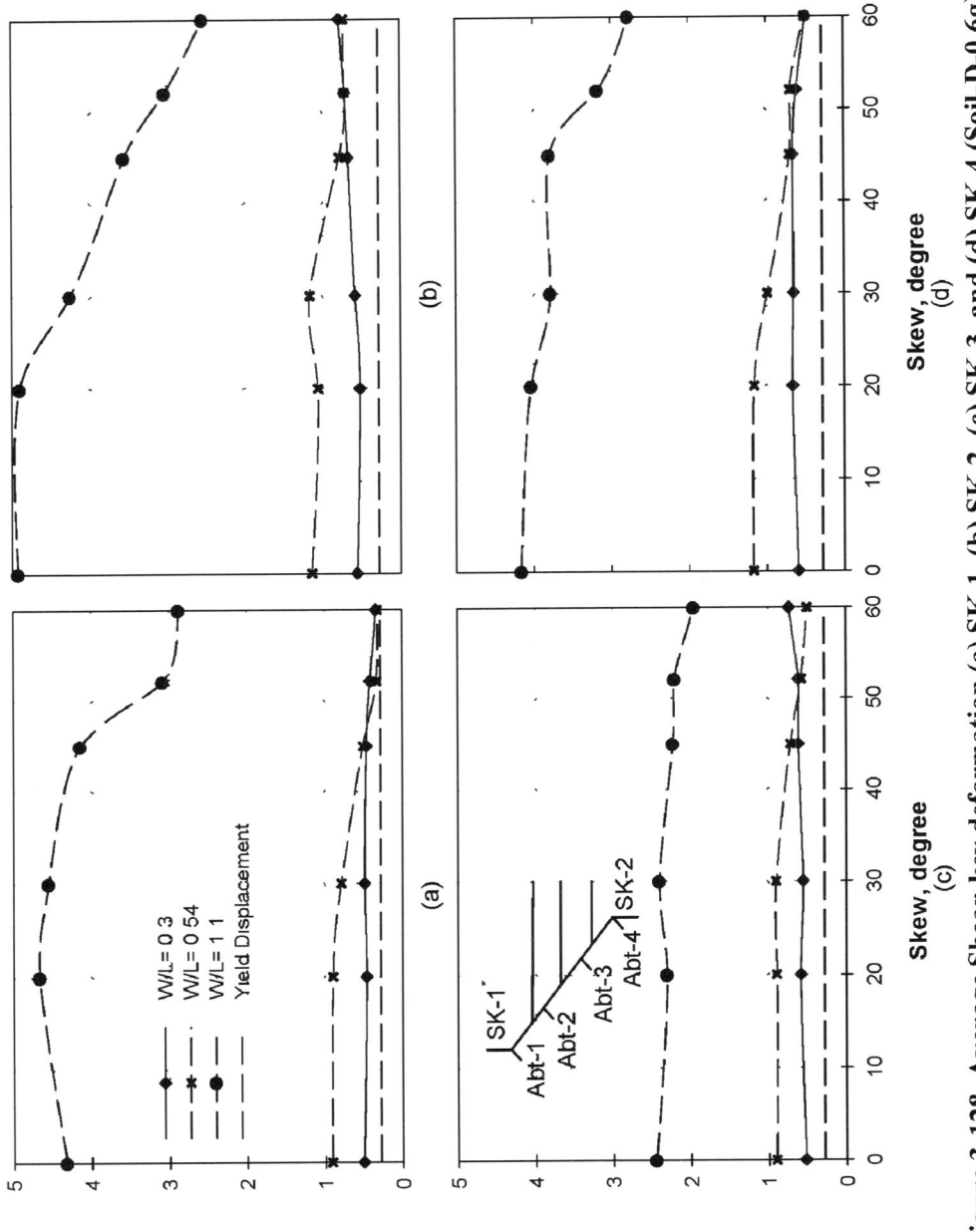

Figure 3-128. Average Shear-key deformation (a) SK-1, (b) SK-2, (c) SK-3, and (d) SK-4 (Soil-D-0.6g)

*SK-3 and SK-4 are at the other abutment

272

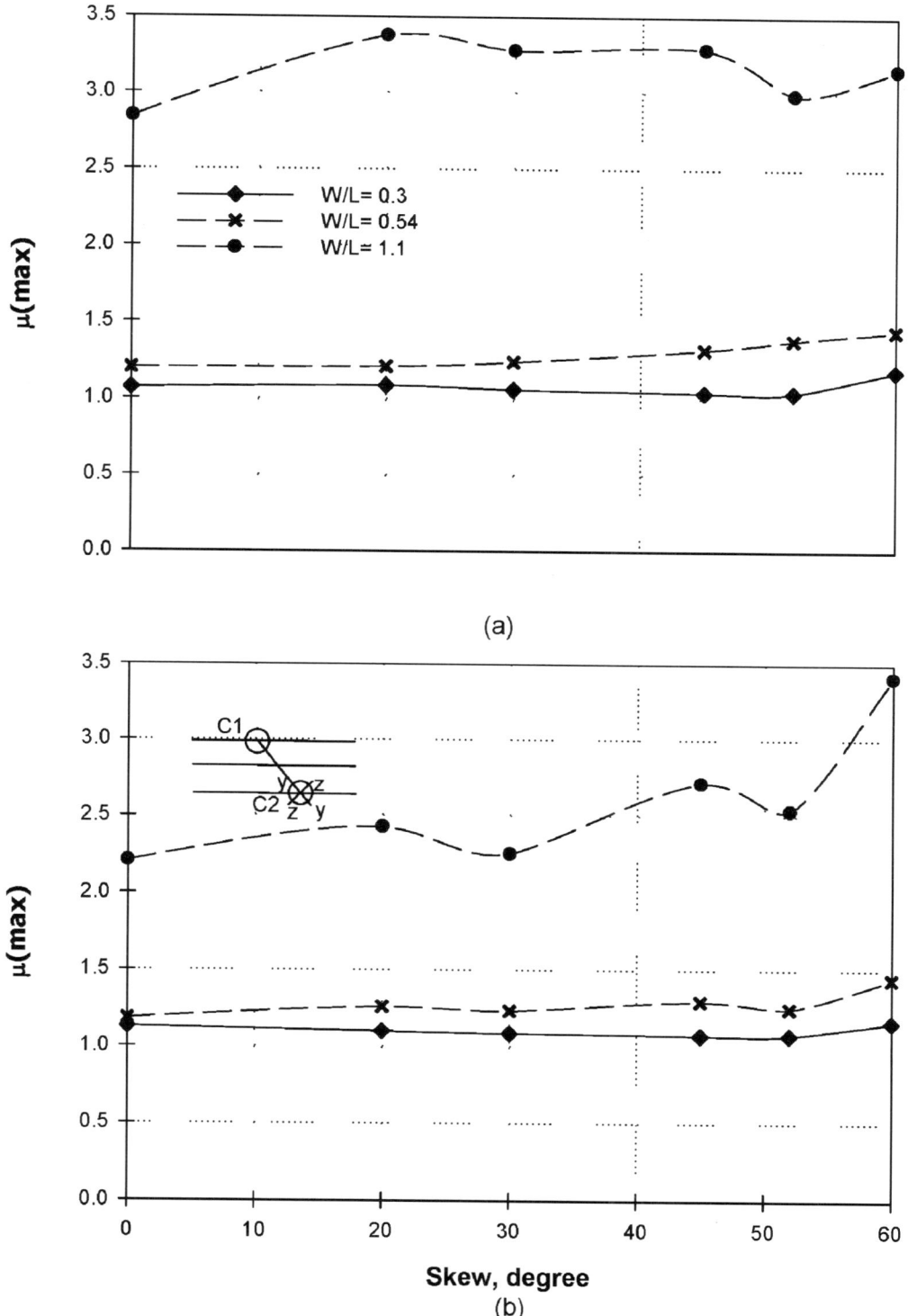

(a)

(b)

Skew, degree

Figure 3-129. Maximum Average Ductility Ratio of (a) C1 and (b) C2 (Soil-D-0.6g)

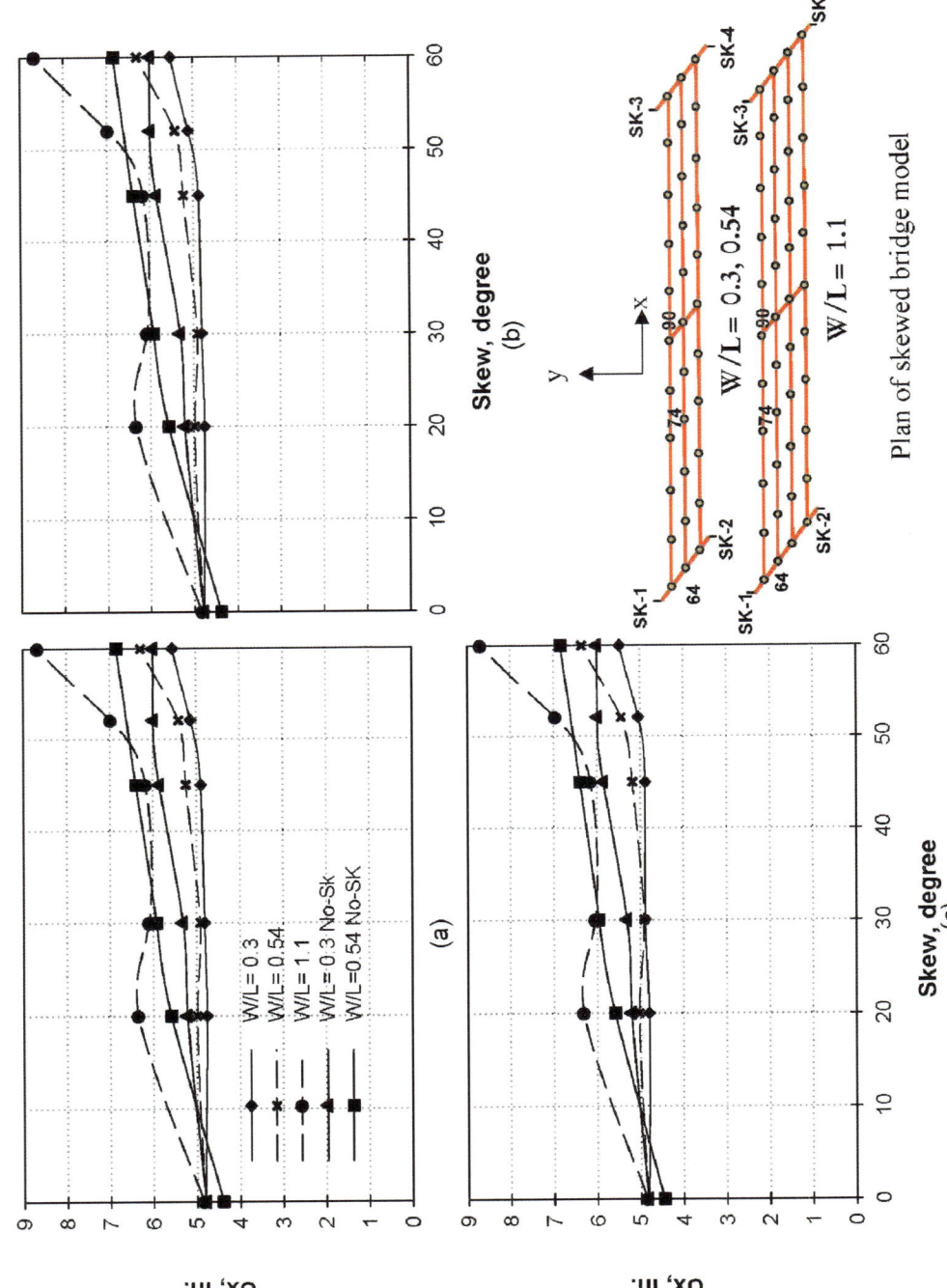

Figure 3-130. Average Displacement in X-direction for nodes (a) 90, (b) 74, and (c) 64 (Soil-D-0.6g)

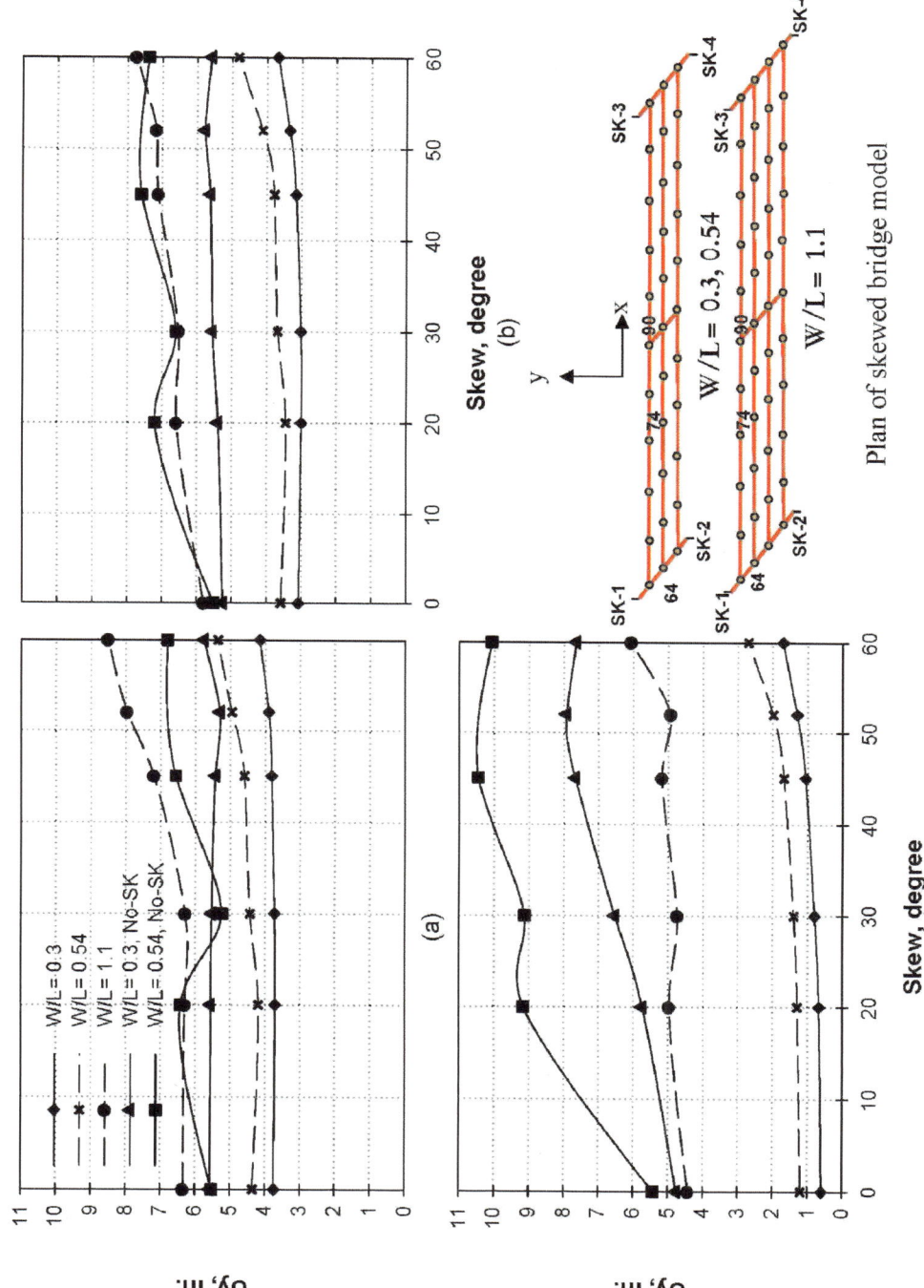

Figure 3-131. Average Displacement in Y-direction for nodes (a) 90, (b) 74, and (c) 64 (Soil-D-0.6g)

275

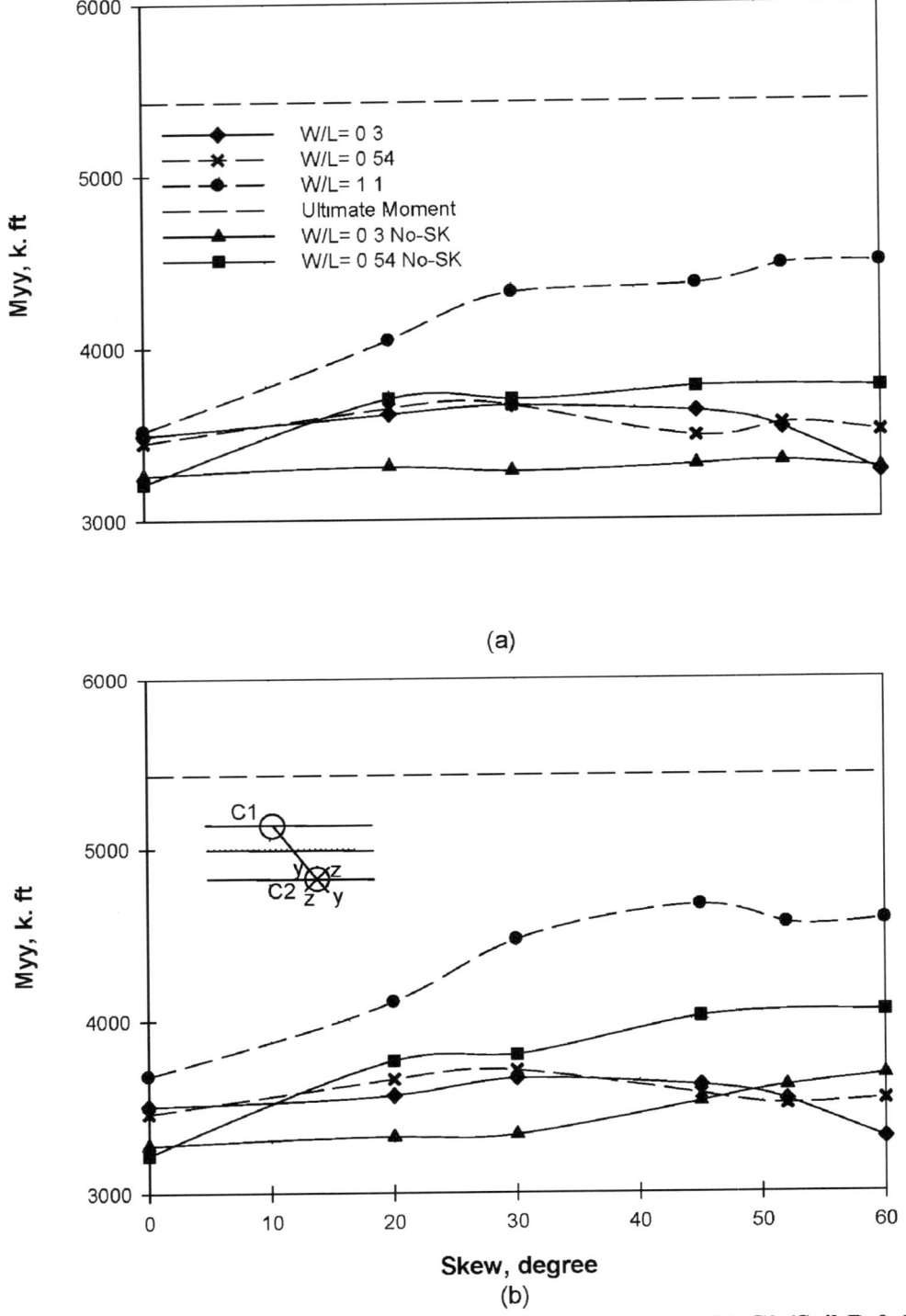

(a)

(b)

Figure 3-132. Average Moment in Y-Directions of (a) C1, and (b) C2 (Soil-D-0.6g)

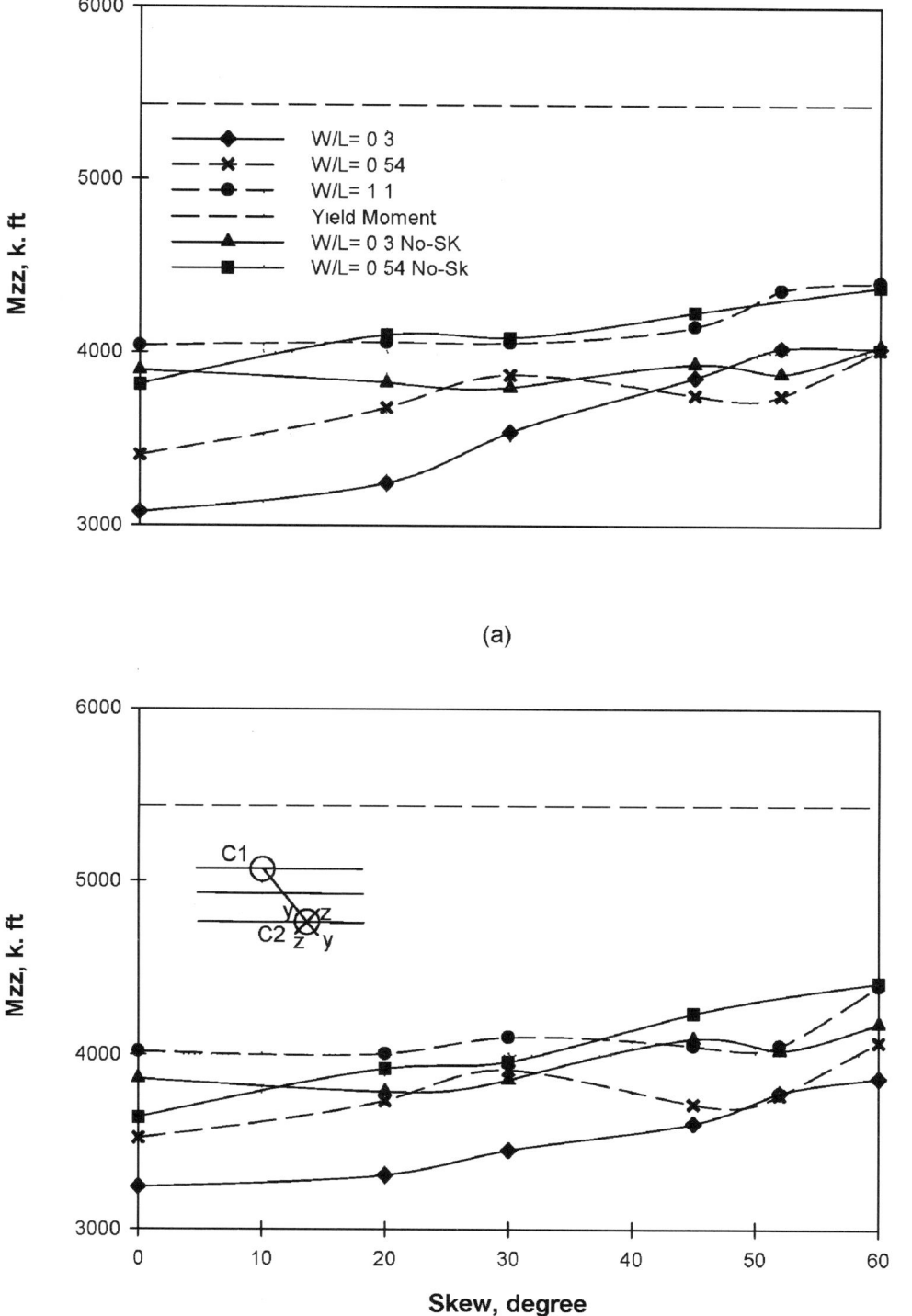

(a)

(b)

Figure 3-133. Average Moment in Z-Directions of (a) C1, and (b) C2 (Soil-D-0.6g)

278

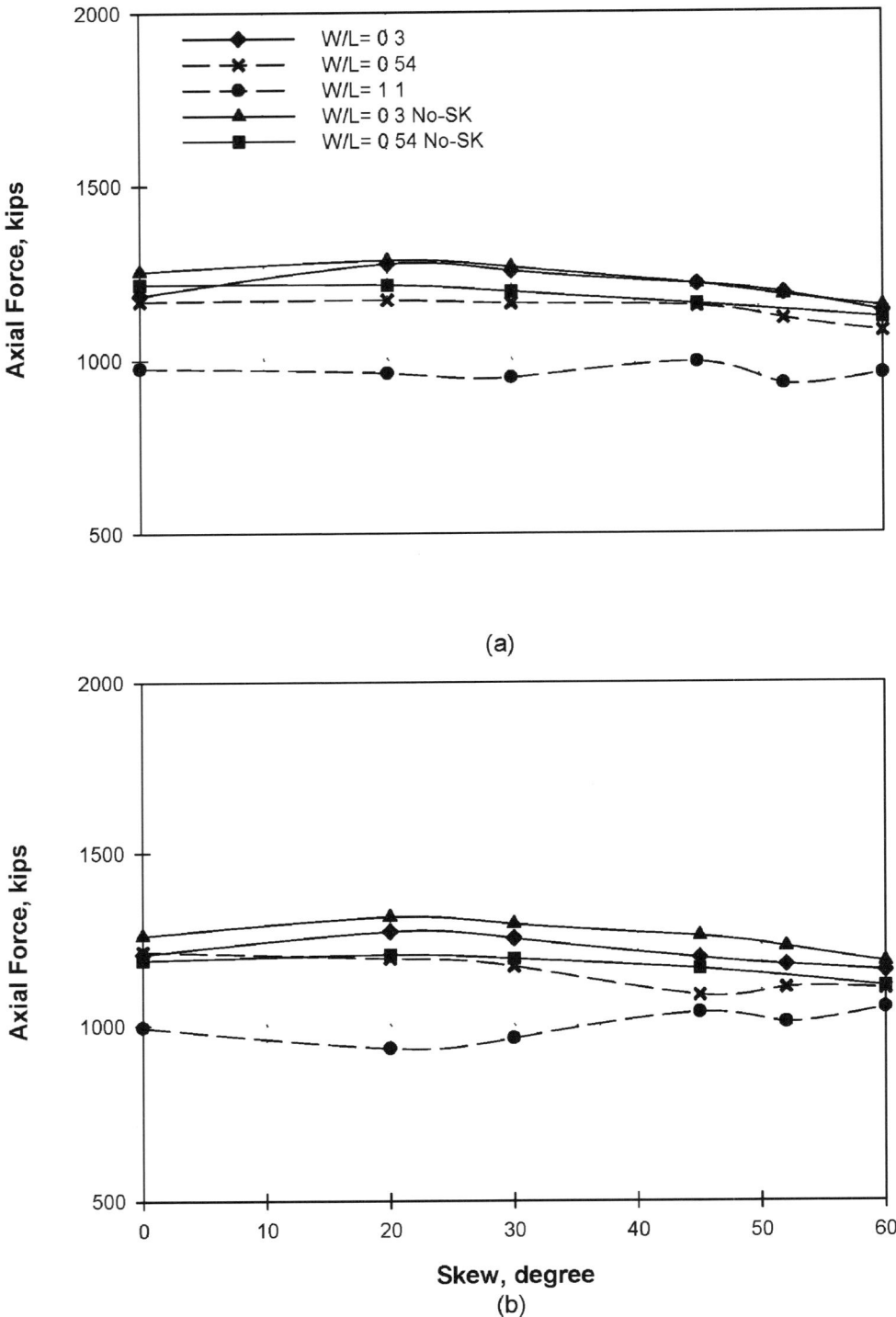

(a)

(b)

Figure 3-134. Average Axial Force in (a) C1, and (b) C2 (Soil-D-0.6g)

279

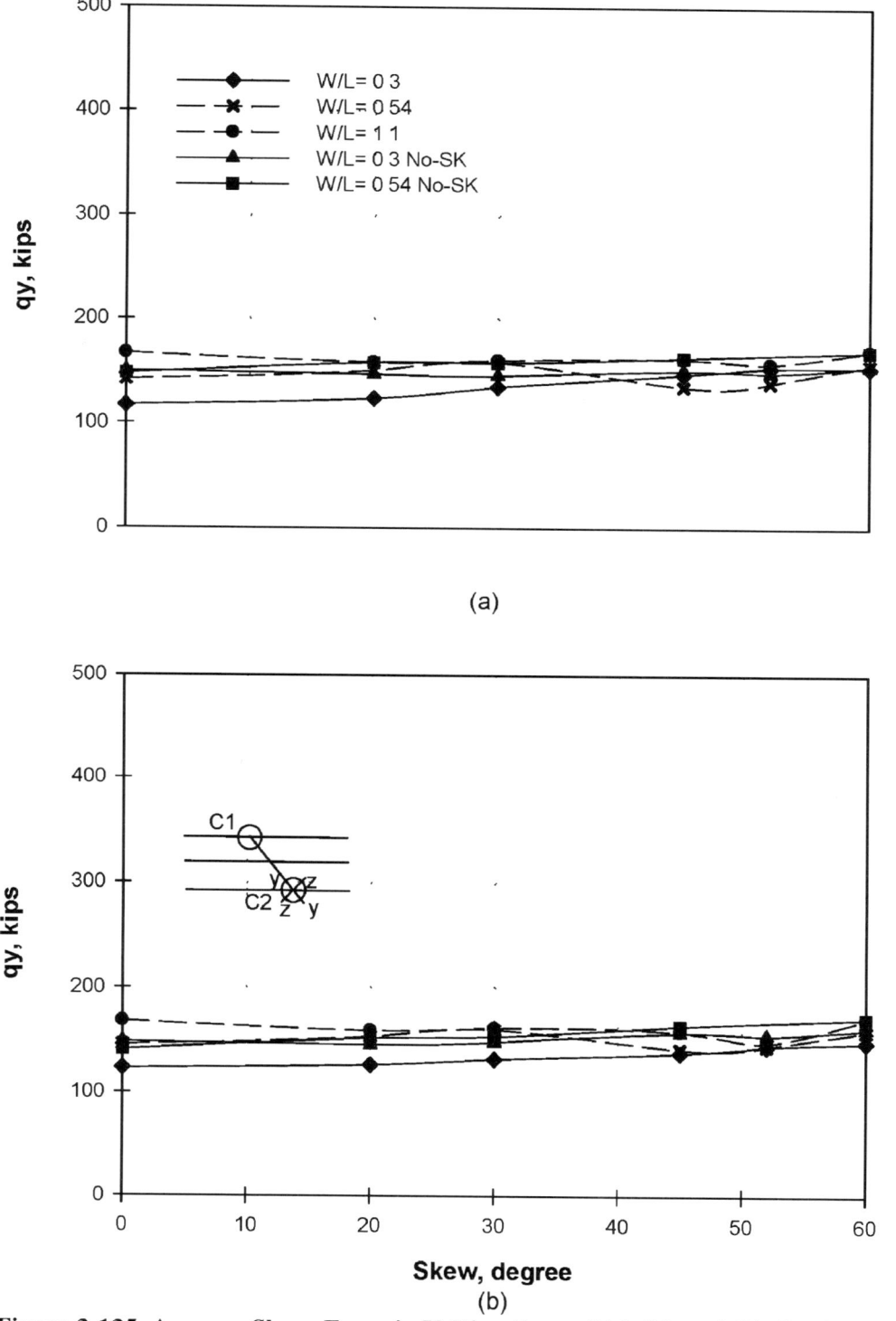

(a)

(b)

Figure 3-135. Average Shear Force in Y-Directions of (a) C1, and (b) C2 (Soil-D-0.6g)

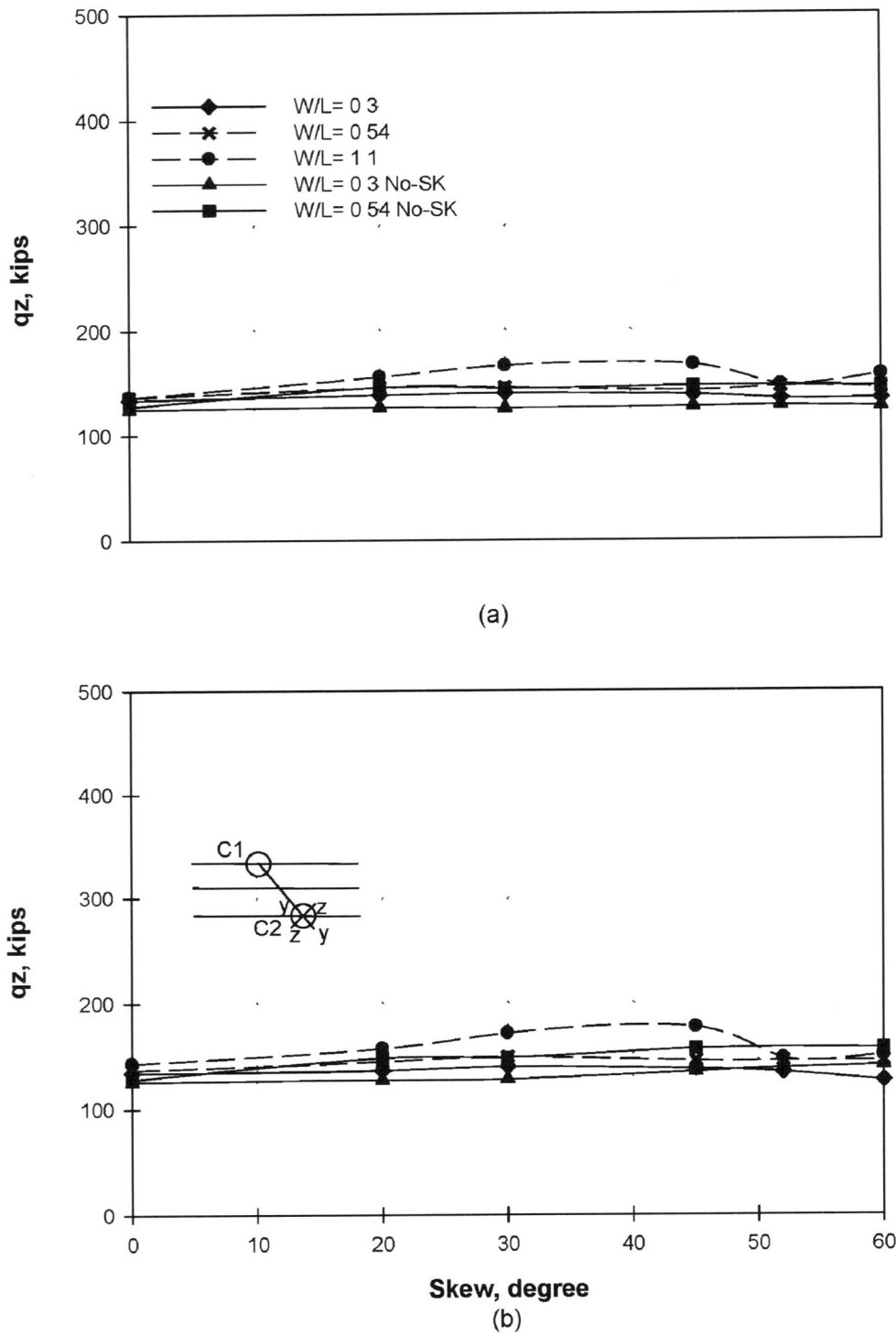

(a)

(b)

Figure 3-136. Average Shear Force in Z-Directions of (a) C1, and (b) C2 (Soil-D-0.6g)

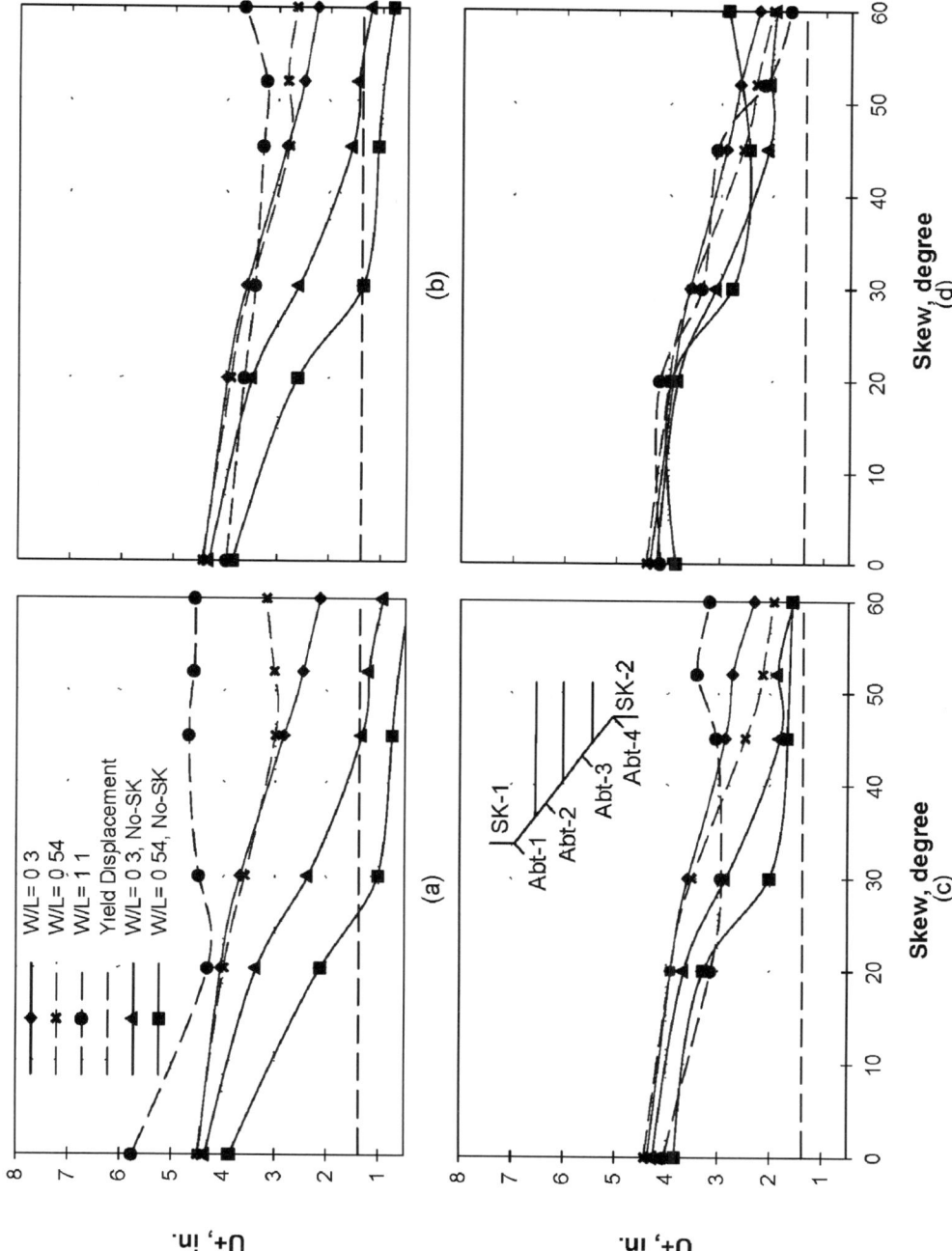

Figure 3-137. Average Displacement "into" Abutment's (a) Abt-1, (b) Abt-2, (c) Abt-3, and (d) Abt-4 (Soil-D-0.6g)

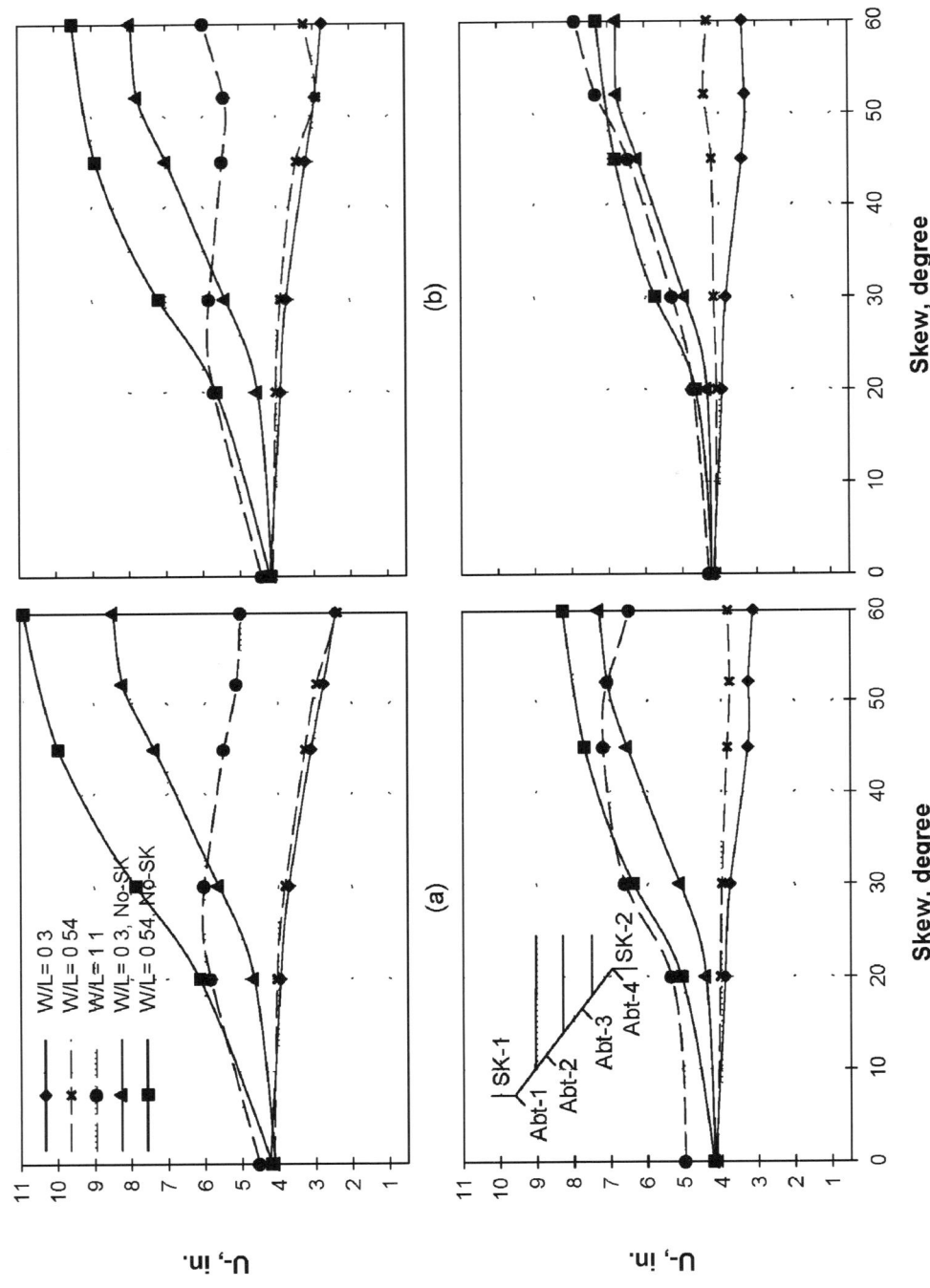

Figure 3-138. Average Displacement "away" from Abutment's (a) Abt-1, (b) Abt-2, (c) Abt-3, and (d) Abt-4 (Soil-D-0.6g)

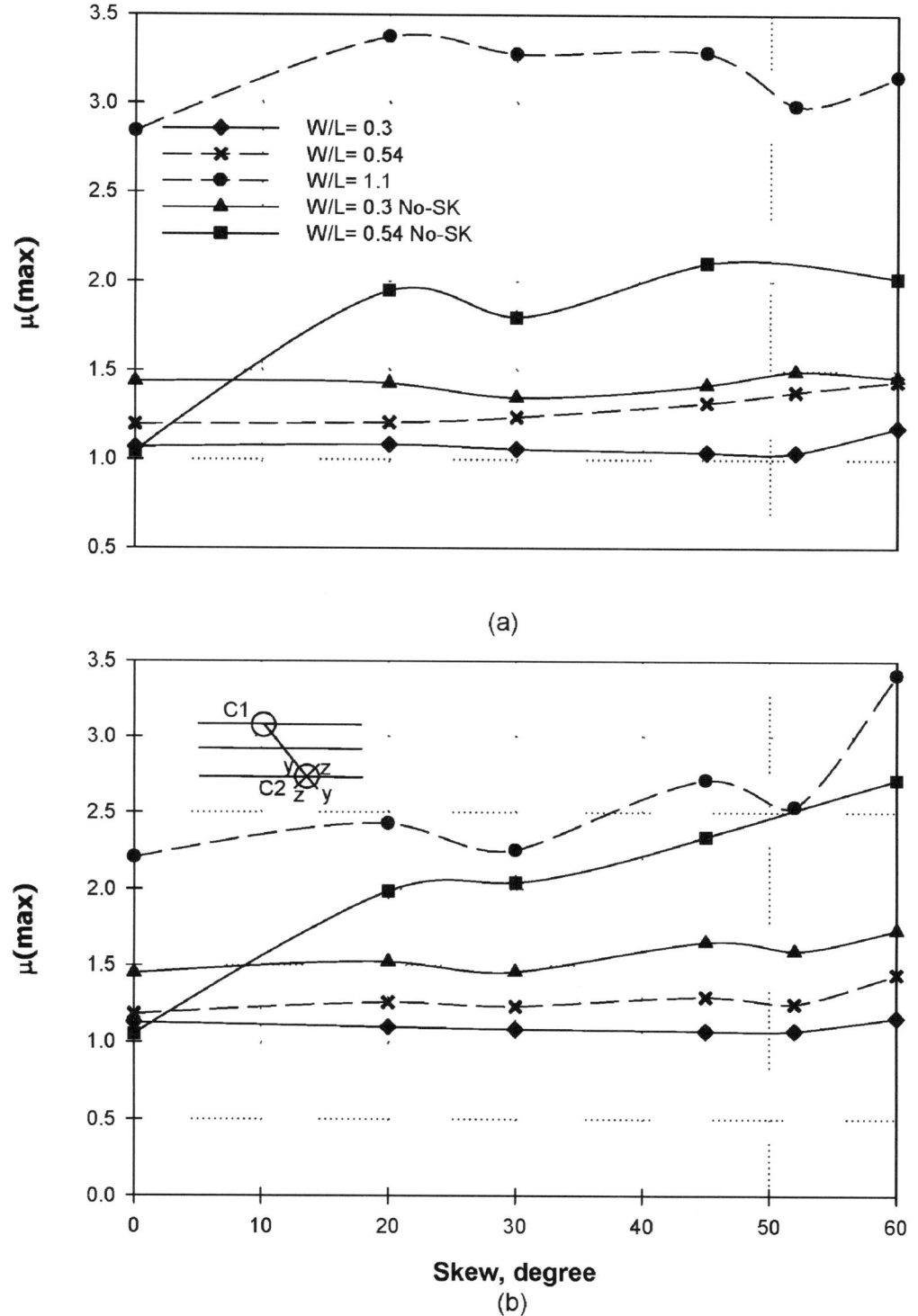

(a)

(b)

Figure 3-139. Maximum Average Ductility Ratio of (a) C1 and (b) C2 (Soil-D-0.6g)

4. Implications for the Seismic Design of Skewed Highway Bridges

4.1. Introduction

A comprehensive set of nonlinear time history analyses was conducted and presented in chapter 3. The analyses covered a range of skewed bridges with various configurations and conditions. However, nonlinear time history analysis is time expensive method and requires complex modeling techniques. Also, the accurate estimation of the seismic demand requires the use of a large suite of ground motions which increase the complexity of conducting nonlinear time history analysis. It is noted that simple methods and design tools are desired to reduce the computational and analytical effort needed to perform nonlinear dynamic analysis. In what follows, methods of analysis used by available guidelines are discussed, and potential tools to facilitate the design of bridges with skew are presented. Also, the applicability of the use of pushover analyses for skewed highway bridges is investigated.

4.2. Current Analysis and Design Guidelines

Caltrans Seismic Design Criteria (SDC, 2006) recommends a system analysis for bridges with irregular geometries. According to SDC, the bridges with irregular geometry, in particular curved and skewed bridges, multiple transverse expansion joints, massive substructure components, and foundations supported by soil can exhibit dynamic response characteristics that are not necessarily obvious and may not be captured in a separate nonlinear subsystem analysis. According to SDC (2006), the seismic analysis is needed to assess the force and deformation demand on structural system. Various analytical methods are presented in section 5 of SDC. The methods are equivalent static analysis (ESA) and elastic dynamic analysis (EDA) which are used to estimate the displacement demand for bridges. Also, inelastic static analysis (ISA) is a type of analysis which is used to estimate the displacement capacity. According to SDC (2006), ESA is used to estimate the displacement demand for ordinary standard bridges with well balanced spans and uniformly distributed stiffness. In this method, the seismic load is assumed as an equivalent static horizontal force applied in proportion to the tributary weight. Hence, the applied force is the product of amplified response spectrum (ARS) and the tributary weight. It is important to note that fundamental period of bridges is about 0.6 seconds. EDA should be used to estimate the demand for structures when ESA is not adequate. In this method, a linear elastic multi-mode spectral analysis using the appropriate response spectrum is conducted. It is recommended to include sufficient number of modes to capture at least 90% mass participation. Also, a minimum of three elements per column and four elements per span are required to perform this analysis and it is recommended to use the complete quadratic combination (CQC) to combine EDA modal results. ISA is commonly known as "pushover" analysis. This method is used by Caltrans to determine the displacement capacity of the structure. The ISA is an incremental analysis which can capture the nonlinear behavior of the elements.

AASHTO-LRFD recommends that the selection of the method of analysis to be based on the seismic zone, regularity, and importance of the bridge. The regular bridge is defined by AASHTO as a bridge that has less than seven spans; no abrupt or unusual change in weight, stiffness or geometry; and no change in these parameters from span to span and support to support. Table 4-1 shows the minimum analysis requirements for seismic analysis (Table 4.7.4.3.1-1 from AASHTO-LRFD). AASHTO-LRFD recommends the use of multi-mode spectral analysis. According to AASHTO, as a minimum, linear dynamic multi-mode analysis should be used with a 3-D model and the number of modes included in the analysis should be at least three times the number of spans of the model. Also, it is recommended to use the complete quadratic combination (CQC) method to combine the modal results.

Selection of a method of analyses is based on the type of the bridge and its characteristics. As can be seen in the tables, the use of multi-mode spectral analysis is recommended for irregular bridges since a great computational effort is needed to perform time history analyses. Accordingly, multi-mode spectral analyses were conducted on all of the bridge models included in this study and presented in chapter 3. As was discussed in chapter 1, multi-mode spectral analysis method is the most accurate spectral analysis method which can be used for irregular bridges since sufficiently large number of mode shapes can be considered. In what follows, a comparison between gap opening results from nonlinear time history analysis, and available guidelines for seat width is presented. Also, a comparison between recommendations of Caltrans displacement ductility and results of displacement ductility from time history analysis and multi-mode response spectra analysis is presented. Based on the findings of this study, design equations are proposed with limited application to highway bridges of similar configurations and geometries studied.

Table 4-1. Minimum Analysis Requirements for Seismic Effect (AASHTO-LRFD)

Seismic Zone	Single-Span Bridges	Other Bridges		Essential Bridges		Critical Bridges	
		Regular	Irregular	Regular	Irregular	Regular	Irregular
1	No seismic analysis required	*	*	*	*	*	*
2		SM^2/UL^1	SM	SM/UL	MM^3	MM	MM
3		SM/UL	MM	MM	MM	MM	TH^4
4		SM/UL	MM	MM	MM	TH	TH

*refers to no seismic analysis required
[1] UL refers to uniform load elastic method
[2] SM refers to single-mode elastic method
[3] MM refers to multi-mode elastic method
[4] TH refers to time history method

4.3. Minimum Seat Width Requirements

Gap opening results from time history analyses presented in chapter 3 are compared to the available guidelines for the recommended minimum seat width. The definition of seat width (N), following AASHTO-LRFD definition which is measured normal to the centerline of the bearing from the edge of the bridge deck to the edge of the abutment (Figure 4-1), is used in this study. It is noted that sufficient seat width must be provided to prevent unseating of bridge decks during major seismic events. As depicted in Figure 4-1, the seat width definations in Caltrans seismic design criteria (SDC) and AASHTO-LRFD are significantly different. Caltrans SDC (2006) requires a minimum width (N_{SDC}) of 30 in. On the other hand, AASHTO-LRFD (equation 4.7.4.4-1) provides the following equation for the determination of minimum seat width for zones 1 and 2:

$$N_{AASHTO} = (8 + 0.02 L + 0.08 H)(1 + 0.000125 S^2) \qquad \text{(4-1)}$$

Where: the minimum seat width is measured normal to the centerline of the bearing, L is the length of the bridge deck to the adjacent expansion joint, or to the end of the bridge deck; for hinges within a span, L is the sum of the distances to either side of the hinge; for single-span bridges; L equals the length of the bridge deck (ft), H is average height of columns supporting the bridge deck to the next expansion joint (ft), and S is the skew of support measured from line normal to span. In the AASHTO-LRFD equation "0.02L" term accounts for the spatial variation in ground motion, "0.08H" term accounts for the base rotation at the bent foundation, and the remaining terms in brackets is to account for the effect of the skew angle.

The envelopes of the gap openings at the abutments based on the comprehensive time history analyses (THA) and response spectrum analyses (RSA) are presented in comparison to Caltrans and AASHTO-LRFD guidelines in Figure 4-2. In the figure, time history responses for different configurations are presented as scatter points. It is important to note that the multi-mode spectral analysis was performed by applying Caltrans ARS for Soil-D biaxially to the bridge models. The bridge models have fundamental periods of about 0.6 second and the second mode of vibration has a period of about 0.4 seconds. Figure 3-3 shows that the average response spectra of ground motions match the Caltrans ARS at the period of interest which is about 0.6 seconds; however, for smaller periods than 0.6 seconds, the average response spectrum of ground motions is much larger than Caltrans ARS. Therefore, the application of transverse mode of vibration and higher modes will lead to smaller deformations when used in the RSA since the Caltrans ARS shows smaller values of response spectrum for smaller periods. It can be observed that the abutment gap openings of the analytical results are not exceeding the seat width required by the two guidelines. It was found that the abutment seat width as specified in the Caltrans seismic design criteria (SDC) was adequate. In the absence of a complete system study Caltrans equation can be deemed appropriate until the level of conservatism is quantified.

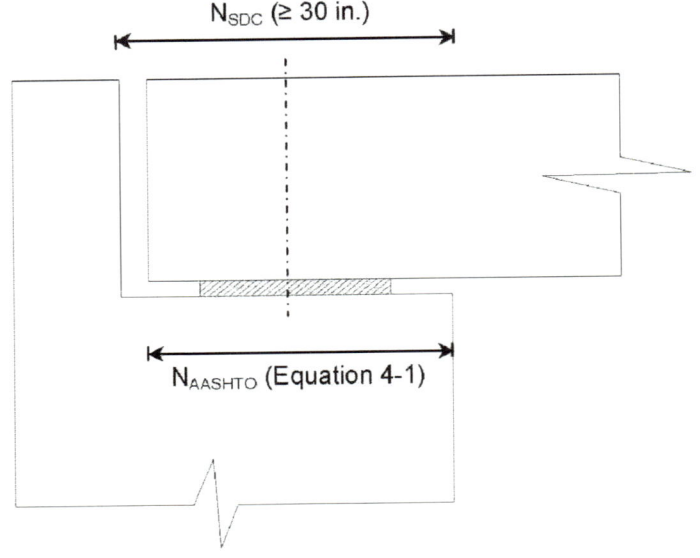

Figure 4-1. Abutment Seat Width

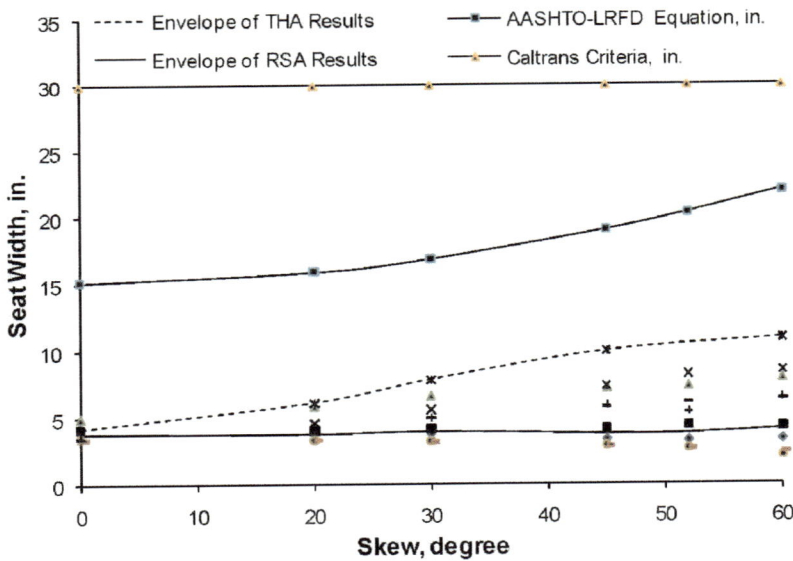

Figure 4-2. Seat Width Comparison

4.4. Displacement Ductility

It was also of interest to assess the adequacy of Caltrans methods of estimating ductility demands. Hence pushover analysis on the entire bent of the bridge models was conducted and the yield displacement of the columns was obtained to be 1.25 in. and 2.5 in. for fixed and pinned base cases, respectively. The ductility was calculated following Caltrans SDC equation (2.2). The equation is presented below and it is a function of the global displacement demand (Δ_D) and the yield displacement of the component (Δ_Y) which is presented above. Caltrans SDC states that the elastic dynamic analysis (EDA) can be used to determine the global displacement. The total displacement is the vector sum of the two horizontal displacement components, including the displacement due to base rotation for the pinned base case. The total displacements were extracted from results of multi-mode spectral analyses (RSA) and the comprehensive time history analyses (THA). It should be noted that the average response spectrum of ground motions matched Caltrans ARS which is used for RSA at the period of interest which is about 0.6 seconds, otherwise, Caltrans ARS has smaller spectra values.

$$\mu_D = \frac{\Delta_D}{\Delta_Y} \tag{4-2}$$

As mentioned before, the ductility factors were obtained using equation (4-2) that from Caltrans SDC for the two analysis types and presented with Caltrans ductility recommendations in the following plot (Figure 4-3). Caltrans ductility target values were calibrated to experimental test results (SDC, 2006) and the recommended displacement ductility factors for different components are:

- For single column bents supports on fixed foundation ($\mu_\Delta \leq 4$)

- For multi-column bents supports on fixed or pinned footings ($\mu_\Delta \leq 5$)

- For pier walls (weak direction) supported on fixed or pinned footings ($\mu_\Delta \leq 5$)

- For pier walls (strong direction) supported on fixed or pinned footings ($\mu_\Delta \leq 1$)

Since the bridge models under study have fixed and pinned foundations and multi-columns bents, a ductility demand of 5 was used and presented in Figure 4-3. The scattered points in the figure show the results of the THA and RSA analyses for models with different configurations. It can be concluded that Caltrans recommendation is conservative.

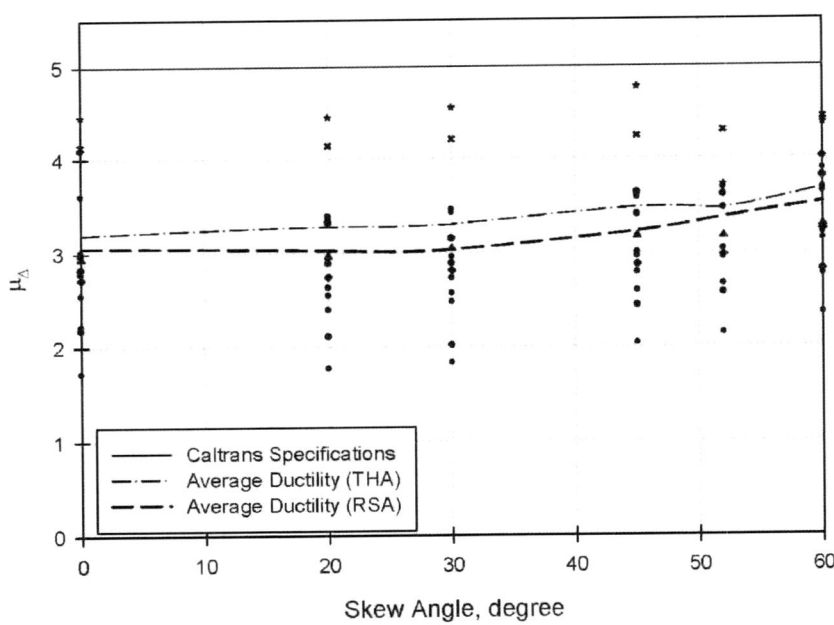

Figure 4-3. Ductility Comparison

4.5. Design Equations for use with RSA

Multi-mode response spectral analysis (RSA) is employed as one of the most commonly used analysis methods for the design of highway bridges. Sufficient number of modes were used to ensure at least 90% mass participation and CQC was used to combine the modal results following the recommendations of Caltrans SDC. Caltrans ARS response spectrum for Soil-D was selected for RSA and presented in chapter 3. Soil-D showed larger response values as presented in chapter 3 therefore it was important to develop the equation for this type of soil. The multi-mode response spectral analysis was conducted on the 3-D improved beam-stick models presented in chapter 2 for all configurations understudy.

(1) Various skew angles (0, 20, 30, 45, 52, 60)

(2) Pinned foundations

(3) Fixed foundations

(4) Various aspect ratios (0.3, 0.5, 1.0)

(5) With shear keys

(6) Without shear keys

The response quantities associated with the parameters of investigation listed below are compared with those obtained from the nonlinear time history analyses (THA) presented in chapter 3 for the corresponding configurations.

(1) Maximum Design Longitudinal Deck Displacement

(2) Maximum Design Transverse Deck Displacement near Abutments

(3) Displacement Demand on Columns

(4) Maximum Design Gap Opening at the Abutment

(5) Maximum Design [External] Shear key Forces

It is important to note that, in general, many sources of nonlinearity can not be captured by linear analysis. According to Caltrans SDC, these sources are the effects of the surrounding soil, yielding of structural components, opening and closing of expansion joints, and nonlinear restrainers and abutment behavior. In the models under study, many sources of nonlinearlity could not be captured because of the nature of the linear analysis such as: cyclic yielding of columns, the opening and closing of abutment gap, the complete hysteresis of bearing pads and shear keys, and the mechanism of shear keys response. Also, response spectra of ground motions matched the target response spectra at periods of interest only. Therefore, the results of RSA are expected to differ considerably from those of THA. Hence, a set of response ratios are developed to utilize the RSA method to obtain accurate results as those obtained from THA.

The response ratio, f, is introduced as the ratio of THA results to RSA results and accordingly plotted versus the skew angle for all response quantities. The response ratio, f, can be used to adjust the results of RSA analysis to obtain more accurate response values that also account for the skew effects. For all of the RSA and bridge models, sufficient number of modes was included in the analyses to ensure a total of 90% or greater mass participation. Certain trends can be observed for various quantities (Figure 4-4 through Figure 4-8). Clearly, the proposed set of plots (or equations) can be used to facilitate the design of highway bridges with various skew and aspect ratios of the similar configurations that were included in the present study. Accordingly, a multi-mode spectral analysis can be conducted and conservative design response values can be obtained by multiplying by the corresponding "f" factors. The design equations for use with RSA are presented (Equation 4-3 through 4-14) to be used to adjust results from RSA. Accordingly, the use of spectral analyses underestimated the response quantities of all the parameters understudy for skewed highway bridges since many sources of nonlinearity can not be captured using linear analysis and the average response spectra of ground motions did not match Caltrans ARS for all periods. From the following equations, skew angle of 30 degree resembled a changing point. Response ratios of bridges with skew angle less than 30 degrees showed different trend than those of bridges with skew angle larger than 30 degrees in most cases.

The following notation was used in the plots and equations: F= Fixed Foundations, P= Pinned Foundations. The superscript is used to denote the limiting applicable aspect ratio (0.3, 0.5, and 1.0), and subscript sk stands for "with shear key" configurations, and no subscript means "without shear key" configurations.

4.5.1. Displacement Demand

Design displacement [demand] is typically associated with either global displacement and/or local member displacement such as; column displacement. Most of the available guidelines use linear analysis to estimate the displacement demand. The global system displacement is the total displacement which includes the effect of foundation flexibility and flexibility of components. Therefore, it is crucial that the analytical model to be used is generic and well detailed as possible. The details include the accurate definition of structural characteristics and boundary conditions. The global displacement is an important parameter since it is used to determine the displacement ductility demand on the system. For the local member deformation, it is defined as the deformations of the member excluding the deformation due to the flexibility of the foundations.

The analytical models used to conduct nonlinear time history analysis were those presented in chapter 2 and the results are presented in chapter 3. It is clear that these models are well detailed since the linear and nonlinear properties of the deck, columns, bearing pads, abutment-soil interaction, and shear keys are defined explicitly and accurately. Also, the commonly used linear analysis is conducted using the multi-mode spectral analysis (RSA). The results of nonlinear time history analyses were compared to those of RSA to develop a set of design equations. These equations are developed to adjust results of RSA to obtain more accurate response quantities for longitudinal deck displacement, transverse deck displacement near the abutments, and displacement demand on columns (local displacement).

The response ratio, $f_{\Delta x}$, (Figure 4-4) is the ratio of the longitudinal deck displacements obtained from time history analyses (THA) results to those obtained from multi-mode response spectral analyses (RSA). It is important to note that the response ratio can be used to adjust the longitudinal deck displacements obtained from RSA for all nodes throughout the bridge deck. The response ratio, $f_{\Delta y}$, (Figure 4-5) is the ratio of the transverse deck displacements near abutments obtained from time history analyses (THA) results to those obtained from multi-mode response spectral analyses (RSA).

In order to determine the displacement demand on columns, the vector sum of the two displacement components (global x and y) of column displacements after subtracting the base rotation contribution for the pinned base case only was determined. The displacements and base rotations were extracted from results of multi-mode spectral analyses (RSA) and the comprehensive time history analyses (THA). A set of factors, $f_{\Delta c}$, (Figure 4-6) were developed for different configurations to be used to adjust the displacements of the columns determined using RSA.

In general, a system analysis was conducted to obtain displacement demand on the bridge following Caltrans recommendations, however, the response quantities obtained using RSA were different from those obtained using THA. This can be attributed to that the average response spectrum of ground motions did not match Caltrans ARS for all periods and the nature of linear analysis which can not capture many sources of nonlinearity, which are accurately presented in the nonlinear models, affecting the performance of the bridge models. The mechanisms such as the abutment gap opening and shear key behavior can not be accounted for in a linear model therefore results became considerably different. Therefore a set of adjustment factors are needed to obtain accurate results comparable to those obtained using THA.

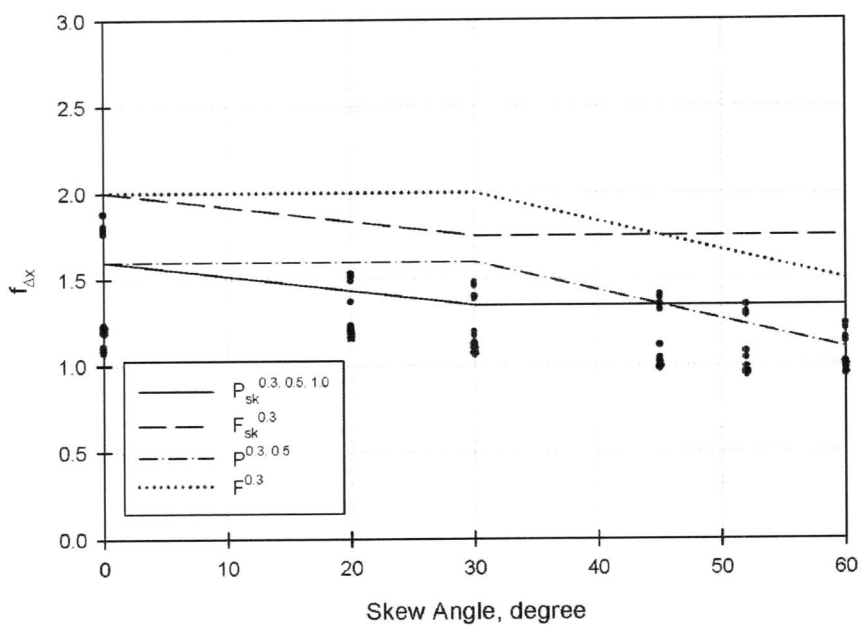

Figure 4-4 Longitudinal Deck Displacement

The following equations are proposed to adjust the longitudinal deck displacements from a multi-mode RSA. It is important to note that θ is the skew angle.

$$f_{\Delta x} = 1.35 - s \times (\theta - 30) \qquad \begin{cases} s = 1/120 & 0 \le \theta \le 30 \\ s = 0 & \theta > 30 \end{cases} \text{ valid for } P_{sk}^{\ 0.3, 0.5, 1.0} \text{ cases} \qquad \textbf{(4-3)}$$

$$f_{\Delta x} = 1.75 - s \times (\theta - 30) \qquad \begin{cases} s = 1/120 & 0 \le \theta \le 30 \\ s = 0 & \theta > 30 \end{cases} \text{ valid for } F_{sk}^{\ 0.3} \text{ case} \qquad \textbf{(4-4)}$$

$$f_{\Delta x} = 1.50 - s \times (\theta - 30) \qquad \begin{cases} s = 0 & 0 \le \theta \le 30 \\ s = 1/60 & \theta > 30 \end{cases} \text{ valid for } P^{0.3, 0.5} \text{ case} \qquad \textbf{(4-5)}$$

$$f_{\Delta x} = 2.00 - s \times (\theta - 30) \qquad \begin{cases} s = 0 & 0 \le \theta \le 30 \\ s = 1/60 & \theta > 30 \end{cases} \text{ valid for } F^{0.3} \text{ case} \qquad \textbf{(4-6)}$$

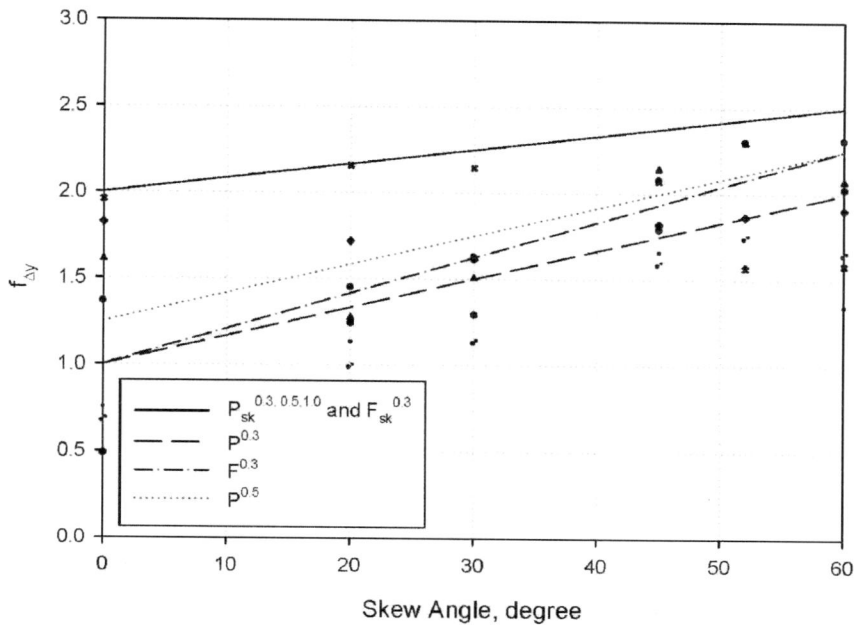

Figure 4-5 Transverse Deck Displacement near Abutments

The following equations are proposed to adjust the transverse deck displacements near abutments from a multi-mode RSA. It is important to note that θ is the skew angle.

$$f_{\Delta y} = \begin{cases} 2.00 + \theta \times 1/120 & \text{valid for } P_{sk}^{0.3,0.5,1.0} \text{and } F_{sk}^{0.3} \text{cases} \\ 1.00 + \theta \times 1/60 & \text{valid for } P^{0.3} \text{case} \\ 1.00 + \theta \times 1/48 & \text{valid for } F^{0.3} \text{ case} \\ 1.25 + \theta \times 1/60 & \text{valid for } P^{0.5} \text{case} \end{cases} \qquad (4\text{-}7)$$

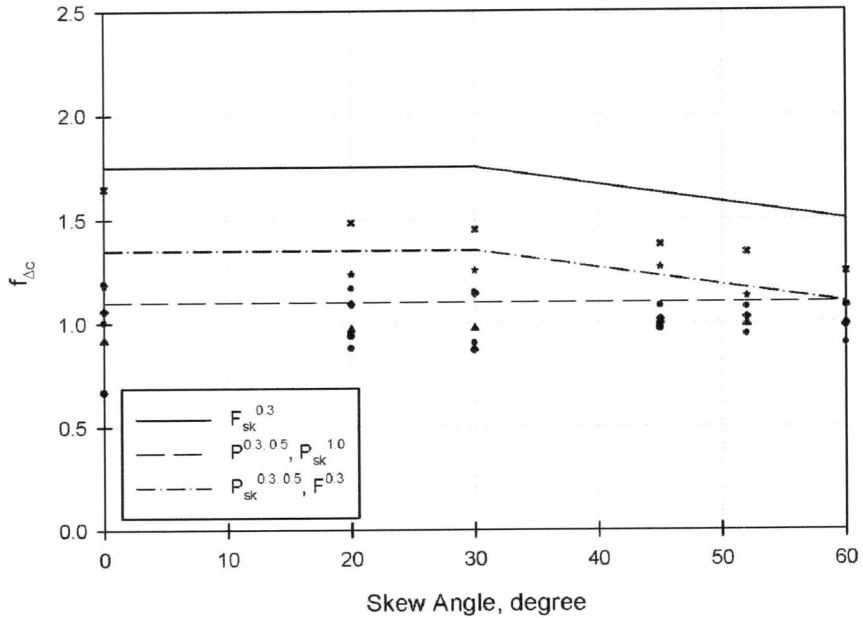

Figure 4-6 Displacement Demand on Columns

The following equations are proposed to adjust the displacement demand on columns from a multi-mode RSA. It is important to note that θ is the skew angle.

$$f_{\Delta c} = 1.75 - s \times (\theta - 30) \begin{cases} s = 0 & 0 \le \theta \le 30 \\ s = 1/120 & \theta > 30 \end{cases} \text{valid for } F_{sk}^{0.3} \text{ cases} \quad \textbf{(4-8)}$$

$$f_{\Delta c} = 1.10 \quad \text{valid for } P^{0.3,0.5} \text{ and } P_{sk}^{1.0} \text{ cases} \quad \textbf{(4-9)}$$

$$f_{\Delta c} = 1.35 - s \times (\theta - 30) \begin{cases} s = 0 & 0 \le \theta \le 30 \\ s = 1/120 & \theta > 30 \end{cases} \text{valid for } P_{sk}^{0.3, 0.5}, F^{0.3} \text{ cases} \textbf{(4-10)}$$

4.5.2. Maximum Design Gap Opening/Closing Perpendicular to the Abutment

In this section, the response ratio, $f_{\Delta a}$, is the ratio of the abutment gap opening/closing perpendicular to the skew obtained from time history analyses (THA) results to those obtained from multi-mode response spectral analyses (RSA). It is important to obtain accurate estimates for the abutment gap opening using a simple analysis method such as RSA. The results can reflect on the abutment sizing and details, and also the provided abutment seat width. As can be seen in Figure 4-7, there is a factor of more than 4 between the two methods of analysis for large aspect ratios. It is important to note that the scattered points in the figure are the actual results of all the configurations used in the study.

It was clear from results of chapter 3 that gap opening increases with the skew angle and there is a potential of unseating under larger level of ground motions. As mentioned earlier, results of RSA are expected to differ considerably from those of THA due to inability of linear methods to captures many sources of nonlinearity including the abutment gap opening and closing. In the nonlinear model, the abutment-soil interaction spring is compression only activated, otherwise, the gap opens. However, in the linear analysis this mechanism can not be modeled which led to smaller deformations at the abutments. Thereby, the use of the following set of factors can adjust the response quantities obtained using RSA to account for the effect of gap opening in the actual model.

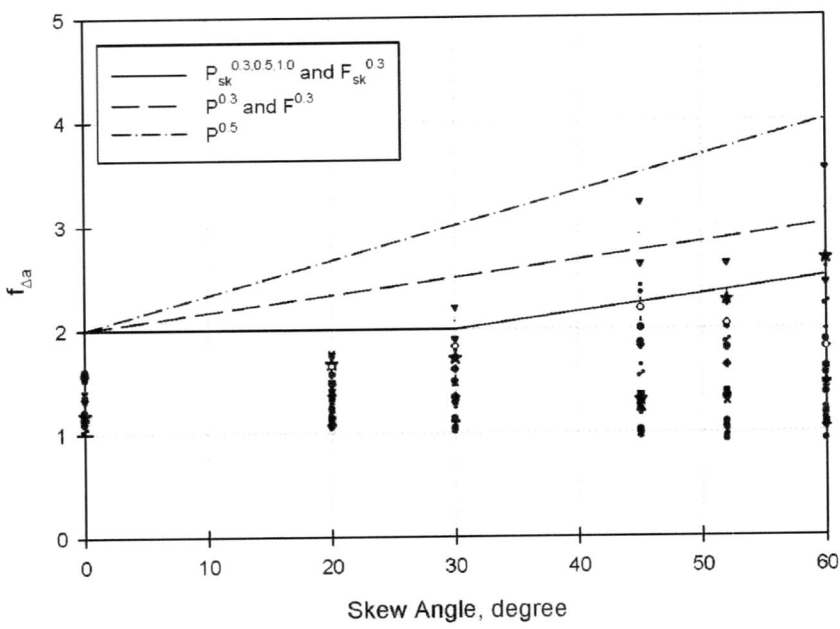

Figure 4-7 Abutment's Gap Opening

The following equations are proposed to adjust the gap opening at the abutments from a multi-mode RSA. It is important to note that θ is the skew angle.

$$f_{\Delta a} = 2.00 + s \times (\theta - 30) \begin{cases} s = 0 & 0 \leq \theta \leq 30 \\ s = 1/60 & \theta > 30 \end{cases} \text{valid for } P_{sk}^{0.3, 0.5, 1.0} \text{and } F_{sk}^{0.3} \text{cases} \quad \textbf{(4-11)}$$

$$f_{\Delta a} = 2.00 + s \times \theta \qquad \begin{cases} s = 1/60 & \text{valid for } P^{0.3} \text{and } F^{0.3} \text{cases} \\ s = 1/30 & \text{valid for } P^{0.5} \text{case} \end{cases} \qquad \textbf{(4-12)}$$

4.5.3. Maximum Design [External] Shear key Force

In this section, the response ratio, $f_{\Delta sk}$, is the ratio of the shear key forces obtained from time history analyses (THA) results to those obtained from multi-mode response spectral analyses (RSA). The design shear keys capacity is limited by the lateral pile capacity and the axial dead load reactions at the abutments. However, it is important for the designers to have a clear knowledge about the accurate amount of forces which will be developed in shear keys during the seismic event.

Chapter 3 presents clearly the importance of shear keys and the role which they play during the seismic event. It was clear that the removal of shear keys led to significant gap opening which increases with the skew. Also, removal of shear keys led to significant increase in the transverse deck displacement. As expected, the linear model failed to predict the accurate response of the shear keys since the abutment-soil and shear keys mechanism can not be modeled accurately in a linear model. The shear key is a compression activated only spring which is not possible to capture in a linear analysis therefore the results using RSA are different from those using THA. The complete hysteresis and the accurate mechanism of shear keys are modeled in the nonlinear modeled which can not be accomplished in a linear model. Therefore, the use of the following set of factors can adjust the response quantities obtained using RSA to account for the difference in modeling of shear keys.

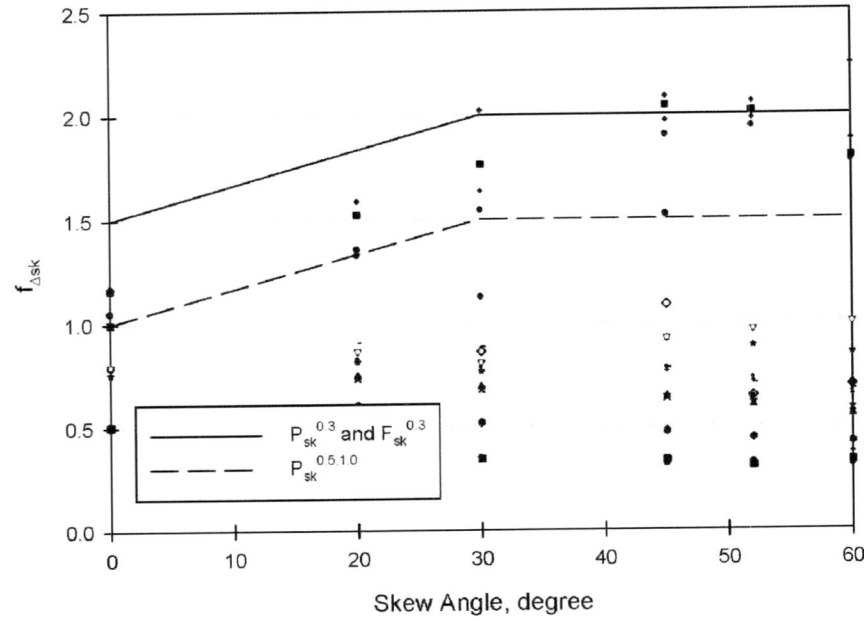

Figure 4-8 Shear key Force

The following equations are proposed to adjust the shear key force at the abutments from a multi-mode RSA. It is important to note that θ is the skew angle.

$$f_{\Delta sk} = 2.00 + s \times (\theta - 30) \quad \begin{cases} s = 1/60 & 0 \le \theta \le 30 \\ s = 0 & \theta > 30 \end{cases} \text{valid for } P_{sk}^{\ 0.3}, F_{sk}^{\ 0.3} \text{ cases} \quad \textbf{(4-13)}$$

$$f_{\Delta sk} = 1.50 + s \times (\theta - 30) \quad \begin{cases} s = 1/60 & 0 \le \theta \le 30 \\ s = 0 & \theta > 30 \end{cases} \text{valid for } P_{sk}^{\ 0.5, 1.0} \text{ case} \quad \textbf{(4-14)}$$

4.6. Comparison between Time History Analysis and Pushover Analyses

Nonlinear static pushover analyses are performed on beam stick models developed with the skew angles under consideration (0°, 20°, 30°, 45°, 52°, and 60°). The objectives are to examine the applicability of pushover analyses for skewed highway bridges. Since the pushover analysis is expected to be sensitive to the selection to the pushover load profile. Therefore, it was important to select a load profile accounting for the combination between modes. A load profile, which is a combination of transverse and longitudinal modes (Mode 1+2), is used. In order to investigate the accuracy of pushover analyses in predicting the lateral capacity of skewed bridges, individual maximum responses from nonlinear time history analyses are plotted on the pushover curves. The maximum resultant base shear and maximum displacement at the center of the bent (control node) in the transverse direction are plotted for each of the six ground motions of the larger level for soil type D. Two abutment support conditions are considered: Case I (with shear keys) and Case II (without shear keys) with two bridge aspect ratios (Figure 4-9 through Figure 4-12). The aspect ratios under consideration are 0.3 and 0.54. It is important to note that nonlinear time history analyses were preceded by nonlinear static analyses for the dead load. Similarly, the nonlinear pushover analyses were preceded by nonlinear static analyses for the dead load.

For Case I and 0.3 aspect ratio (Figure 4-9), the time history results compare well for the 0 degree skew. Nonlinear time history results follow the pushover curves very well;

although, the base shear capacity is slightly underpredicted by the pushover curve. For the 20 degree skew, the pushover curves compare well to the time history results. For the 30 degree skew, the pushover curve continues to compare well to the time history results. For the 45 degree skew, the pushover curve resembles the results of time history analyses adequately since it runs between the individual ground motion results forming an average. For the 52 degree skew, the pushover curve underpredicts the capacity for most of the ground motions. For the 60 degree skew, the time history results compare well with the pushover curve. It can be concluded that the pushover curves compare well to the nonlinear time history analysis and can be used to predict the capacity of skewed bridges.

For Case I and 0.54 aspect ratio (Figure 4-10), the time history results compare well for the 0 degree skew and pushover curve predicts the capacity reasonably. For the 20 degree skew, the pushover curves slightly overpredict the base shear capacity. For the 30 degree skew, the pushover curve continues to overpredict most of time history results. For the 45 degree skew, the pushover curve predicts the base shear capacity adequately since it runs between the individual ground motion results. For the 52 degree skew, the pushover curve overpredicts the capacity for most of the ground motions. For the 60 degree skew, the pushover curve overpredicts the capacity for most of the ground motions. It can be concluded that the pushover curves compare well to the nonlinear time history analysis and can be used to predict the capacity of skewed bridges.

For Case II and 0.3 aspect ratio (Figure 4-11), in this case the shear key is assumed to fail, the pushover curve is not complete due to divergence of the analysis for the 0 degree skew. However, the pushover curve overpredicts the base shear capacity. For the 20 degree skew, the pushover curve underpredicts the capacity. For the 30 degree skew, the pushover curve overpredits the capacity for all of the ground motions. For the 45 degree skew, the pushover curve converges to the individual ground motion results and agrees well with them. For the 52 degree skew, the pushover curve predicts the capacity for two of the ground motions and overpredicts some of them. For the 60 degree skew, the pushover curve runs among the points resembling the results for the individual ground motions showing adequate prediction of the base shear capacity.

For Case II and 0.54 aspect ratio (Figure 4-12), the pushover curve underpredicts the base shear capacity for the 0 degree skew. For the 20 degree skew, the pushover curve overpredicts the capacity for all of the ground motions. For the 30 degree skew, the pushover curve fairly predicts the average capacity due to all of the ground motions. For the 45 degree skew, the pushover curve agrees well with the results due to two of the ground motions and overpredicts the capacity for most of them. For the 52 degree skew, the pushover curve overpredicts the capacity for all of the ground motions. For the 60 degree skew, the pushover curve runs among the scatter resembling the results for the individual ground motions showing adequate prediction of the base shear capacity. In general, it can be concluded that pushover analyses can be used to predict the base shear capacity of skewed highway bridges with and without shear keys and with different aspect ratios as well.

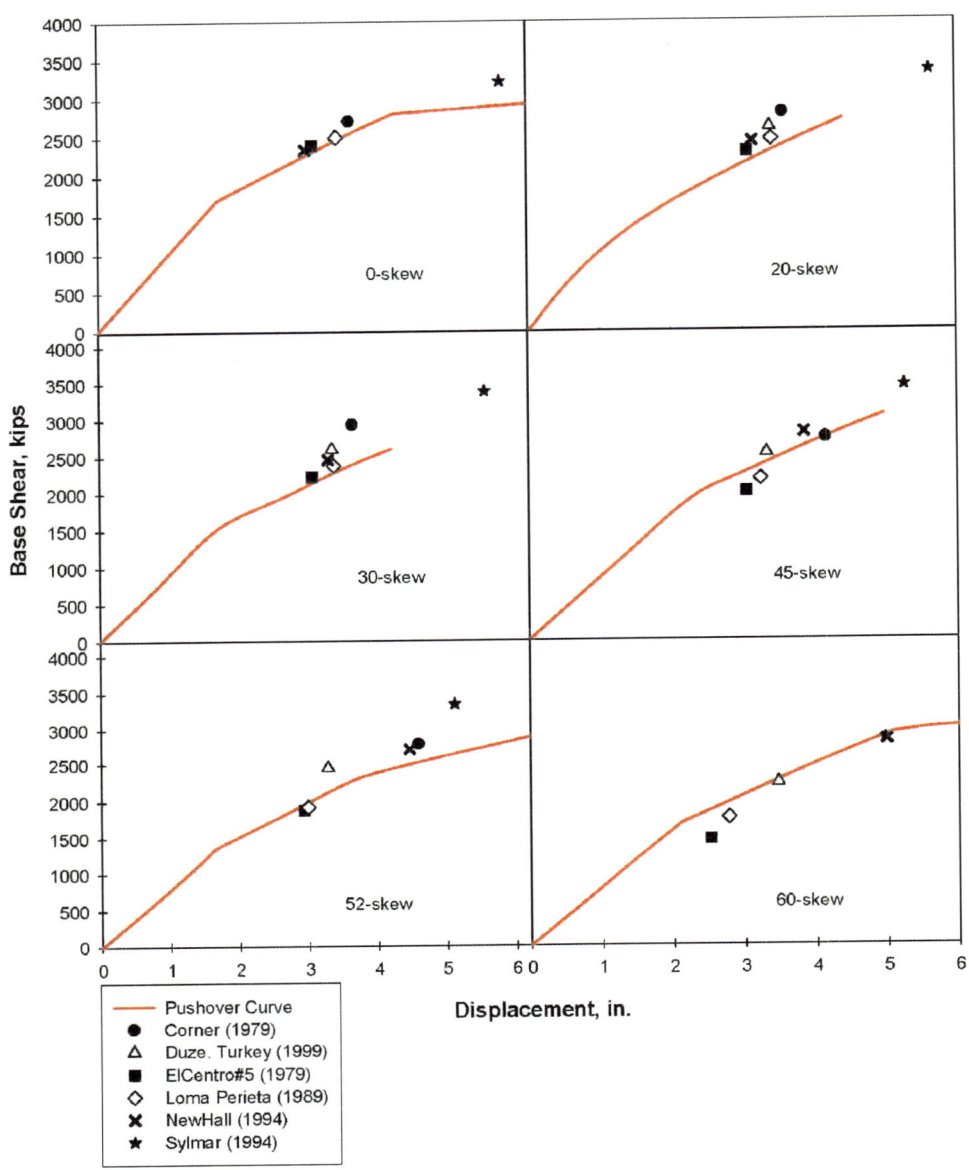

Figure 4-9. Comparison of Time History Results and Pushover Capacities for 0.3 Aspect Ratio, Case I- with shear keys

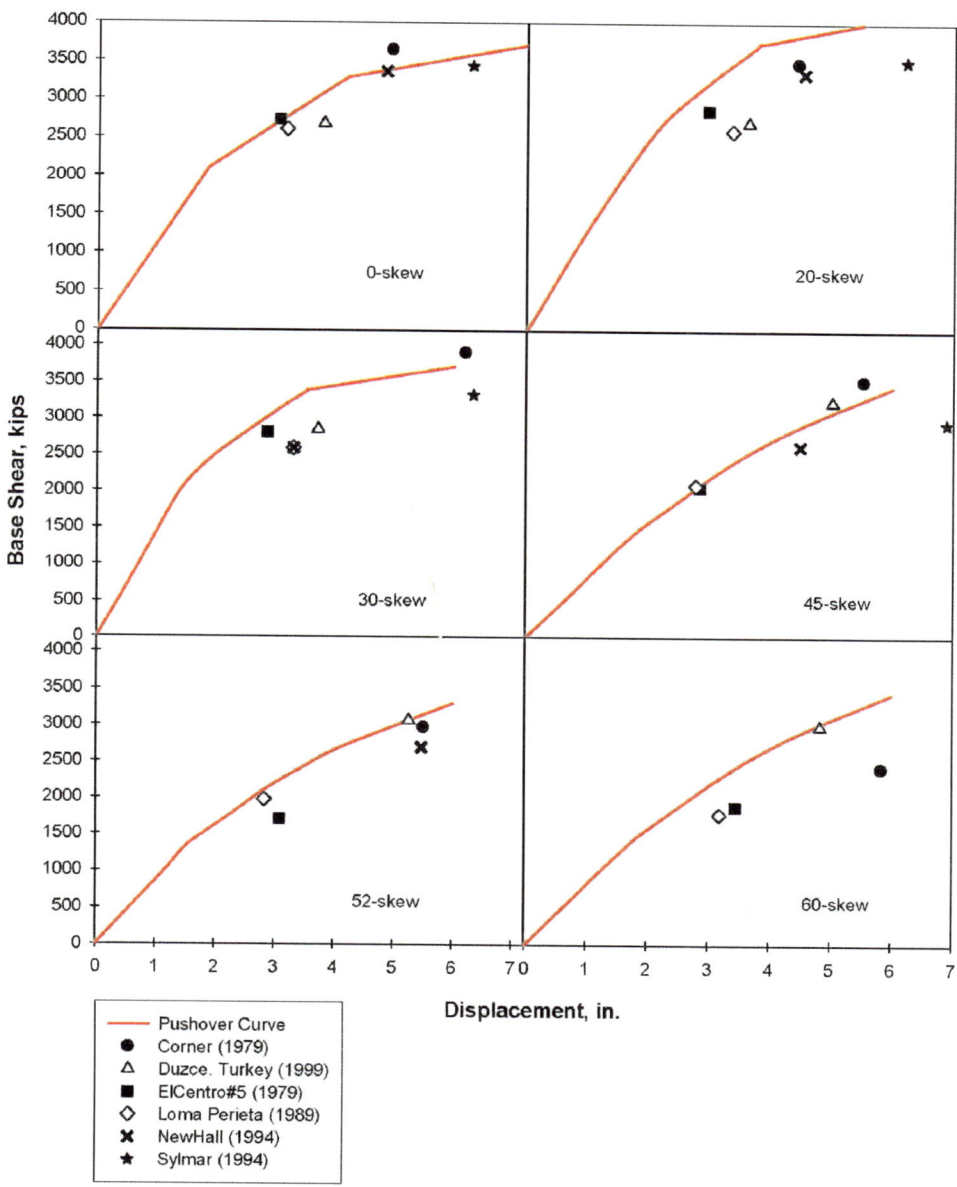

Figure 4-10. Comparison of Time History Results and Pushover Capacities for 0.54 Aspect Ratio, Case I- with shear keys

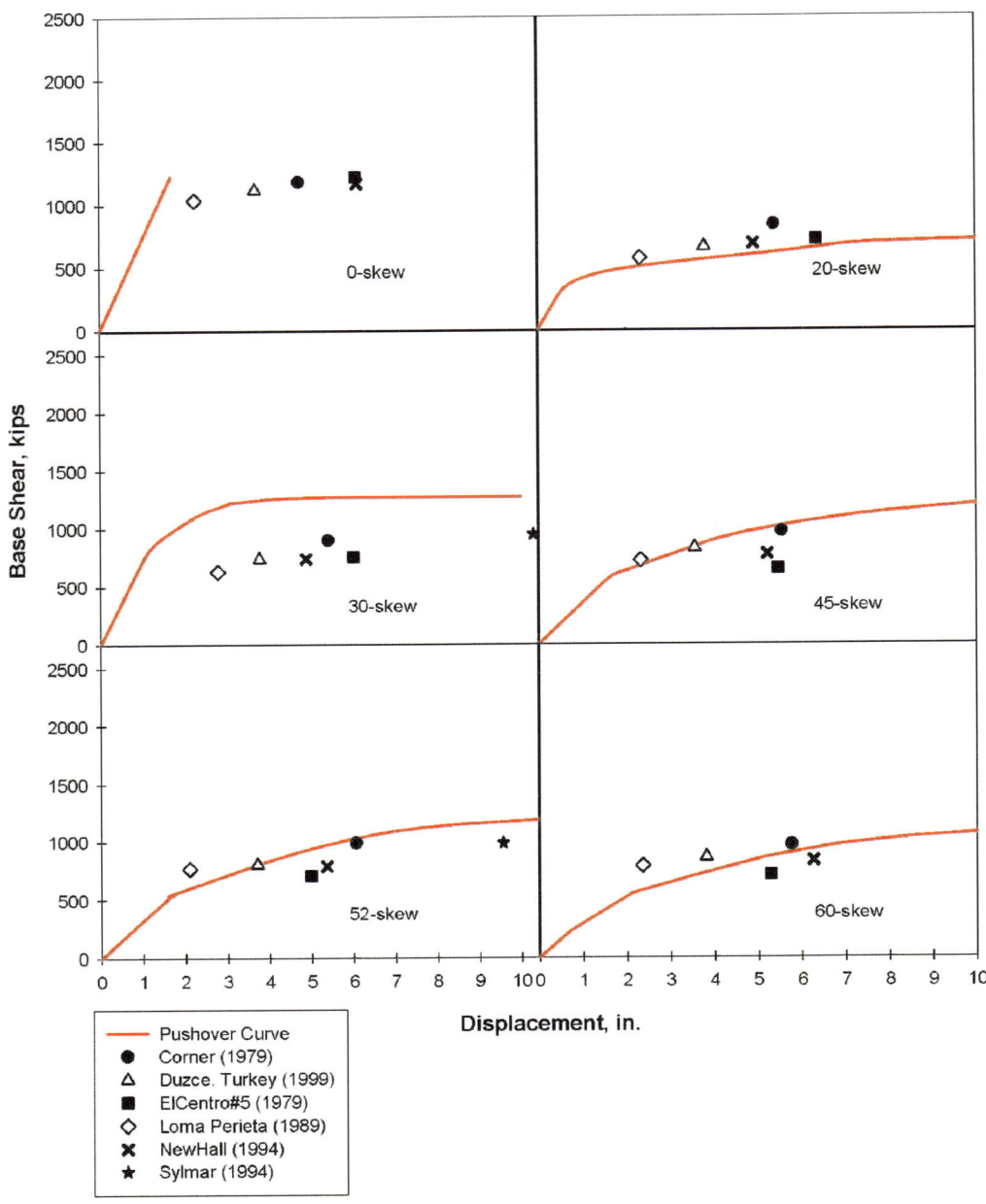

Figure 4-11. Comparison of Time History Results and Pushover Capacities for 0.3 Aspect Ratio, Case II- without shear keys

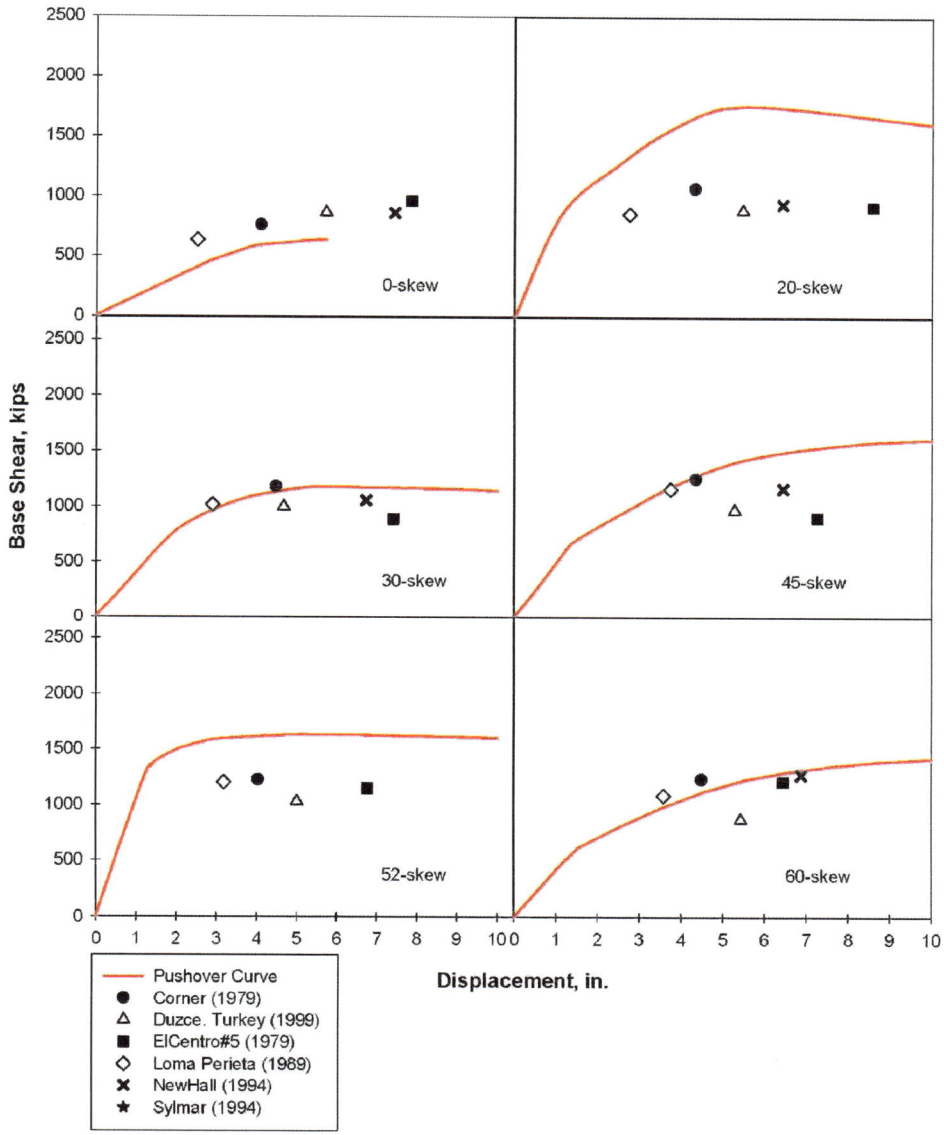

Figure 4-12. Comparison of Time History Results and Pushover Capacities for 0.54 Aspect Ratio, Case II- without shear keys

5. Analytical Fragility Curves for Skewed Highway Bridges

5.1. Introduction

In this section, a brief discussion of the development of the so-called fragility curves (a.k.a. functions) is presented first. Following the recommended improved and calibrated modeling techniques in the earlier chapters, detailed nonlinear models of a three-span highway bridge were developed with three different skew angles, namely 0, 30 and 60. These are selected to represent straight bridge, moderate skew and large skew angles. Incremental dynamic analysis (IDA) procedure was used with a total of 45 ground motion pairs and nonlinear time-history analyses were conducted. Analytically "observed" damage is recorded in view of the established limit states to arrive at analytical fragility curves. These are compared to the fragility curves established by the HAZUS method and also used to demonstrate the effect of the skew angle on the fragility curves for highway bridges.

5.2. Background

An important aspect and component of the so-called Seismic Risk Analysis (SRA) is the assessment of losses (e.g. social and economical) due to hazards (e.g. earthquake) and to develop reliable mitigation scenarios. Recently, SRA is being integrated in many fields that require various levels and stages of hazard (event) emergency management. Essential to the assessment of transportation systems' resilience and seismic performance is the availability of technically-sound methodologies for SRA of an overall transportation system. In this context, seismic resilience of a system is defined and measured by its robustness during and after earthquake and the rapidity of its recovery after an event. With these in mind, a desired tool that implements a SRA methodology should enable (1) pre-earthquake planning, (2) assessment of seismic improvement options, and (3) post-earthquake response including traffic-flows and economic losses; and should incorporate features that assist in the seismic risk reduction decision making.

In recognition of this need, the Federal Highway Administration (FHWA, 1993) has embarked on a multiyear seismic research program and has funded collaborative research studies with an ultimate goal to develop and study SRA methodologies for transportation systems. The culmination of this specific program has led to the development of a new methodology and tool named REDARS (Risks from Earthquake Damage to the Roadway System).

Other methodologies were developed by numerous researchers (ATC, 1985 and 1991; King et al., 1997; Shinzouka et al., 1997; Veneziano et al., 2002). The Applied Technology Council (ATC) led the first efforts towards developing a unified SRA methodology for building structural systems. In order to overcome some of the drawbacks identified in this groundbreaking effort, ATC (1991) developed a more generic document for the conterminous United States on the seismic vulnerability and disruption of lifelines. To further improve the SRA methodologies, the Federal Emergency Management Association (FEMA, 1997) developed a geographical information system (GIS)-based risk assessment tool named HAZUS (HAZards US) which was mainly based on expert opinion. FEMA attempted to improve the estimation methodologies with the recent versions of HAZUS in 1999 and 2003.

All of the aforementioned tools and methodologies allow the use of either simplified or user-defined fragility models (curves) to characterize the probabilistic seismic performance of bridges in the highway system. Clearly, simplified fragility models present limited applicability as they may be suitable for highway bridges that can be categorized as regular only. However, more than 60% of the Nation's highway bridges fall under irregular or *other* bridge category. Although recent studies have contributed to the improvement of the tools by introducing either analytical or empirical [calibrated] fragility curves for other and also retrofitted bridges, there is still a major need to

(1) improve the existing fragility curves based on recent available and additional research. It is noted that fragility curves should be "accurate" to the extent possible for reliable assessment of post-earthquake operational state, repair time, cost and other risks.

(2) introduce new fragility functions suitable for modeling bridges that have unique configurations (e.g. curvature and skew), been retrofitted and response characteristics not well represented by the default models.

5.2.1. Fragility Curves

The fragility curves *predict* the probability of reaching or exceeding specific damage states for a given level of peak earthquake intensities (represented here by the peak ground acceleration). The fragility curves essentially associate the hazard level (intensity or other engineering demand parameter) with structural damage in a probabilistic sense.

It is noted that formulation of fragility curves can be based on one of the following two methods: (1) Numerical simulation where analytical fragility curves are derived via large number of nonlinear time-history analyses of structural systems allowing variations (uncertainty) in properties; and (2) Empirical fragility curves based on observed earthquake damage data. Furthermore, recently alternative methods are also proposed by Nielson and DesRoches (2007), and Padgett and DesRoches (2006). In some cases, good agreement between empirical and analytical fragility curves has been reported (Basoz et al., 1999; Mander and Basoz, 1999; Banerjee and Shinozuka, 2008).

Bridge damage data immediately after an earthquake serves as invaluable *field experimental results* (Banerjee and Shinozuka, 2008). Hence, empirical fragility curves can be developed on the basis of real damage data such as those observed after the 1989 Loma Prieta and 1994 Northridge earthquakes. Empirical fragility curves using data from these two earthquakes were proposed by many researchers (Basoz and Kiremidjian, 1997; Der Kiureghian, 2002; Shinzouka et al. 2003; Elnashai et al., 2004). It is noted, however, that empirical fragility curves are highly specific to a particular seismo-tectonic, geotechnical and built-environment. In addition, the observed damage data tend to be scarce and highly clustered in the low-damage and low-intensity range. And because of the high variability of the observational damage classification, empirical fragility curves introduce significant uncertainties. Hence, they may have limited application and may not represent most of the associated uncertainties.

On the other hand, analytically derived fragility curves can be more generic and diverse. Analytical fragility curves are derived based on the damage distributions simulated from numerical/computational analyses. Clearly, this will reduce the bias and increase reliability of the vulnerability estimate for different structures (Chrysanthopoulos et al., 2000; Mosalam et al., 1997; Reinhorn et al., 2001). They can be developed and used for any class of bridge and/or when real damage data are not available or rare. Development of analytical fragility curves consist of three major steps: (1) simulation or selection of a suitable set of ground motions, (2) nonlinear modeling of bridges, and (3) developing fragility curves from response results of analysis. There are different types of analysis that are commonly used to determine the vulnerability of bridge structures such as (1) nonlinear time history analysis (Shinozuka et al., 2000b; Hwang et al., 2000a; Karim and Yamazaki, 2001), (2) elastic spectral analysis (Hwang et al., 2000b), or (3) nonlinear static analysis (Mander and Basoz, 1999; Shinozuka et al., 2000b). Nonlinear time history analysis is considered to be the most accurate method; however, substantial computational effort is involved. Furthermore, limitations in modeling capabilities as well as the choices of the analysis method, idealization, seismic hazard, and damage models may influence the derived curves and have been seen to cause significant discrepancies in seismic risk assessments. However, the development of fragility curves by analytical means still remains to be the most direct and systematic approach.

In order to develop analytical fragility curves, models with varying complexities are typically developed; and computational models may involve bridges that include substructure and superstructure components including single and multiple-column bents, shear-keys, diaphragms, gap [opening/closing], restrainers, various bearing types and details; abutment, backfill, as well as various retrofit details. These models include ideally individual component characteristics calibrated based on engineering judgment, available past research and observations. Finally, nonlinear time-history analyses are conducted considering various earthquake intensities and mechanisms.

There have been many studies for the derivation of analytical fragility curves. Most, however, have been derived based on the assumption that structures are supported on rigid foundations even though the soil foundation systems have considerable effects on the response of the structure. A recent attempt to include the effect of flexible foundation on bridges by using lumped spring was made (Nielson, 2005), which is the most commonly adopted approach to model soil foundation system.

The fragility curves presented here were generated using the probabilistic seismic demand model (PSDM) using nonlinear time history analysis. The concept is to relate the engineering demand parameter (D_D) to the ground motion intensity (IM) using the scaling approach which is known as the incremental dynamic analysis (IDA) (Zhang et al., 2008). Equation (5-1) represents the probability of being in or exceeding a given limit state:

$$P[D_I \geq L_{si} | I_M] = \frac{n_i}{N} \qquad \text{(5-1)}$$

Nonlinear time history analyses are conducted for each ground motion intensity where the damage probability, P is calculated as the ratio of the number of cases, n_i at which the damage index, D_I exceeds or equals the limit state, L_{si} over the total number of cases, N for a certain intensity level (Karim and Yamazaki, 2001). Figure 5-1 illustrates a typical fragility curve.

5.2.2. Damage States and Damage Limits

The fragility curves *distribute* damage among distinct and descriptive damage states, such as those presented in Table 5-1. Damages states associated with the fragility curves are used along with the corresponding engineering demand and structural response parameters that establish quantifiable damage limits (e.g. drift limits for column, ductility demand of the column, deformation limits at the abutments). Ideally, survey of experimental and/or post-event field observations is necessary to arrive at consistent definitions and limits. Furthermore, available observed damage data may be used to verify, calibrate and improve the definitions of damage states, hence analytical fragility curves (Banerjee and Shinozuka, 2008).

In most of the published research on the development of fragility curves, deformation limits are established and used for certain components that are identified as the most vulnerable to damage and as critical to the function and structural integrity of the bridge structure. These deformation limits may be associated with column ductility, bearings, hinges, etc. (Hwang et al., 2000; Karim and Yamazaki, 2001; Mackie and Stojadinovic,

2001; Nateghi and Shahsavar, 2004). In the presented study the rotational ductility of columns (μ_θ) and abutment unseating potential were used to assess the damage states as per Table 5-2. It is noted that N is the seat width measured normal to the center line of the bearing from the edge of the bridge deck to the edge of the abutment (Figure 5-2). The damage states for column ductility were those proposed by (Choi et al., 2004).

Finally, it is noted that the analytical fragility curves may be highly sensitive to the definitions and quantified limits use to derive them. Furthermore, the so-called system fragility is more recommended over component fragility as it is more generic and more representative of the overall state of damage in bridges (Nielson and DesRoches, 2006). It was demonstrated that the use of any of the bridge component alone to assess the vulnerability of the bridge may lead to significant underestimation of the overall bridge fragility.

5.3. Benchmark Bridge

To facilitate the objectives of this part of the study, a highway bridge was chosen from Federal Highway Administration's (FHWA) Seismic Design of Bridges Series (Design Example No.4). The bridge is a continuous three-span box-girder bridge with 320 ft total length, spans of 100, 120, and 100 ft, and 30° skew angle (Figure 5-3; FHWA, 1996). The superstructure is a cast-in-place concrete box-girder with two interior webs and has a width-to-span ratio (W/L) of 0.43 for the end spans and 0.358 for the middle span. The intermediate bents have a cap beam integral with the box-girder and two reinforced concrete circular columns. Reinforced concrete columns of the bents are 4 ft in diameter supported on spread footings. The longitudinal reinforcing steel ratio of the column is approximately 3% and the volumetric steel ratio of the spirals is approximately 0.8%. Also, the axial load ratio of the column is 14%. The abutments are seat-type with elastomeric bearings under the web of each box girder. In the longitudinal direction, movement of the superstructure is limited by the gap (6 in) between the superstructure and the abutment back-wall. In the transverse direction, interior shear keys prevent the movement. This bridge was designed to be built in the USA in a zone with an acceleration coefficient of 0.3g following 1995 AASHTO guidelines (AASHTO, 1995).

5.3.1. Modeling of Nonlinearity

Simplified models for fragility analysis were developed and presented in chapter 2. The models will be calibrated against 3D FE models of the benchmark bridge as per the next section. In the models, the superstructure was assumed to be linear-elastic, and all the nonlinearity was assumed to take place in the substructure elements, external shear keys, bearings, and abutment-soil springs. The nonlinearity is assumed to take place in the form of localized plastic hinges at the top and bottom of columns. Figure 5-4 presents the moment-rotation properties of lumped plastic hinges. The footings were assumed to present fixed conditions. At the abutment locations, springs were assigned to the deck element. Figure 5-5 shows the properties of the nonlinear lumped bearing pads. The plastic (Wen) link element available in SAP2000 was used to model the hysteresis of bearing pads. The bearing links were assigned in the longitudinal and transverse directions of the bridge. A combination of nonlinear link elements in SAP2000 can simulate the behavior of abutment-soil interaction as discussed in chapter 2. Figure 5-6a represents the backbone curve of the force displacement relationship of the soil. The hysteretic properties of shear keys were determined after the experimental study reported by Megally et al. (2002). Figure 5-6b shows the backbone of force-displacement relationship of the shear keys. The shear keys are external and four were assigned, one at each corner of the bridge. The shear keys were aligned along the transverse direction of the bridge. A combination of nonlinear link element was used in SAP2000 to simulate the hysteretic behavior of shear keys (chapter 2). It is important to note that models with skew angles of 0, 30, and 60 degrees were developed.

5.4. Preliminary Comparative Time-History Analysis

The simplified models of the bridges with three skew angles, namely 0, 30 (benchmark), and 60, were developed. The model of the benchmark bridge with 30° skew angle is compared against the more complex finite element model to measure its accuracy. 1940 El Centro ground motion acceleration record was chosen in order to conduct the preliminary time-history analyses to verify the accuracy of the single spine model. A nonlinear static analysis including dead load preceded the time history analyses. Figure 5-7 through Figure 5-9 show comparison of response of the models at the abutments and bents. The comparison is in terms of the longitudinal and transverse deck displacements and accelerations. Also, the bent force displacement normal to the skew and along the skew were presented and compared for the two models. A very good agreement between the two models can be seen.

5.5. Selection of Ground Motions

The benchmark bridge (30° skew) was designed to be built in the western area of USA in a zone with an acceleration coefficient of 0.3g. The PEER Strong Motion Database (http://peer.berkeley.edu/smcat) was searched for ground motions with epicenteral distances of up to 15 kilometers, Soil-D, and PGA up to 1.0g. A total of 45 ground motions were found and selected for the study, with PGAs ranging from 0.1 to 1.0g. The stronger component of each ground motion was applied in the transverse direction of the three models, while the weaker component was applied in the longitudinal direction. It is noted that no significant effect of orientation of excitation with respect to the skew angle is expected based on the recent preliminary study (Schroeder, 2006). To conduct the incremental dynamic analyses described earlier, each ground motion pair was scaled to PGAs of 0.1 to 1.0g with increments of 0.1g. It is noted that the stronger component was consistently scaled to the target level and weak component was scaled accordingly. Therefore, a total of 1350 nonlinear time history analyses were conducted using SAP2000 version 12 (CSI, 2008). The fragility curves developed here are based on the analyses of the three dimensional (3-D) bridge systems.

5.6. Skewed Bridge Fragility Curves and Comparison

Following the equation (5-1), fragility curves for the three bridges were generated (Figure 5-10 and Figure 5-11). The figures present the fragility curves of the three bridges using the limit states for column ductility and abutment unseating potential which are presented in Table 5-2. It is important to note that 100%N means complete unseating at the abutment. N has a value of 42 in. Response of two components of the bridges were monitored, reported, and used in assessing the damage states of the bridges. These are the column rotations and unseating potential at the abutments. Previous research showed that in most of the cases, fragility curves can be represented by a normal or log-normal distribution function (Zhang et al., 2008). In the presented study both functions are employed and presented (Figure 5-10 and Figure 5-11). The comparisons suggest a better representation by log-normal distribution.

5.6.1. Analytical Fragility Curves

As was mentioned above, in the present study, the system fragility was evaluated on the basis of "damage" to two major components/mechanisms; namely, column rotational ductility and unseating potential at the abutments. The composite damage states, DS1 through DS4 were quantified as presented in Table 5-2 for the class and type of bridge studied herein. Accordingly, reaching to the limiting rotational ductility in any one of the four bent columns or to the limiting unseating potential at the abutments leads to the assignment of the corresponding damage state. Fragility curves of the overall system (Figure 5-11) suggest that skew results in increased vulnerability. This effect is clearly more pronounced for bridges with large skew angles. In other words, skewed bridges with 60° skew showed the highest vulnerability to damage while 30° skew bridges showed slight vulnerability of damage in comparison to the no skew case, especially, at lower levels of PGA. It is important to note that skewed bridges experienced higher potential for unseating under large PGAs (> 0.5g) while unseating preceded by complete loss of shear keys. This observation suggests that a more refined definition of quantified damage states may be necessary.

5.6.2. Median PGA Comparison

Table 5-3 and Figure 5-12 show a comparison among the results of skewed highway bridges in terms of median PGAs. It can be observed that the increase of skew angle lead to lower median PGAs which confirms the vulnerability of skewed bridges. However, at the collapse damage level, close values of median PGAs were observed which indicate, along with Figure 5-11, the failure of all the bridges under large level of PGAs. At the collapse damage level, Figure 5-11 shows that lower median PGAs were achieved by bridges with skew rather than the straight bridge since deck of bridges with skew have higher chances to unseat at abutments than the straight bridge.

5.6.3. HAZUS Fragility Curves Comparison

It is important to compare the generated set of fragility curves to the existing ones. National Bridge Inventory (NBI, 1999) produced a document summarizes steps to develop HAZUS fragility functions. HAZUS deals with skewed bridges as a standard bridge, however, a modification factor called K_{skew} is applied.

$$K_{skew} = \sqrt{\sin(\alpha)} \tag{5-2}$$

where α is 90 degree for a straight bridge and the skew angle is (90-α). For the four damage states, median PGAs of 0.91, 0.91, 1.05, and 1.38 were used in generating HAZUS fragility curves as recommended in Table 18 in NBI (1999) and Table 7.7 of FEMA (1999) document. Also, soil amplification factors where those presented in Table 4.10 of FEMA (1999) document. The recommended dispersion values, β, for all bridges, is 0.6 based on the study by Mander and Basoz (1999). However, this value is later modified to 0.4 in HAZUS (FEMA, 2003). Table 5-3 shows the dispersion values (β) of the analytical results. Table 5-4 shows the calculated median PGAs using HAZUS method.

Figure 5-13 through Figure 5-15 present the generated fragility functions for bridges using HAZUS method and the recommendations accompanied with analytical fragility curves produced earlier for 0, 30, and 60 degree skews, respectively.

Figure 5-16 presents a comparison of the median PGAs for the bridges with 0, 30, and 60 degree skew angles between analytical and HAZUS results with the dispersion value (β). In the figure, HAZUS median and median$\pm\beta$ PGAs are presented in comparison to the analytical median and median$\pm\beta$ PGAs. All presented figures in Figure 5-16 together show that HAZUS underestimated significantly the vulnerability of bridges with skew. In other words, the analytical fragility curves are more fragile than those obtained from HAZUS. Difference between the analytical and HAZUS results may be attributed to the fact that HAZUS methodology is developed using nonlinear static methods. The analysis like that used to develop HAZUS fragility function does not account for the dynamic effect on the response. Also, mechanisms such as; gap opening and closing which are expected to affect the dynamic response of the complete system (the bridge) can not be accounted for in this analysis. The mechanism at the abutments may lead to larger demand on the columns which can not be accounted for in such an elastic analysis as that used to develop HAZUS fragility curves. Together, all these reasons led to the discrepancy between HAZUS fragility functions and the analytical ones. However, looking at Figure 5-16, HAZUS "median-β" predicted the analytical median PGAs for the three bridges with 0, 30, and 60 degree skew angles fairly. Also, overlapping occurred between curves for analytical median$\pm\beta$ PGAs and HAZUS median$\pm\beta$ PGAs.

5.6.4. Generated Fragility Curves of Skewed Bridges from those for Straight Bridge

It was of interest to generate fragility functions based on those developed analytically for the no skew case using the skew equation (5-2) that accounts for the skew angle effect. Table 5-5 presents the median PGAs for the no skew case and the developed PGAs for the 30 and 60 degree skew cases. The results for the two skew angles were compared to those calculated analytically and the percent of error was reported. It was found that relatively close agreements were achieved for the 30 degree skew while; however, the percent of error increases as the skew angle increases. Figure 5-17 and Figure 5-18 show a comparison between analytical and generated fragility functions using HAZUS's dispersion value (β) of 0.4 and the skew equation for the 30 and 60 degree skew, respectively. It can be observed that however percent of error increases with the skew, the generated fragility functions are overly conservative. Looking at Figure 5-11, it can be suggested that the skew angle equation (5-2) can be function of the damage level or the PGA intensity since the difference between median PGAs changes when moving from slight damage to collapse.

Table 5-1. Damage States and Descriptions (FEMA, 2003)

Damage State	Description
DS1 (None)	None
DS2 (Slight/Minor)	minor cracking and spalling to the abutment, cracks in shear keys at abutments, minor spalling and cracks at hinges, minor spalling at the column (damage requires no more than cosmetic repair) or minor cracking to the deck
DS3 (Moderate)	[any] column experiencing moderate (shear cracks) cracking and spalling (column structurally still sound), moderate movement of the abutment (<2"), extensive cracking and spalling of shear keys, any connection having cracked shear keys or bent bolts, keeper bar failure without unseating, rocker bearing failure or moderate settlement of the approach.
DS4 (Extensive)	[any] column degrading without collapse – shear failure – (column structurally unsafe), significant residual movement at connections, or major settlement approach, vertical offset of the abutment, differential settlement at connections, shear key failure at abutments.
DS5 (Collapse)	[any] column collapsing and connection losing all bearing support, which may lead to imminent deck collapse, tilting of substructure due to foundation failure.

Table 5-2. Definition of Damage Indicies

Component \ Damage Level	Column Ductility, μ_θ	Abutment Unseating Potential
Slight	>1	>10% N[*]
Moderate	>2	>30%N
Extensive	>4	>50%N
Collapse	>7	>100%N

[*] N is the abutment seat width

Table 5-3. Median PGA's of Skewed Bridge from Analytical Results

Damage Level	0-Skew	β	30-Skew	β	60-Skew	β
Slight	0.144	0.260	0.138	0.265	0.124	0.290
Moderate	0.266	0.290	0.262	0.291	0.245	0.320
Extensive	0.520	0.300	0.507	0.310	0.481	0.320
Collapse	0.598	0.320	0.595	0.320	0.592	0.350

Table 5-4. Median PGA for Skewed Bridges using HAZUS

Damage Level	0-Skew	30-Skew	60-Skew
Slight	0.569	0.569	0.569
Moderate	0.663	0.617	0.469
Extensive	0.765	0.711	0.541
Collapse	1.005	0.935	0.710

Table 5-5. Median PGA's for Skewed Bridges using Skew Equation

Damage Level	Straight 0-Skew	30-Skew	60-Skew	30-Skew, %error	60-Skew, %error
Slight	0.144	0.134	0.102	3.24	18.12
Moderate	0.266	0.247	0.188	5.62	23.27
Extensive	0.520	0.484	0.368	4.48	23.49
Collapse	0.598	0.556	0.423	6.54	28.54

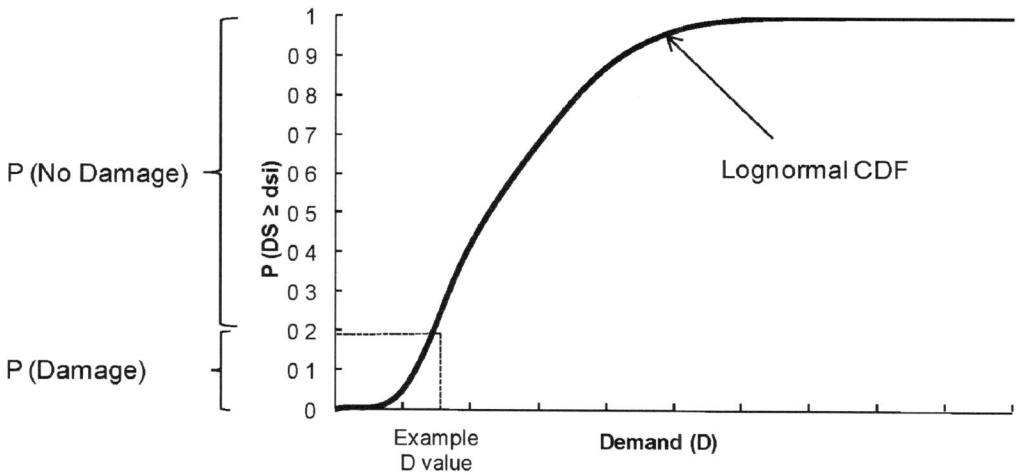

Figure 5-1. Fragility Curve for Damage State, ds_i (Baker, 2008)

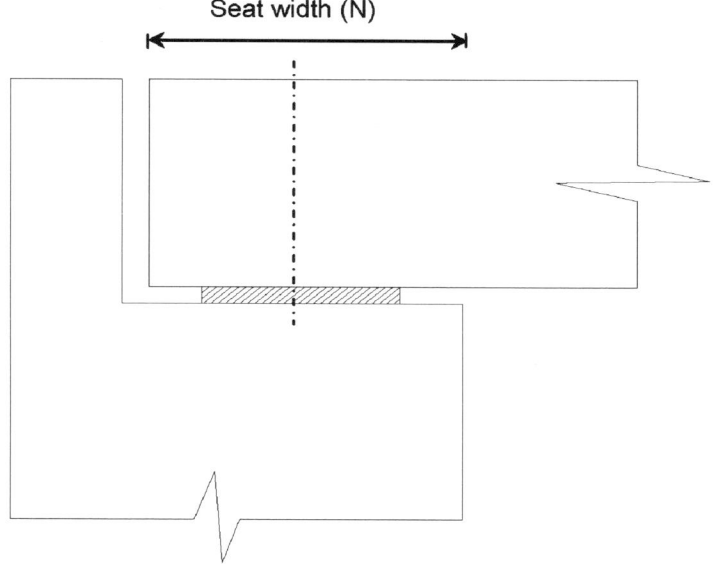

Figure 5-2. Abutment Seat Width

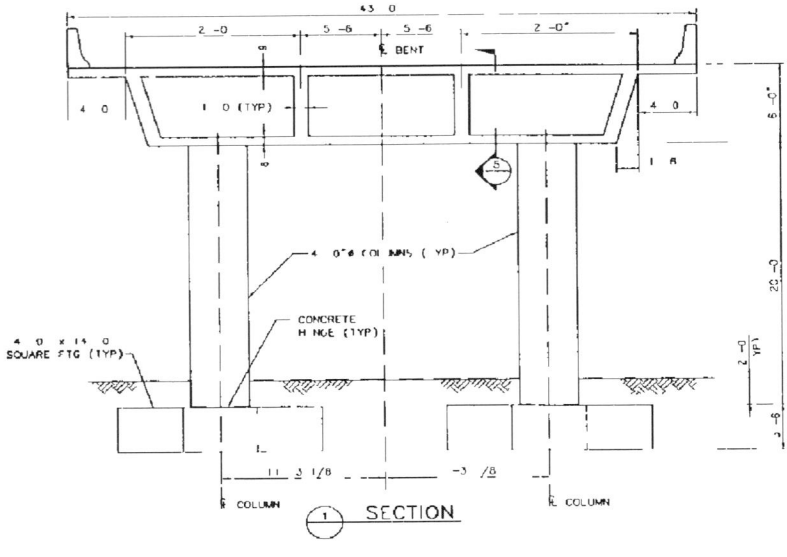

(a)Bent Elevation

PLAN

ELEVATION
(LOOKING PARALLEL TO BENTS)

(b) Plan and Elevation Views

Figure 5-3. Benchmark Bridge Geometry

331

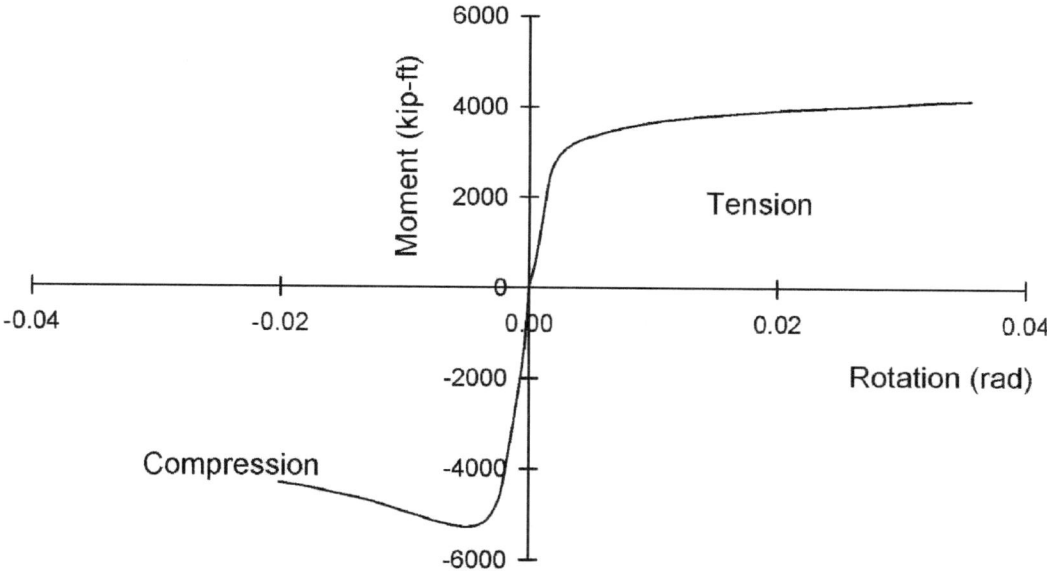

Figure 5-4. Lumped Plastic Hinge Properties

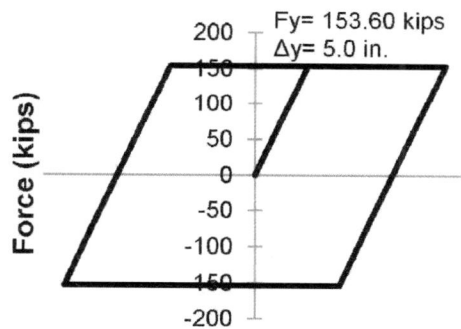

Figure 5-5. Hysteresis of Typical Bearing Pad

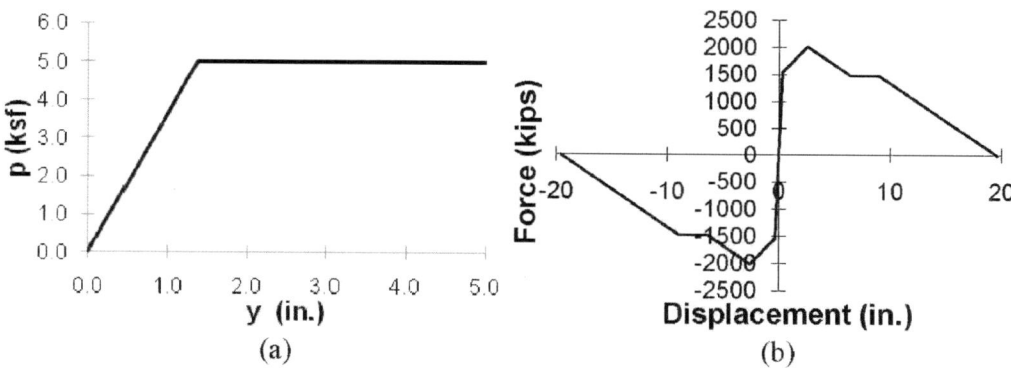

(a)   (b)

Figure 5-6. Force-Displacement Relation of (a) Abutment-soil and (b) Shear Key

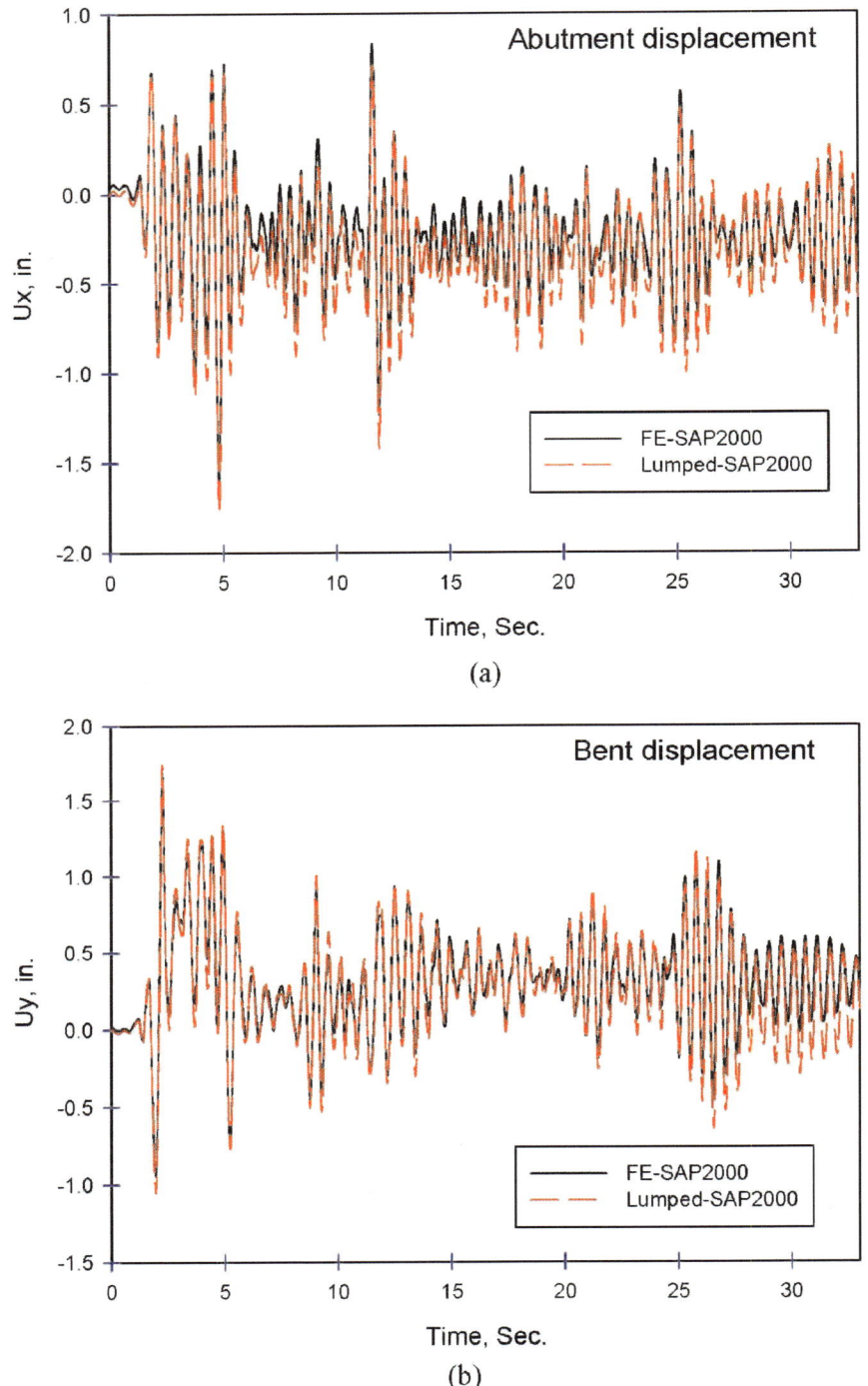

(a)

(b)

Figure 5-7. Comparison of Deck Displacement (a) longitudinal direction at the abutment and (b) transverse direction at the bent

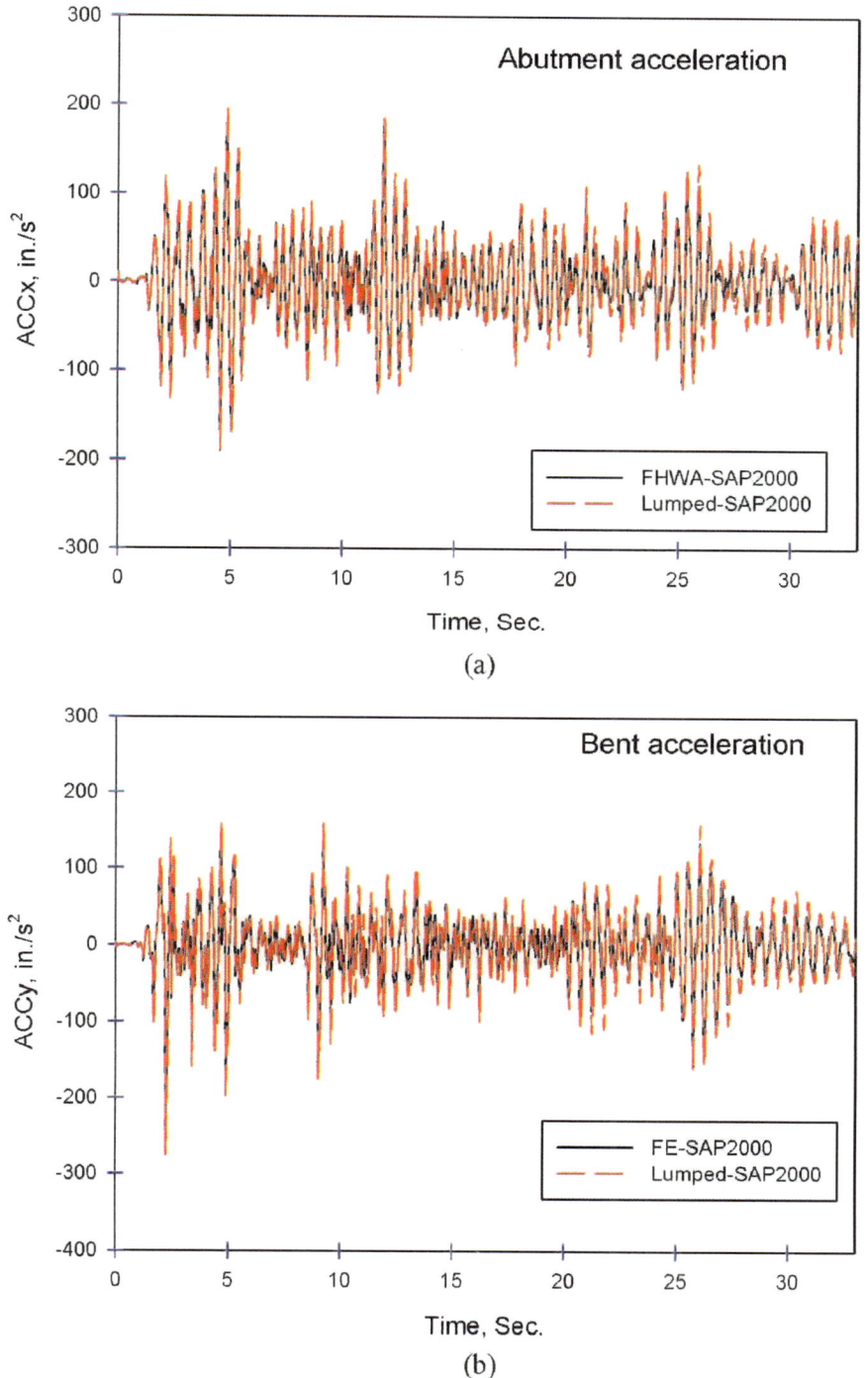

Figure 5-8. Comparison of Deck Acceleration (a) longitudinal direction at the abutment and (b) transverse direction at the bent

334

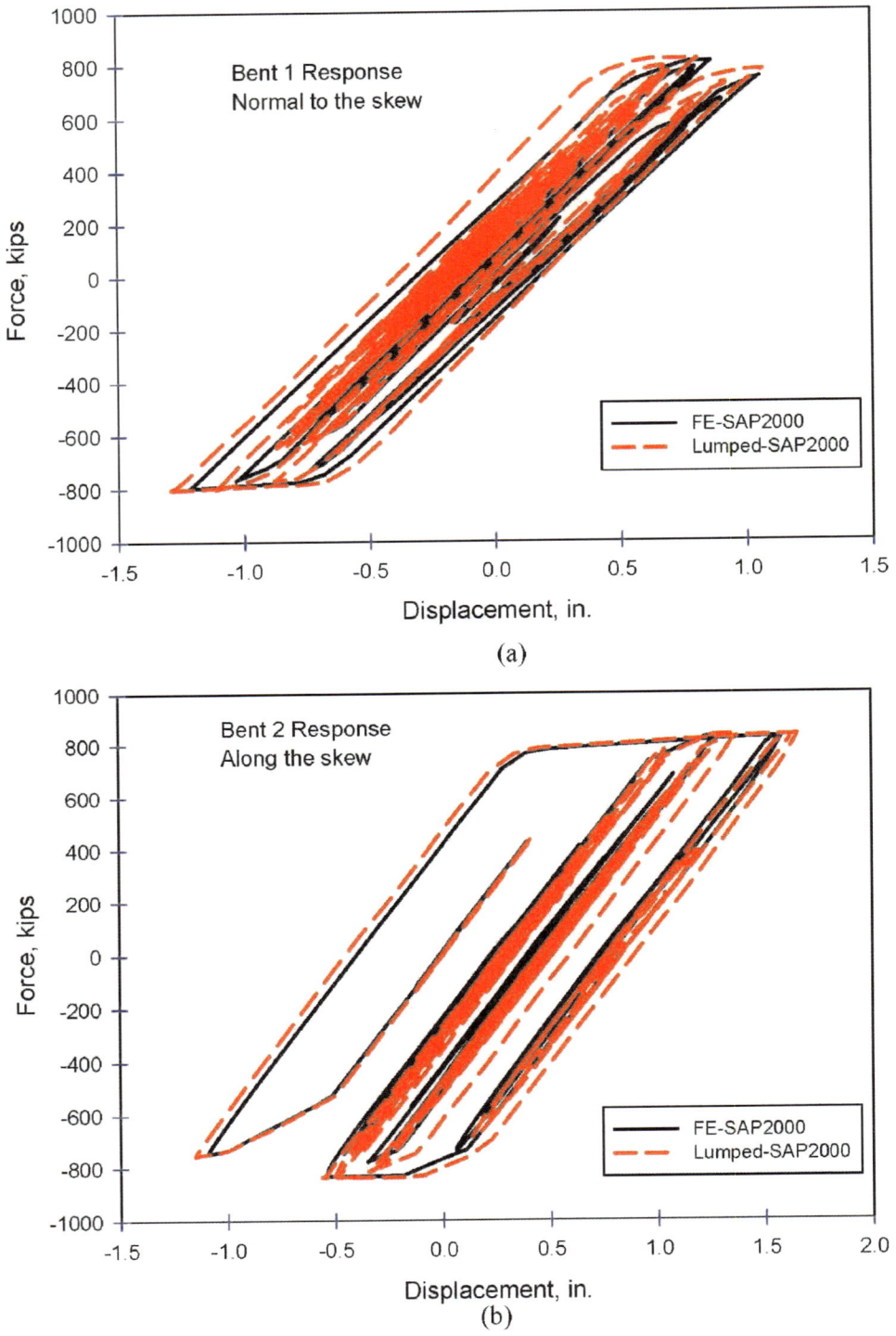

Figure 5-9. Bent Force-Displacement (a) Normal to the Skew and (b) Along the skew

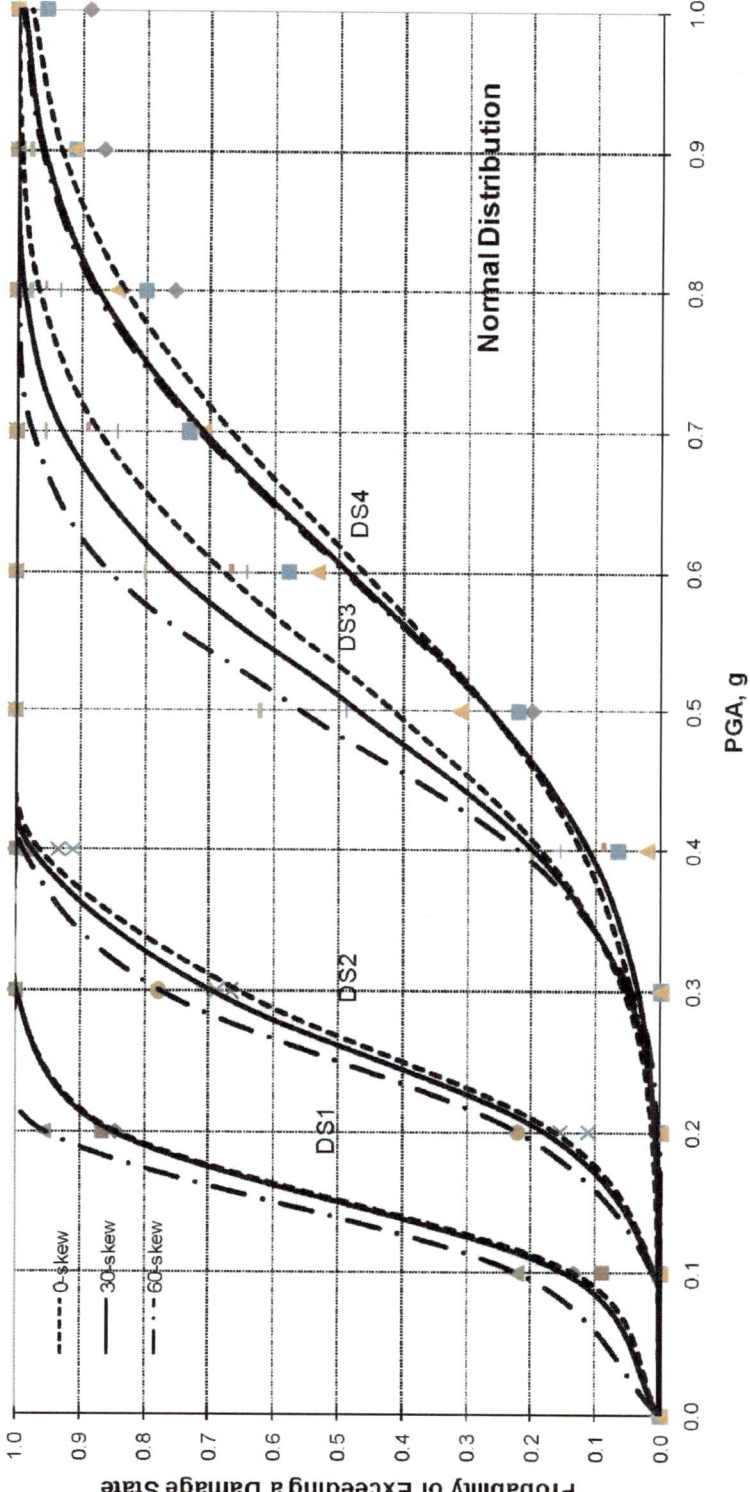

Figure 5-10. Fragility Curves of Skewed Highway Bridges

335

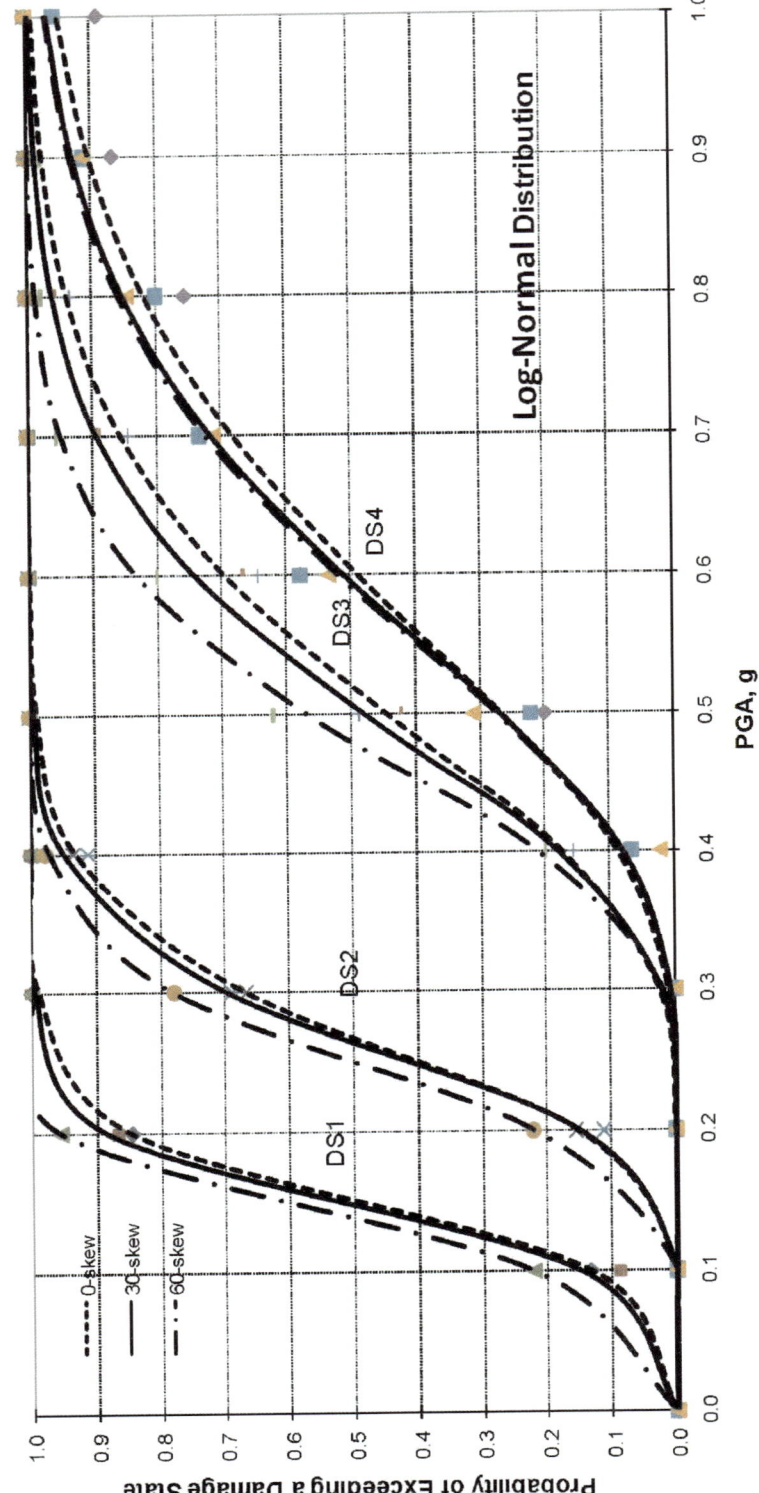

Figure 5-11. Fragility Curves of Skewed Highway Bridges

336

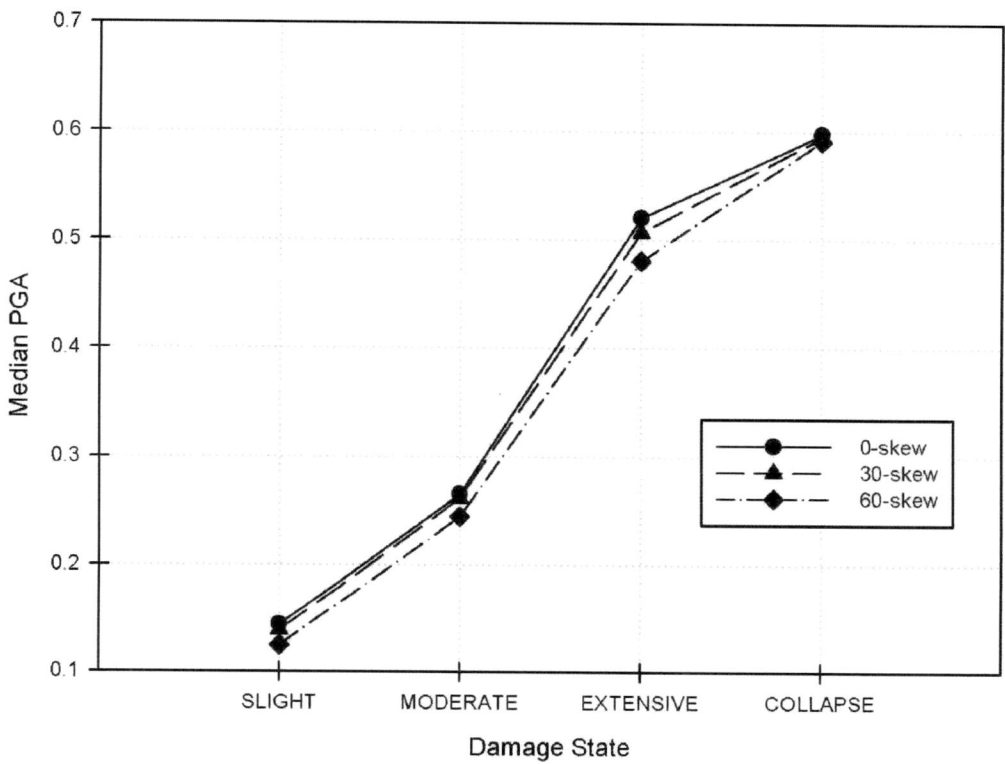

Figure 5-12. Comparison of Analytical Median PGAs

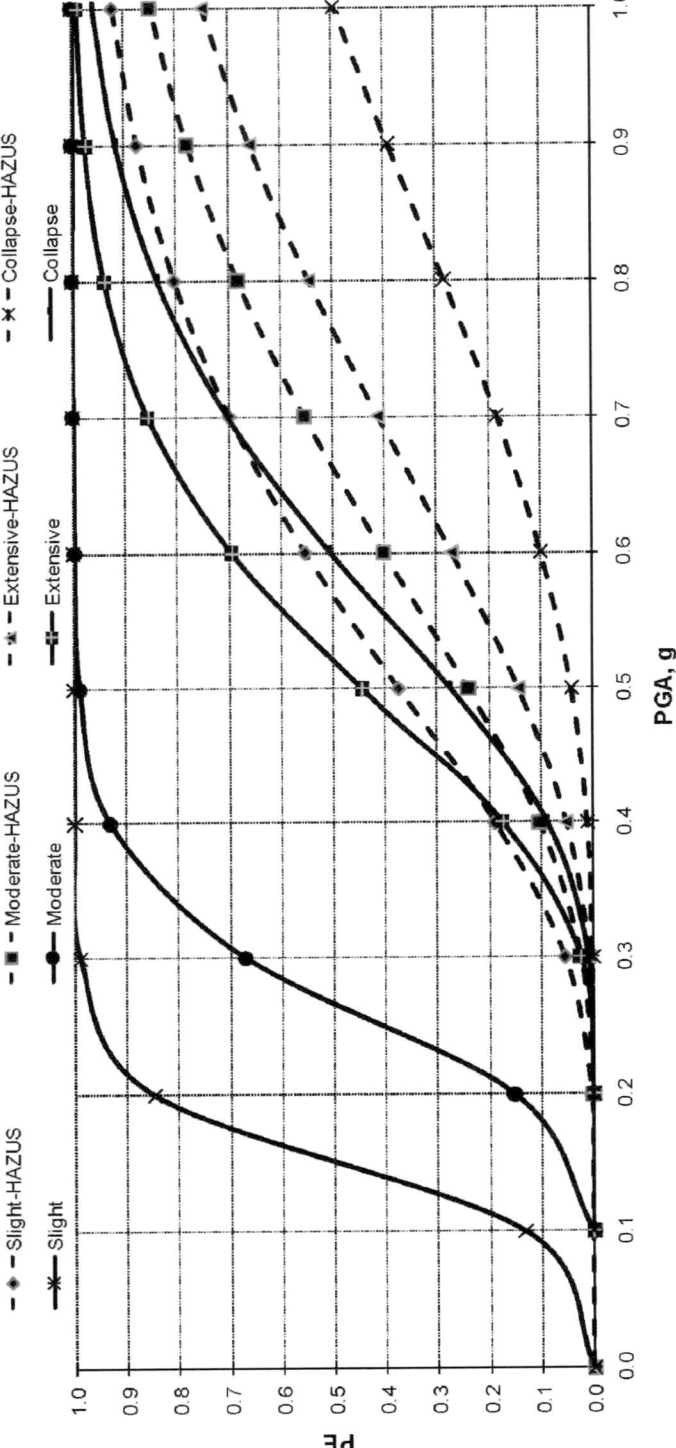

Figure 5-13. Comparison of Analytical Fragility Curves and HAZUS for 0 degree Skew

338

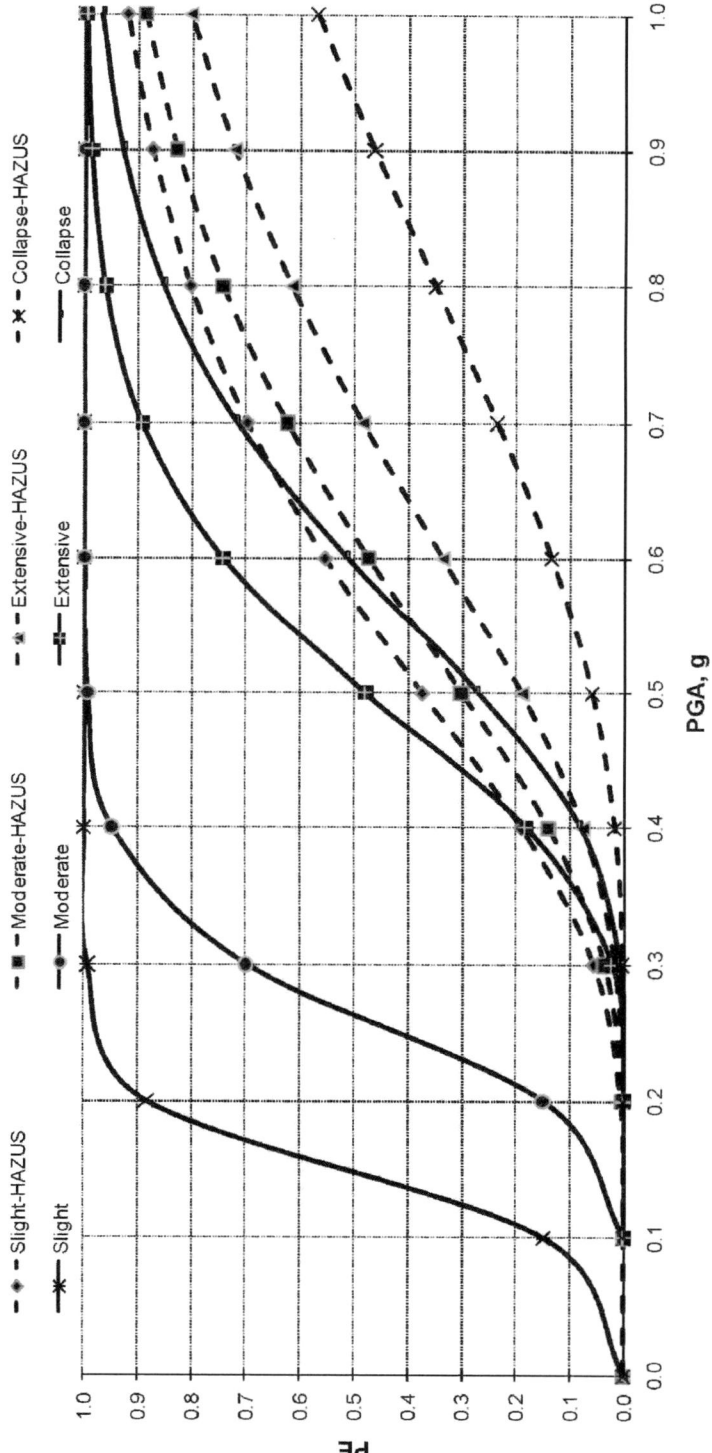

Figure 5-14. Comparison of Analytical Fragility Curves and HAZUS for 30 degree Skew

339

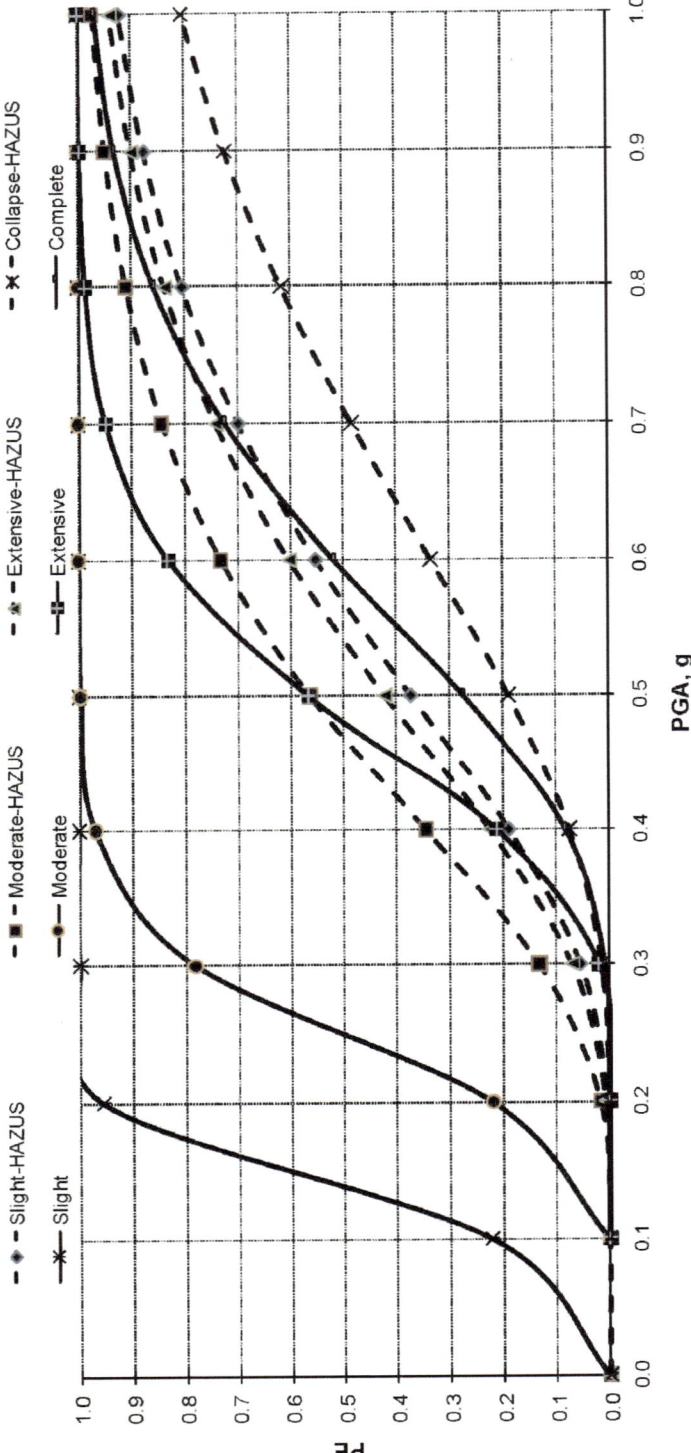

Figure 5-15. Comparison of Analytical Fragility Curves and HAZUS for 60 degree Skew

340

341

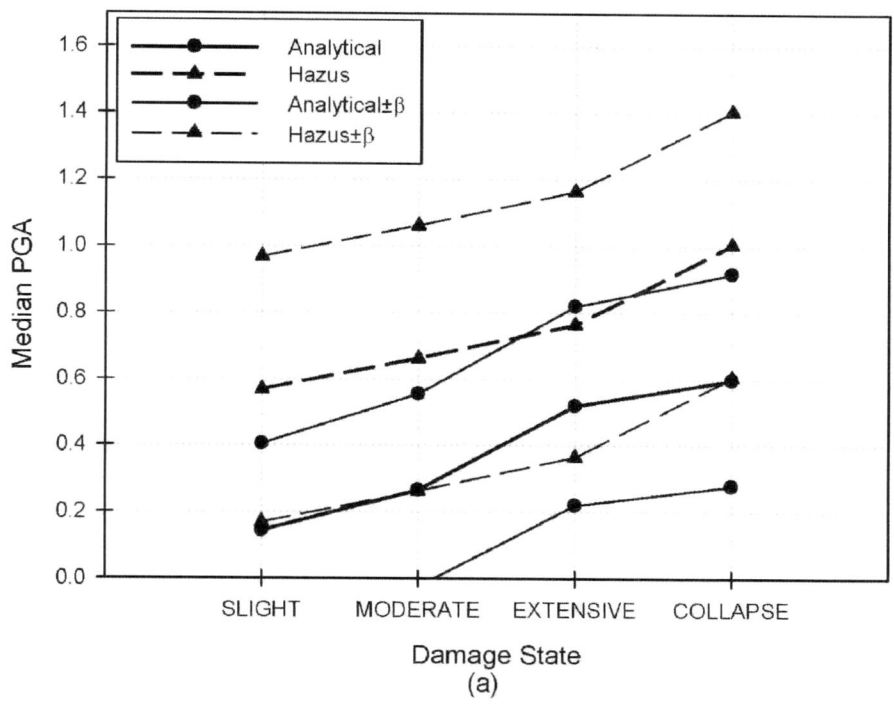

(a)

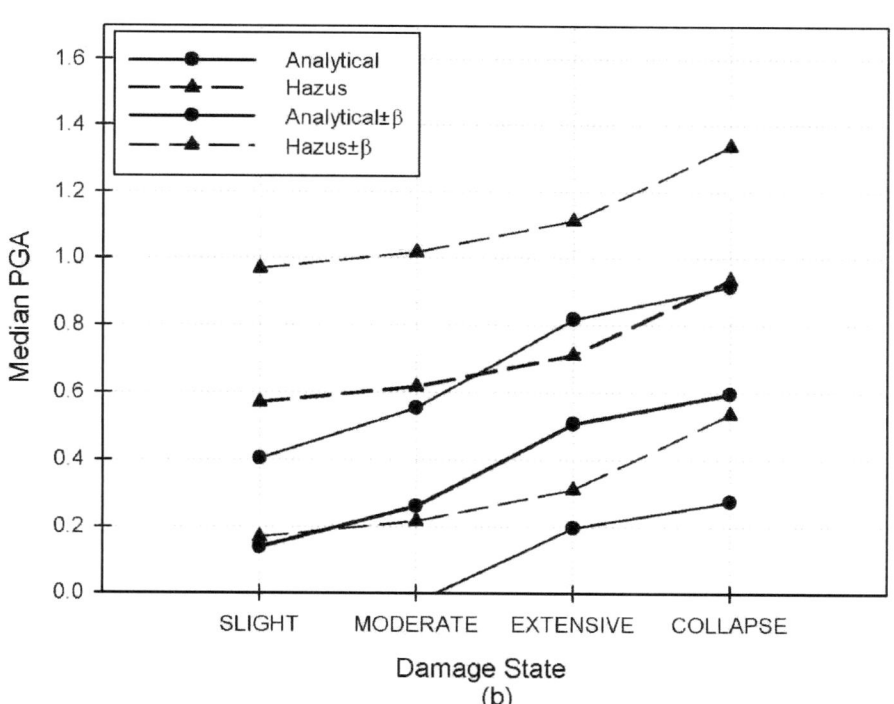

(b)

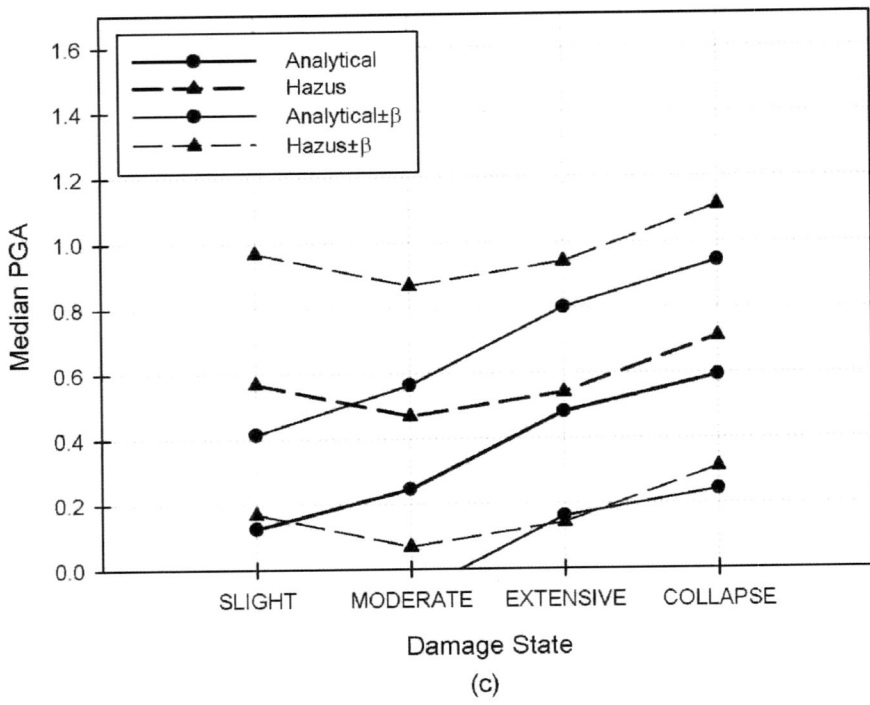

**Figure 5-16. Comparison of Median PGAs of Analytical Results and HAZUS (a) 0°
Skew, (b) 30° Skew, and (c) 60° Skew**

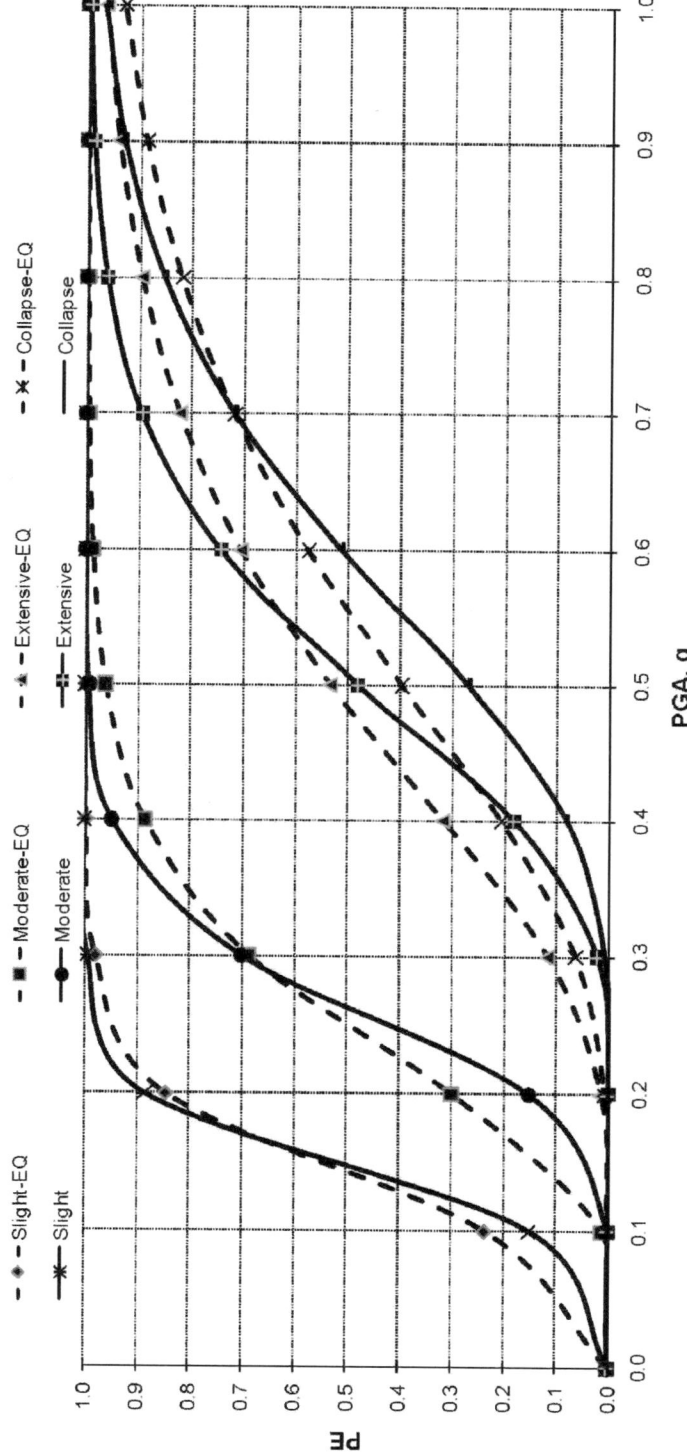

Figure 5-17. Comparison of Analytical Fragility Curves and Generated Curves using Skew Equation and HAZUS Standard Deviation for 30 degree Skew

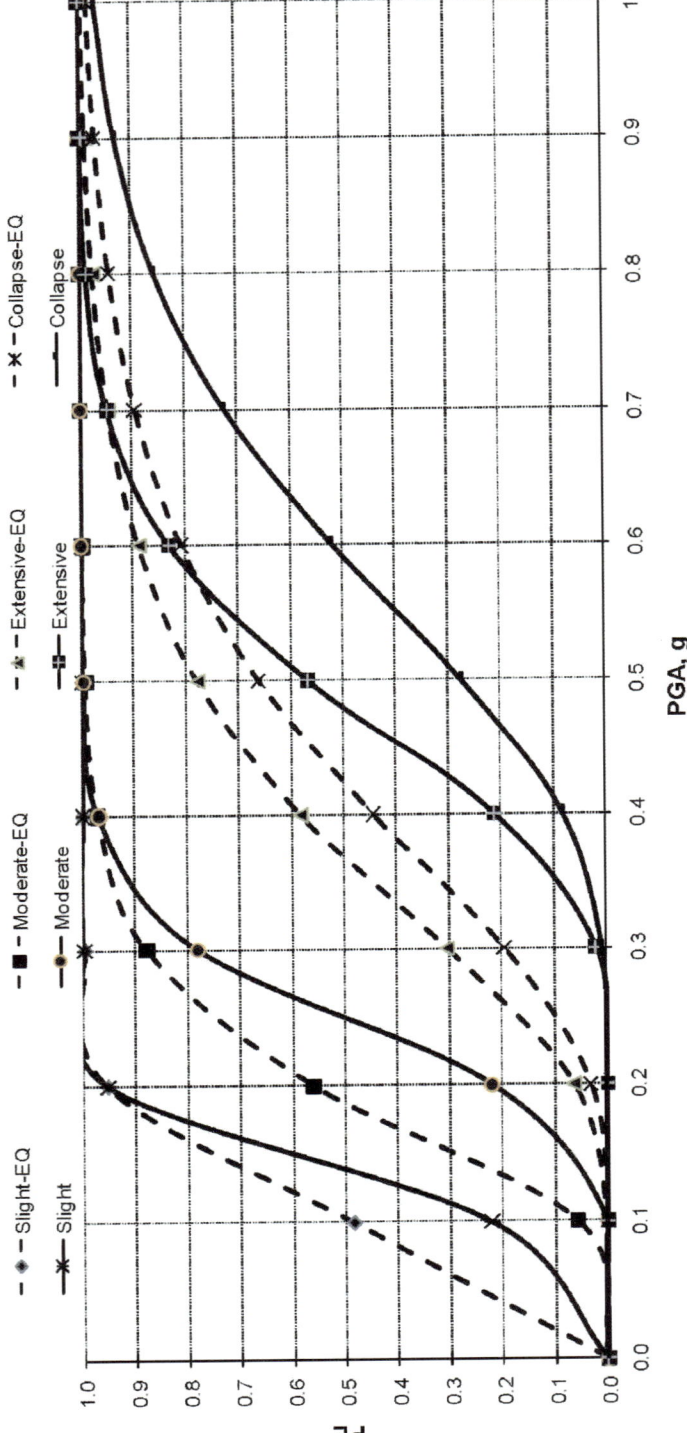

Figure 5-18. Comparison of Analytical Fragility Curves and Generated Curves using Skew Equation and HAZUS Standard Deviation for 60 degree Skew

344

6. Summary, Conclusions and Recommendations

6.1. Summary

Overall response behavior of skewed highway bridges under both service and seismic loads is inherently complex. Several parameters interact with the skew of the bridge and lead to this complex response behavior, such as: superstructure flexibility, boundary conditions, support details at the abutments and intermediate supports, shear keys, soil-foundation-structure interaction, in-span hinges (if any) and restrainers along with sub and super-structure interactions, width-to-span ratios, and mass and stiffness eccentricity.

Accordingly, the primary objective of the present study was to investigate the effect of skew angle on the overall seismic response characteristics by analytical and computational means. The study reported herein consisted of three major parts: (1) investigation of improved but at the same time simplified modeling techniques, (2) seismic response of assessment of a series of bridges with various skew angles, and (3) effect of skew on the probabilistic fragility curves for bridges.

6.1.1. Analytical Modeling of Bridges with Skew

Needless to say, one very important aspect of nonlinear analysis is the need to identify and model accurately the inelastic response characteristics of individual components. For the present study, bent columns, shear-keys, and abutment-soil interaction were considered as the most important components to model with full nonlinearity in that respect. Various parametric studies were conducted to establish rules and techniques for the accurate simulation of the nonlinear response behavior of columns, abutment-soil interaction, and shear keys. In order to arrive at a consistent and accurate modeling approach, a parametric study was performed to determine the proper number of fibers and the proper distribution and the proper number of segments of a PMM element in DRAIN3DX and also to determine the suitable location of plastic hinges and the proper plastic hinge length. Another study was conducted to model the complex nonlinear hysteretic behavior due to abutment-soil interaction and of the shear keys. For these parametric studies, components (columns and shear-keys) were modeled using both SAP2000 and DRAIN3DX. These components were selected from previously published experimental studies that involved quasi-static testing of subassemblies. Hence, analytical responses were compared against experimentally recorded response from past experiments and models were calibrated.

6.1.2. Overall Seismic Response Assessment

A comprehensive analytical investigation of the seismic response characteristics of a series of generic highway bridges with typical details was presented. Since the main focus of the investigation was to study the effect of skew angle along with other interacting conditions at the superstructure, abutments, and foundation, a benchmark bridge was first selected and then it was altered to have skew angles 0 to 60 degrees (0, 20, 30, 45, 52, and 60). It should be noted that the selection of the benchmark bridge was intentionally constrained to California bridges.

Following the parametric studies and improved modeling techniques, in the second phase, various detailed models of the bridges were developed and a comprehensive analytical investigation was carried out by means of modal and nonlinear time history analyses. This investigation was conducted in view of some of the details which were anticipated to affect the seismic performance of skewed highway bridges. Various parameters were investigated such as: the effect of intensity of ground motion, the soil type, the abutment support and foundation support conditions, aspect ratio, and number of columns per bent interacting with skew angle on the seismic response of two-span California skewed highway bridges. A complete analysis matrix was presented earlier in chapter 2. Various implications for the seismic design of skewed highway bridges are discussed in chapter 4.

6.1.3. Development of Preliminary Fragility Curves

The last part of the study investigated the vulnerability of skewed bridges to induced damage subjected to various levels of ground motion intensity. The system fragility was assessed and a set of fragility curves were developed analytically. Damage states were presented and quantified based on column rotational ductility and abutment deformations. Analytical fragility curves were compared to HAZUS fragility curves.

6.2. Conclusions

The important conclusions drawn and observations made are listed under three important areas. Firstly, conclusions pertaining to the analytical part of the study and accompanying computational modeling are given. Secondly, implications for the design and practical applications are summarized. Finally, concluding remarks are made regarding the preliminary fragility curves and effect of skew on the vulnerability of bridges.

6.2.1. Analytical Modeling of Bridges with Skew

The following conclusions are drawn

(1) The number of segments used to divide the distributed plasticity element is important and it may increase the execution time substantially. DRAIN3DX has the distributed plasticity element namely, fiber beam-column element (element 15). It is recommended to divide the fiber column into 10 segments (at least) in order to obtain relatively accurate results,

(2) The use of 32 fiber elements by lumping the longitudinal steel at the corners may lead to accurate nonlinear modeling for the PMM fiber hinge in both of SAP2000 and DRAIN3DX. Approximately, 12 concrete fibers can be used to represent the concrete core and another 12 fibers as close to the concrete cover in addition to 8 steel lumped fibers. The column cross section has to be modeled using a sufficient number of fibers. However, increasing the number of fibers can increase the execution time drastically. The recommended fiber distribution can be used for columns with different dimensions; however, a sufficient number of fibers has to be used to obtain cross section properties within 5% of those of the actual column properties,

(3) Elasto-plastic spring elements can be used to model the bearing pad hysteresis adequately. Examples for this type of element are the plastic (Wen) link in SAP2000 and inelastic truss bar element "element 01" in DRAIN3DX,

(4) A combination of nonlinear link elements has been demonstrated to model very accurately the complex hysteretic response due to abutment-soil interaction and of the shear keys in SAP2000 resulting in reliable results as discussed in detail in chapter 2, and

(5) DRAIN3DX has the capability of modeling hysteresis of abutment-soil interaction and internal and external shear keys accurately using single spring "element 09".

Early in this phase of the study, even though detailed Finite Element (FE) models of the bridges were developed and used in the preliminary analyses, it was noted that the time required to complete only one set of analyses was in the order of several weeks. Therefore, a simplified three-dimensional (3D) modeling approach was investigated in which the complex 3D system interactions could still be represented without losing the accuracy. This was verified with a series of modal analysis and limited nonlinear time history analyses of selected bridges (with small to large skew) modeled using both the simplified as well as the FE approach. Accordingly,

(6) Improved beam stick (BS) models are preferred to conduct both modal analyses and nonlinear time history analyses of skewed bridges. BS models predicted periods of all models for all skews well in comparison to counterpart finite element (FE) models. Improved BS models are also capable of capturing coupling of higher modes. Also the accuracy of BS models to capture nonlinear time history response of skewed highway bridges is demonstrated.

6.2.2. Overall Seismic Response Assessment and Design Implications

A series of comprehensive nonlinear time history analyses was conducted to investigate the effect of the various parameters listed and their interaction with the skew angle as it is reflected by the seismic performance of skewed highway bridges. A range of skew angles from 0 to 60 degrees was examined. Two levels of ground motions (sets of six) with PGAs of 0.3g and 0.6g for two different soil conditions (Soil-D and Soil-B) were considered. The effect of shear keys (with full nonlinear hysteresis and without), foundation boundary conditions (pinned and fixed), and effect of aspect ratios (0.3, 0.54, and 1.1) were examined thoroughly and abutment-soil structure interaction was considered explicitly. Accordingly, the following conclusions are drawn and recommendations are made:

(1) The presence as well as the level of post-tensioning in concrete box-girder bridges with or without skew does not have any significant effect on the overall seismic response as demonstrated earlier with the preliminary analytical investigation,

(2) Larger response quantities (i.e. displacements, forces, etc.) were observed with larger levels of ground motions in general. Nonetheless, bent columns as well as shear keys remained linear-elastic with negligible abutment-soil interaction during the design level excitations,

(3) Relatively stiff soil conditions (Soil-B) versus soft soil (Soil-D), lead to smaller response for the comparable levels of excitations, regardless the skew angle,

(4) It was found in general that shear keys have a predominant effect on the overall seismic response of the bridges studied herein. Shear keys are commonly used at the abutments to provide resistance to lateral loads. It is typically assumed that

shear keys cease to provide resistance once their capacity is reached. Therefore, bridge columns and bents are designed to provide capacity to resist the full seismic [deformation] demand. Caltrans bridge design specifications state that damage to abutments during a major seismic event is acceptable provided that collapse or unseating is prevented. This study suggest that however, these assumptions may lead to an overly conservative design of the bridge columns and bents, as well as the provided design deformation capacities of the components particularly at the abutments,

(5) Clearly, this study suggests that failure of shear keys is followed by elevated transverse displacements and increased demand on columns. The "absence" of shear keys may lead to increased abutment-soil structure interactions. In other words, gap opening is larger and increases with skew. It was also noted that the relative "effectiveness" of shear keys in controlling the seismic response of bridges diminish as the skew angle becomes larger,

(6) It was noted that shear keys were found to be particularly effective in reducing the torsional response of columns especially for large skew angles,

(7) An overall comparison of pinned versus fixed foundation cases suggests that the former results in significantly larger deformations whereas larger force demand on the bent component was introduced in the later case,

(8) Introducing larger aspect ratios slightly increased displacement in the longitudinal and significantly increased displacement in the transverse direction of the bridge deck. The bent forces on columns increased which led to the increased demand on columns. However, complex response behavior was observed in bridges with

larger aspect ratios. On the other hand, gap opening and closing at the abutments become more pronounced and the demand on shear keys increased, and increased in-plane deck rotations with skew was evident. Uneven distribution of abutment forces may take place which can lead to progressive failure, and complex global system behavior,

(9) However, an interesting observation was related to the general trend for the gap opening in bridges with and without shear keys. Bridges modeled with explicit shear key elements experienced larger gap openings at the obtuse corner that became smaller toward the acute corner. This behavior was observed regardless of the aspect ratio, however, only in bridges with a skew angle 45 degree or larger. For smaller skew angles, a rather uniform gap opening can be seen in the associated figures of chapter 3. It should be noted that although the shear key elements yielded and provided only marginal resistant in the transverse direction in most of the cases, the observation regarding the gap opening was still valid. On the other hand, skewed bridges that were modeled without the shear keys experienced larger gap openings at the acute corner that became smaller toward the obtuse corner. And the latter observation was valid for all skew angles and for all aspect ratios as well. And finally, the gap openings were significantly smaller in bridges modeled with the shear key elements,

(10) It was found that the abutment seat width as specified in the Caltrans SDC (Seismic Design Criteria) and AASHTO-LRFD equation 4.7.4.4-1 was adequate. It was observed that the abutment gap openings of the analytical results are small in comparison to the seat width provided by the two guidelines. In the AASHTO-

LRFD equation there is no near-fault effect hence a potential large displacement effect exist. In the absence of a complete system study, Caltrans equation can be deemed appropriate until the level of conservatism is quantified, and

(11) Caltrans recommendation for displacement ductility is conservative for skewed bridges.

It was important to analyze skewed bridges using a simple model such as response spectra analysis (RSA). The response quantities associated with maximum design deck displacement, maximum design transverse deck displacement near the abutment, displacement demand on columns, maximum design gap opening at abutments, and maximum design shear key force were compared with those from the nonlinear time history analyses (THA). Design equations for the use with RSA were presented in chapter 4 to be used to adjust results from RSA. Also, the applicability of pushover analysis was investigated to be used to predict the base shear capacity of skewed highway bridges Accordingly,

(12) The use of spectral analyses underestimated the response quantities of all the parameters understudy for skewed highway bridges since the average response spectrum of ground motions did not match Caltrans ARS, which is used for RSA for all periods, and naturally many source of nonlinearity can not be captured using linear analysis and

(13) Skew angle of 30 degree resembled a changing point. Response ratios of bridges with skew angle less than 30 degrees showed different trend than those of bridges with skew angle larger than 30 degrees in most cases.

(14) The pushover analyses may be used to predict the base shear capacity of skewed highway bridges with and without shear keys with different aspect ratios as well.

6.2.3. Development of Preliminary Fragility Curves

Three skewed highway bridges with 0, 30, and 60 degree skew were modeled with shear key and abutment-soil interaction springs. 45 pairs of ground motions were applied biaxially for each level of intensity. The bridges were exposed to 10 levels of intensities ranging from 0.1 to 1.0g with 0.1g increments. Damage states and limits are discussed in detail in chapter 5. The following conclusions are drawn.

(1) As the skew angle increases, the bridges become more vulnerable, especially, those with skew angles larger than 30 degree,

(2) However, bridges with 30 degree skew did not show a significant increase in vulnerability in comparison to no skew case. Particularly bridges skew angle greater than 30 degree may be deemed more vulnerable,

(3) Skewed bridges showed higher tendency to unseat rather than straight bridges. It is important to note that unseating was preceded by a complete loss of shear keys,

(4) Fragility curves derived based on the HAZUS recommendations underestimate significantly the vulnerability and skew effects. This can be attributed to that fragility functions of HAZUS developed using nonlinear static analysis which does not account for the entire dynamic effects in the response. However, subtracting the dispersion value recommended by HAZUS from HAZUS median PGAs led to fair predication of the analytical median PGAs, and

(5) The use of the HAZUS-MH equation to account for the skew angle effect resulted in conservative results. The use of the equation to drive fragility curves for skewed bridges using fragility information of straight bridge with similar geometry may be reliable.

6.3. Recommendations for Future Research

Analytical and computational approaches to study the seismic response characteristics of bridges are the most economically feasible methods. However, analytical models are as accurate as the adopted assumptions in modeling response characteristics of components. Hence, significant effort is necessary to arrive at consistent inelastic modeling assumptions for various structural details. In order to facilitate efficient analytical investigations, attention should be given to ensure that the analytical models are simple but general enough to capture the global bridge response and at the same time detailed enough to allow accurate estimation of component-level seismic response both in linear and nonlinear range. This is particularly important for analytical fragility studies. The work presented here should be extended through:

(1) Investigate the performance of skewed bridges due to vertical ground motions. The study has to be comprehensive through including all the parameters interacting with the skew and for a wide range of skew angles. This study can be extended for bridges with irregular geometries such as curved bridges. The current study did not address the effect of vertical ground motions,

(2) Develop generalized modeling guidelines that can be used to model nonlinearity of bridges with irregular geometry. The modeling guidelines can be developed

through a comprehensive study of a range of bridges with different number of spans, aspect ratios, column heights, and soil conditions. The current study presented modeling guidelines for skewed bridges,

(3) Investigate the effect of the application of incoherent ground motions on the seismic performance of skewed highway bridges. Also, the effect of the change in bridge columns heights on the bridge performance need to be studied,

(4) Investigate the effect of soil-structure interaction (SSI) on the seismic performance of skewed highway bridges,

(5) Improve the existing fragility functions based on recent available research findings and results of the proposed research program. It is noted that fragility curves should be "accurate" to the extent possible for reliable assessment of post-earthquake operational state, repair time, cost and other risks,

(6) Fragility functions need to be generated for bridges with different geometries and for a wider range of skew angles. It is important to assess the vulnerability of bridges with in-span hinges and restrainers along with sub and super-structure interactions and those with mass and stiffness eccentricity. The current study investigated the fragility functions of bridges with reported geometry and a range of skew angles while there were no in-span hinges and restrainers,

(7) Introduce new fragility functions suitable for modeling skewed and curved bridges, been retrofitted and response characteristics not well represented by the default models. When numerical simulation is used, analytical fragility functions are derived via large number of nonlinear time-history analyses of structural systems allowing variations (uncertainty) in properties,

(8) Verify the use of the nonlinear static methods to develop fragility functions for bridges with special geometry such as skewed and curved bridges. Nonlinear static methods are simple and quick methods and have been used to develop fragility functions of regular bridges by many researchers. The use of these methods needs to be calibrated, validated, and used for skewed and curved bridges. The current study used nonlinear time history analyses to develop the fragility functions,

(9) Develop fragility functions for a range of bridges with unique configurations (e.g. curve and skew) designed using old codes and guidelines. It is important redesign those bridges following the latest codes and guidelines. Finally, a comparison should be conducted between the two sets of fragility functions and recommendations can be developed. The findings of these studies will be important to designers and departments of transportations as it will asses in developing maintenance plan and designing new bridges,

(10) Investigate the fragility of skewed bridges with soil-structure interaction (SSI). The presence of soft soil leads to ground motions amplification, structural period elongation, hysteretic damping, and permanent soil deformations. Developing fragility curves for bridges with SSI can be an important tool for a regional seismic risk assessment,

(11) HAZUS-MH has an equation to modify fragility median PGA to account for the skew effect. The equation was developed based on static analysis therefore a wide range of fragility analysis using complex nonlinear analysis needs to be

conducted. Also, an equation to modify fragility median PGA for curved bridges can be developed,

(12) The effect of liquefaction on the fragility functions needs to be assessed. Also, methodologies to account for the liquefactions in fragility curves have to be developed. The current study did not take in to account this type of hazard, and

(13) A simplified method to generate fragility functions can be developed. The simplified method has to incorporate all the bridge components vulnerability and not just the effect of a single element. A large effort is needed to validate the use of this method to develop fragility functions of bridges with irregular geometries.

7. References

American Association of State Highway and Transportation Officials (AASHTO), 1998, "LRFD Bridge Design Specifications," 2nd Edition, Washington D.C.

American Association of State Highway and Transportation Officials (AASHTO), 1995, "Standard Specifications for Highway Bridges, Division I-A: Seisimic Design, American Assoication of the State Highway and Transportation Officials, Inc.," 15th Edition, as amended by the Interim Specification-Bridges, Washington D.C.

Applied Technology Council (ATC), 1996, "Seismic Evaluation and Retrofit of Concrete Buildings," SSC 96-01: ATC-40, 1, Redwood City, CA.

ATC, 1985, "Earthquake Damage Evaluation Data for California," Report ATC-13, Applied Technology Council.

ATC, 1991, "Seismic Vulnerability and Impact of Disruption of Lifelines in the Conterminous United States," Report ATC-25, Applied Technology Council.

Aydinoglu, M.N, 2003, "An Incremental Response Spectrum Analysis Procedure Based on Inelastic Spectral Deformation for Multi-mode Seismic Performance Evaluation," Bulletin of Earthquake Engineering, Vol. 1, No. 1, pp. 2-26.

Aydinoglu, M.N., 2004, "An Improved Pushover Procedure for Engineering Practice: Incremental Response Spectrum Analysis IRSA," Proceedings of the International Workshop Performance-Based Seismic Design Concepts and Implementation, Bled, Slovenia, pp. 345–356.

Aviram, A., Mackie, K.R., and Stojadinovic, B., 2008, "Guidelines for Nonlinear Analysis of Bridge Structures in California," Peer Report 2008/03. http://peer.berkeley.edu/publications/peer_reports/reports_2008/reports_2008.html

Baker, J.W., 2008, "Introducing Correlation among Fragility Functions for Multiple Components," 14th World Conference on Earthquake Engineering, Beijing, China.

Banerjee, S., and Shinozuka, M., 2008, "Experimental verification of bridge seismic damage states quantified by calibrating analytical models with empirical field data," Journal Earthquake Engineering and Engineering Vibration, IEM, Vol. 7, No. 4, pp. 383-393.

Banerjee, S., and Shinozuka, M., 2008, "Mechanistic quantification of RC bridge damage states under earthquake through fragility analysis," Probabilistic Engineering Mechanics, Vol. 23, pp. 12-22.

Basoz, N., and Kiremidjian, A.S., 1997, "Evaluation of bridge damage data from the Loma Prieta and Northridge, CA Earthquakes," The John A. Blume Earthquake Engineering Center, Department of Civil Engineering, Stanford University, Report No. 127.

Basoz, N., and Kiremidjian, A.S., 1999, "Development of Empirical Fragility Curves for Bridges," 5th U.S. Conference on Lifeline Earthquake Engineering, *ASCE*, Seattle, WA.

Basoz, N.I., Kiremidjian, A.S., King, S.A., and Law, K.H., 1999, "Statistical Analysis of Bridge Damage Data from the 1994 Northridge, CA earthquake," *Earthquake Spectra*, Vol. 15, No. 1, pp. 25–53.

Bentz, E.C., and Collins, M.P., 2000, Response-2000, Version 1.0.5, University of Toronto, Canada.

Bjornsson, S., Stanton, J., and Eberhard, M., 1997, "Seismic Response of Skew Bridges," 6th U.S. National Conference on Earthquake Engineering, pp. 1-12.

Bracci, J.M., Kunnath, S.K., and Reinhorn, A.M., 1997, "Seismic Performance and Retrofit Evaluation for Reinforced Concrete Structures," ASCE *Journal of Structural Engineering*, No. 123, pp. 3-10.

Bruneau, M., Wilson, JC., and Tremblay, R., 1996, "Performance of steel bridges during the 1995 Hyogo-Ken Nanbu (Kobe, Japan) earthquake," *Canadian Journal of Civil Engineering*, Vol. 23, No. 3, pp. 678–713.

Buckle, I.G., editor, 1994, "The Northridge California earthquake of January 17, 1994: performance of highway bridges," NCEER-94-0008, National Center for Earthquake Engineering Research, Buffalo (NY), March 24, 1994.

Buckle, I.G., Mayes, R.L., and Button M.R., 1987, "Sesimic Design and Retrofit Manaual for Highway Bridges," FHWA-IP-87-6 U.S. Department of Transportation, Federal Highway Administration.

Button, M.R., Cronin, C.J., and Mayes, R.L., 1999, "Effect of Vertical Ground Motions on the Structural Response of Highway Bridges," MCEER-99-0007, Multidisciplinary Center for Earthquake Engineering Research, State University of New York at Buffalo, NY.

Caltrans, 1992, "Seismic Design References," California Department of Transportation, Sacramento, CA.

Caltrans, 2004, "Seismic Design Criteria." Version 1.3, California Department of Transportation, Sacramento, CA.

Caltrans, 2006, "Seismic Design Criteria." Version 1.4, California Department of Transportation, Sacramento, CA.

Chan, M.C., and Penzien, J., 1975, "Analytical investigation of seismic response of short, single or multiple-span highway bridges," UCB/EERC 75-4, Earthquake Engineering Research Center, University of California, Berkeley.

Chang, KC., 2000, "Seismic Performance of Highway Bridges," *Earthquake Engineering and Engineering Seismology*, Vol. 2, No. 1, pp. 55 –77.

Cheok, G.S., and Stone, W.C., 1990, "Behavior of 1/6-Scale Model Bridge Columns Subjected to Inelastic Cyclic Loading," *Journal of Structural Enigneering*, ACI, Vol. 87, No. 6, pp. 630-638.

Choi, E., DesRoches, R., and Nielson, B., 2004, "Seismic Fragility of Typical Bridges in Moderate Seismic Zones," *Engineering Structures*, Vol. 26, pp. 187–199.

Chopra, A.K., 2001, *"Dynamics of Structures: Theory and Applications to Earthquake Engineering,"* 2nd Edition, Prentice-Hall, Inc., New Jersey.

Chopra, A.K., and Goel, R.K., 2002, "A Modal Pushover Analysis Procedure for Estimating Seismic Demands for Buildings," *Earthquake Engineering and Structural Dynamics*, Vol. 31, pp. 561–582.

Chopra, A.K., and Goel, R.K., 2004, "Evaluation of Modal and FEMA Pushover Analyses: SAC buildings," *Earthquake Spectra*, Vol. 20, No. 1, pp. 225–254.

Chryssanthopoulos, M. K., Dymiotis, C., and Kappos, A. J., 2000, "Probabilistic evaluation of behaviour factors in EC8-designed R/C frames," *Engineering Structures*, Vol. 22, No. 8, pp. 1028-1041.

Comartin, C., Green, M., and Tubbesing, S., 1995, "The Hyogo-Ken Nanbu earthquake," Preliminary Reconnaissance Report, Earthquake Engineering Research Institute, Oakland , CA, Feb. 1995.

Computers and Structures, Inc., 2005, SAP2000, Version 10.0.5, Integrated Structural Analysis and Design Software, Berkeley, CA.

Computers and Structures, Inc., 2008, SAP2000, Version 12.0.1, Integrated Structural Analysis and Design Software, Berkeley, CA.

CSI, 2005, "CSI Analysis Reference Manual," Computers and Structures, Inc., Berkeley, CA.

De Rue, G.M., 1998, "Nonlinear Static Procedure Analysis of 3D Structures for Design Applications," Master Thesis, University of New York at Buffalo, August 1998.

Der Kiureghian, A., 2002, "Bayesian Methods for Seismic Fragility Assessment of Lifeline Components," *Acceptable Risk Processes: Lifelines and Natural Hazards, Monograph No. 21*, A. D. Kiureghian, ed., Technical Council on Lifeline Earthquake Engineering, ASCE, Reston VA.

DesRoches, R., and Muthukumar, S., 2002, "Effect of Pounding and Restrainers on Seismic Response of Multiple-frame Bridges," ASCE *Journal of Structural Engineering*, No. 128, pp. 860-869.

Elnashai, A., Borzi, B., and Vlachos, S., 2004, "Deformation-Based Vulnerability Functions for RC Bridges," *Structural Engineering and Mechanics*, Vol. 17, No. 2, pp. 215–244.

Elnashai, A.S., 2001, "Advanced Inelastic Static (Pushover) Analysis for Earthquake Applications," *Structural Engineering and Mechanics*, Vol. 12, No. 1, pp. 51-69.

Engineer Manual, 1999, "Response Spectra and Seismic Analysis for Concrete Hydraulic Structures U.S. Army Corps of Engineers," Washington, DC.

Esmaeily, G.A., and Xiao, Y., 2002, "Seismic Behavior of Bridge Columns Subjected to Various Loading Batterns," Peer Report 2002/5.

Federal Emergency Management Agency (FEMA), 1997, "NEHRP guidelines for the seismic rehabilitation of buildings," FEMA-273, Washington, D.C.

Federal Emergency Management Agency (FEMA), 2000, "Prestandard and Commentary for the Seismic Rehabilitation of Buildings," FEMA-356, Washington, D.C.

Federal Emergency Management Agency (FEMA), 2004, "Improvement of Nonlinear Static Seismic Analysis Procedures," FEMA-440, Washington, D.C.

Federal Highway Administration (FHWA), 1996, "Seismic Design of Bridges, Design Example No. 4 – Three-span Continuous CIP Concrete Bridge," Publication No. FHWA-SA-97-009, October 1996.

Federal Highway Administration (FHWA),1996, "Seismic design of bridges, Design example No. 4 – Three-span continuous CIP concrete bridge," Publication No. FHWA-SA-97-009, October 1996.

FEMA, 1999, *HAZUS 99: Technical Manual.* Federal Emergency Management Agency, Washington DC.

FEMA, 2003, *HAZUS-MH MR1: Technical Manual*, Vol. Earthquake Model. Federal Emergency Management Agency, Washington DC.

Fu, C.C., and Alayed, H., 2003, "Seismic Analysis of Bridges using Displacement-based Approach," (03-3175) Transportation Research Board, Washington, D.C., Jan. 2003.

Ghobarah, A.A., and Tso, W.K., 1974, "Seismic Analysis of Skewed Highway Bridges with Intermediate Supports," *Earthquake Engineering and Structural Dynamics*, Vol. 2, pp. 235-248.

Gupta, B., and Kunnath, S.K., 2000, "Adaptive Spectra-based Pushover Procedure for Seismic Evaluation of Structures," *Earthquake Spectra*, Vol. 16, pp. 367-392.

Hwang, H., and Huo, J.R., 1998, "Probabilistic Seismic Damage Assessment of Highway Bridges," Proceedings of the 6th U.S. National Conference on Earthquake Engineering, Seattle, WA, May-Jun., 1998.

Hwang, H., Jernigan, J.B., and Lin, Y.-W., 2000a, "Seismic Fragility Analysis of Highway Bridges." MAEC RR-4, Center for Earthquake Research Informaion, Memphis, TN.

Hwang, H., Jernigan, J.B., and Lin, Y.-W., 2000b, "Evaluation of Seismic Damage to Memphis Bridges and Highway Systems," *Journal of Bridge Engineering*, Vol. 5, No. 4, pp. 322-330.

Hwang, H., Liu, J.B., and Chiu, Y.-H., 2001, "Mid-America Earthquake Center Technical Report MAEC RR-4 Project Center for Earthquake Research and Information," The University of Memphis.

Japan Society of Civil Engineers (JSCE), "The 1999 Ji-Ji earthquake, Taiwan—investigation into the damage to civil engineering structures," Available from: http://www.jsce.or.jp/e/index.html.

Kappos, A.J., Paraskeva, T.S., and Sextos, A.G., 2004, "Seismic Assessment of a Major Bridge Using Modal Pushover Analysis and Dynamic Time-history Analysis," Proceedings of the International Conference on Computational and Experimental Engineering and Science, Madeira, Portugal.

Karim, K.R., and Yamazaki, F., 2001, "Effect of Earthquake Ground Motions on Fragility Curves of Highway Bridge Piers Based on Numerical Simulation," *Earthquake Engineering and Structural Dynamics*, Vol. 30, pp. 1839–1856.

Karim, K.R., and Yamazaki, F., 2003, "A Simplified Method of Constructing Fragility Curves for Highway Bridges," *Earthquake Engineering and Structural Dynamics*, Vol. 32, pp. 1603–1626.

King, S.A., Kiremidjian, A.S., Basoz, N., Law, K., Vucetic, M., Doroudian, M., Olson, R.A., Eidinger, J.M., Goettel, K.A., and Horner, G., 1997, "Methodologies for Evaluating the Socio-Economic Consequences of Large Earthquakes," *Earthquake Spectra*, Vol. 13, No. 4, pp. 565–584.

Kircher, CA., Nassar, AA., Kustu, O., Holmes, WT., 1997, "Development of building damage functions for earthquake loss estimation," *Earthquake Spectra*, Vol. 13, No. 4, pp. 663–682.

Mackie, K., and Stojadinovic, B., 2001, "Probabilistic Seismic Demand Model for California Bridges," *Journal of Bridge Engineering*, Vol. 6, No. 6, pp. 468–480.

Mackie, K., and Stojadinovic, B., 2004, "Fragility Curves for Reinforced Concrete Highway Overpass Bridges," 13th World Conference on Earthquake Engineering, Vancouver, BC, Canada.

Maleki, S., 2002, "Deck Modeling for Seismic Analysis of Skewed Slab-girder Bridges," *Engineering Structures*, Vol. 24, pp. 1315-1326.

Mander, J.B., and Basoz, N., 1999, "Seismic Fragility Curve Theory for Highway Bridges," Proceeding of the 5th U.S. Conference on Lifeline Earthquake Engineering, Seattle, WA, August 12-14, 1999, pp. 31-40.

Maragakis, E., 1984, "A Model for the Rigid Body Motions of Skew Bridges," Doctoral Thesis, California Institute of Technology, Pasadena, CA.

Maragakis, E., Douglas, B., and Vrontinos, S., 1991, ''Classical Formulation of the Impact Between Bridge Deck and Abutments During Strong Earthquakes.'' Proceedings of the 6th Canadian Conference on Earthquake Engineering, University of Toronto Press, Canada, pp. 205–212.

Megally, S.H., Silva, P.F., and Seible, F., 2002, "Seismic Response of Sacrificial Shear Keys in Bridge Abutments," SSRP-2001/23, Department of Structural Engineering, University of California, San Diego.

Meng, J.Y., and Lui, E.M., 2000, "Seismic Analysis and Assessment of a Skew Highway Bridge," *Engineering Structures*, Vol. 22, pp. 1433-1452.

Meng, J.Y., and Lui, E.M., 2002, "Refined Stick Model for Dynamic Analysis of Skew Highway Bridges," *Journal of Bridge Engineering*, Vol. 7, No. 3, pp. 184-194.

Moehle, JP., 1994, "Northridge earthquake of January 17, 1994," Reconnaissance Report, volume 1—Highway Bridges and Traffic Management, *Earthquake Spectra*, Vol. 11, No. 3, pp. 287–372.

Mosalam, K. M., Ayala, G., White, R. N., and Roth, C., 1997, "Seismic fragility of LRC frames with and without masonry infill walls," *Journal of Earthquake Engineering*, Vol. 1, No. 4, pp. 693-720.

Nateghi, F., and Shahsavar, V.L., 2004, "Development of Fragility and Reliability Curves for Seismic Evaluation of a Major Prestressed Concrete Bridge," 13th World Conference on Eartquake Engineering, Vancouver, BC, Canada.

National Bridge Inventory (NBI), 1999, "Enhancement of the Highway Transportation Lifeline Module in HAZUS," Draft 7, March 1999.

Nielson, B.G., and DesRoches, R., 2004, "Improved Methodology for Generation of Analytical Fragility Curves for Highway Bridges," 9th ASCE Specialty Conference on Probabilistic Mechanics and Structural Reliability, *ASCE*, Albuquerque, NM.

Nielson, B.G., and DesRoches, R., 2006, "Seismic Fragility Methodology for Highway Bridges using a Component Level Approach," *Earthquake Engineering and Structural Dynamics*, Vol. 36, pp. 823–839

PEER Strong Motion Database, 2000, Pacific Earthquake Engineering Research Center, Berkeley, CA.

Prakash, V., Powell, G.H., and Campbell, S., 1994, "Drain-3DX: Static and Dynamic Analysis of Inelastic 3D Structures," Department of Civil Engineering, University of California, Berkeley.

Priestley, M.J.N., Seible, F., and Calvi, G.M., 1996, "*Seismic Design and Retrofit of Bridges*," 1st Edition, John Wiley & Sons, Inc., 1996.

Reinhorn, A. M., Barron-Corvera, R., and Ayala, A.G., 2001, "Spectral evaluation of seismic fragility of structures, Structural safety and reliability," (ICOSSAR 2001).

Reinhorn, A.M., 1997, "Inelastic Analysis Techniques in Seismic Evaluations," Proceedings of the International Workshop on Seismic Design Methodologies for the Next Generation of Codes, Bled, Slovenia, pp. 277–288.

Saiidi, M., and Orie, D., 1992, "Earthquake Design Forces in Regular Highway Bridges," Computers & Structures, Vol. 44, No. 5, pp. 1047-1054.

Schroeder, B.L., 2006, "Seismic Response Assessment of Skew Highway Bridges," Masters Thesis, Department of Civil and Environmental Engineering, University of Nevada, Reno.

Shamsabadi, A., Yan, L., and Martin, G., 2004, "Three Dimensional Nonlinear Seismic Soil Foundation-Structure Interaction Analysis Of A Skewed Bridge Considering Near Fault Effects,"

Shinozuka, M., Chang, S., Eguchi, R.T., Abrams, D.P., Hwang, H., and Rose, A., 1997, "Advances in Earthquake Loss Estimation and Application to Memphis, Tennessee," *Earthquake Spectra*, Vol. 13, No. 4, pp. 739–758.

Shinozuka, M., Feng, M.Q., Kim, H., Uzawa, T., and Ueda, T., 2003, "Statistical Analysis of Fragility Curves," MCEER-03-0002, MCEER.

Shinozuka, M., Feng, M.Q., Kim, H.-K., and Kim, S.-H., 2000b, "Nonlinear Static Procedure for Fragility Curve Development," *Journal of Engineering Mechanics*, Vol. 126, No. 12, pp. 1287-1296.

Shinozuka, M., Feng, M.Q., Lee, J., and Naganuma, T., 2000a, "Statistical Analysis of Fragility Curves," *Journal of Engineering Mechanics*, Vol. 126, pp. 1224–1231.

SIMQKE-2, 1999, Version 2.0, Massachusetts Institute of Technology, Cambridge, Massachusetts.

Spacone, E., Martino, R., and Kingsley, G., 1999, "Nonlinear Pushover Analysis of Reinforced Concrete Structures," Colorado Advanced Software Institute.

Sucuoglu, M., and Darek, B., 1999, "Circular High-Strength Concrete Columns Under Simulated Seismic Loading," *Journal of Structural Enigneering*, ACI, Vol. 125, No. 3, pp. 272-280.

Tirasit, P., and Kawashima, K., 2005, "Seismic Torsion Response of Skewed Bridge Piers," JSCE *Journal of Earthquake Engineering*.

Veneziano, D., Sussman, J.M., Gupta, U., and Kunnumkal, S.M., 2002, "Eartquake Loss Under Limited Transportation Capacity: Assessment, Sensitivity and Remediation," 7th U.S. National Conference on Earthquake Engineering, Boston, Mass. EERI.

Wakefield, R.R., Nazmy, A.S., and Billington, D.P., 1991, "Analysis of Seismic Failure in Skew RC Bridge," ASCE *Journal of Structural Engineering*, Vol. 117, No. 3, March 1991.

Werner, S.D., Lavoie, J.-P., Eitzel, C., Cho, S., Huyck, C., Ghosh, S., Eguchi, R.T., Taylor, C.E., and Moore II, J.E., 2003, "REDARS 1: Demonstration Software for Seismic Risk Analysis of Highway Systems" http://mceer.buffalo.edu/publications/resaccom/03-SP01/02werner.pdf.

Werner, S.D., Taylor, C., and Moore, J., 1997, "Loss Estimation Due to Seismic Risks to Highway Systems," *Earthquake Spectra*, Vol. 13, No. 4, pp. 585–604.

Whitman, R.V., Biggs, J.M., Brennan III, J.E., Cornell, C.A., de Neufville, R.L., and Vanmarcke, E.H., 1975, "Seismic Design Decision Analysis," *Journal of the Structural Division, ASCE*, Vol. 101(ST5), pp. 1067–1084.

Yamazaki, F., Motomura, H., and Hamada., T., 2000, "Damage assessment of expressway networks in Japan based on seismic monitoring," 12th World Conference on Earthquake Engineering. CD-ROM, Paper No. 0551.

Zhang, J., Huo, Y., Brandenberg, S.J., and Kashighandi, P., 2008, "Effects of Structural Characterizations on Fragility Functions of Bridges Subject to Seismic Shaking and Lateral Spreading," *Earthquake Engineering and Engineering Viberation*, Vo.7, No.4, pp. 369-382.

Appendix I. Calculation of the Properties of the Benchmark Bridge

AI.1. Deck

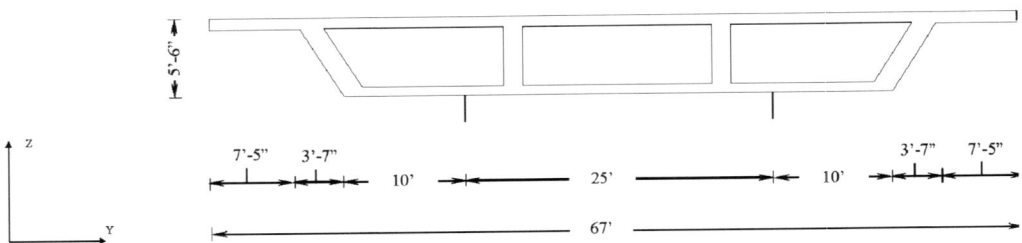

Figure AI-1. Cross-Section of Benchmark Bridge

Average span length = 134'

Width = 52' 2'' (in the horizontal direction)

2-span bridge

A= 9.52E3 in.^{2}

I_y = 6.157E6 in.^{4}

I_z = 162.136E6 in.^{4}

Skew angle = 52^0

Weight = 150* 9516.96/144 = 9.91E3 plf

Weight = 9913.5 * 134 = 1.33E6 lb (one span of body of the bridge)

AI.1.1. Calculation of Polar Moment of Inertia

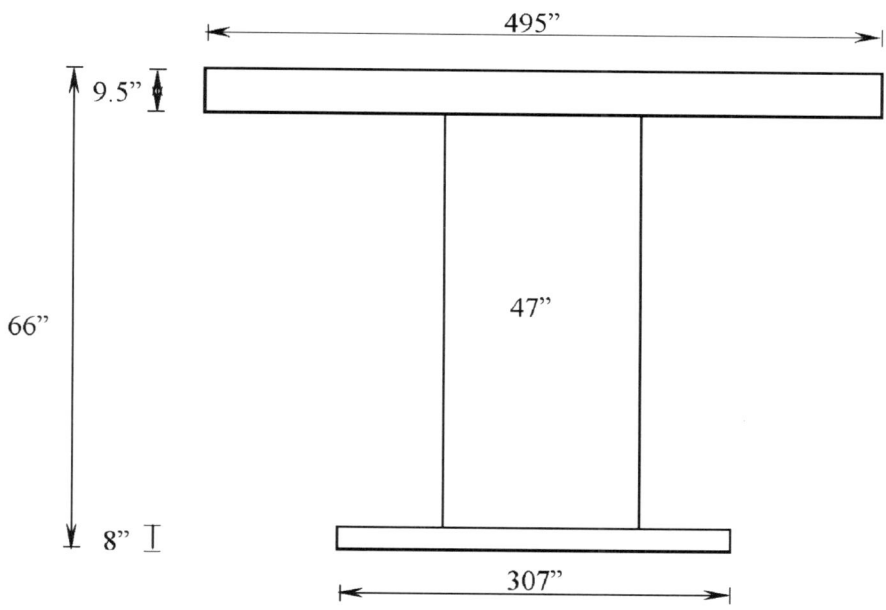

Figure AI-2. Equivalent Transformed Cross-Section

$$J = \frac{1}{3}\sum t^3 S$$

$$= 1.895E6 \text{ in.}^4$$

Table AI-1. Properties of uncracked section:

Weight per unit length	9.91E3 Plf
A	9.52E3 in.^{2}
I_y	6.157E6 in.^{4}
I_z	162.136E6 in.^{4}
J	1.895E6 in.^{4}

AI.2. Bent Cap (Cracked)

z

11 # 36*

5 # 22*(typ.)

x   x

72" x 66"

11 # 36*

z

*Bars Id in millimeters

$$I_x = \frac{72 \times 66^3}{12} = 1\ 724E6\ in^4$$

$$I_z = \frac{72^3 \times 66}{12} = 2\ 052E6\ in^4$$

$$f_c = 4000 psi$$

$$E_c = 57000\sqrt{f_c}\ psi$$
$$E_c = 3,604,996\ 5\ psi$$
$$E_s = 29 \times 10^6\ psi$$

$$n = 8\ 05$$

Calculation in X-direction

$$nA_s = 8\ 05 \times 11 \times \frac{\pi \times 1\ 42^2}{4} = 140\ 23 in^2$$

$$\therefore 72 \times y \times (\frac{y}{2}) = 140\ 23 \times (61 - y)$$

$\therefore y = 13.6"$ (*location of neutral axis*)

$$I_{crx} = \frac{1}{3} \times 72 \times (13.6)^3 + 140.23 \times 47.4^2 = 375.434E3 \ in.^4$$

$$I_{ex} = 1.050E6 \ in.^4$$

Calculation in Z-direction:

$$nA_s = 8.05 \times 5 \times \frac{\pi \times 0.87^2}{4} = 23.93 \ in^2$$

$$\therefore 66 \times y \times (\frac{y}{2}) = 23.93 \times (67 - y)$$

$\therefore y = 6.6"$ (*location of neutral axis*)

$$I_{crz} = \frac{1}{3} \times 66 \times (6.6)^3 + 23.93 \times 60.4^2 = 93.625E3 \ in.^4$$

$$I_{ez} = 1.073E6 \ in.^4$$

As inertia of Cap Beam in both directions is much larger than that of the columns, it can be modeled using rigid 3-D frame elements

AI.3. Columns (Cracked)

$$I = I_x = I_y = \frac{\pi \times 48^4}{64} = 260.6E3 \ in^4$$

$$n = 8.05$$

$$nA_s = 8.05 \times 12 \times \frac{\pi \times 1.42^2}{4} = 152.98 \ in^2$$

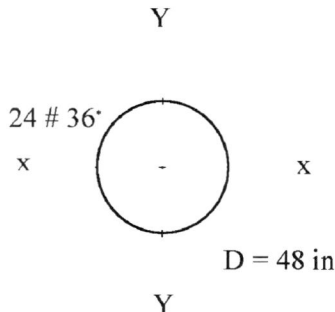

24 # 36·

D = 48 in

$$\therefore \frac{2}{3} \times 2 \times \sqrt{y(48-y)} \times (\frac{2}{5}y) = 152.98 \times (34.19 - y)$$

$$\therefore y = 30.7 \quad (\textit{location of neutral axis})$$

$$I_{cr} = 80505.1 + 152.98 \times 3.49^2 = 82.368E3 \ in.^4$$

$$I_e = 171.472E3 \ in.^4$$

Appendix II. Calculation of the Properties of Beam-Stick Models

AII.1. Calculation of Girder Spacing for Beam-Stick Model, S

$$A_s = \frac{A}{3} = \frac{9516\ 96}{3} = 3\ 17E3 \text{ in }^2 \text{ (Meng and Lui, 2002)}$$

$W_S = 3\ 3E3\ plf$

$$I_{sy(middle)} = \frac{92\ 5}{357\ 5} \times 6,157,516\ 65 = 1\ 593E6 \text{ in }^4$$

$$I_{sy(edge)} = \frac{132\ 5}{357\ 5} \times 6,157,516\ 65 = 2\ 282E6 \text{ in }^4$$

$$J_{s(edge)} = \frac{J}{2} = \frac{1,895,314\ 6}{2} = 947\ 657E3 \text{ in }^4$$

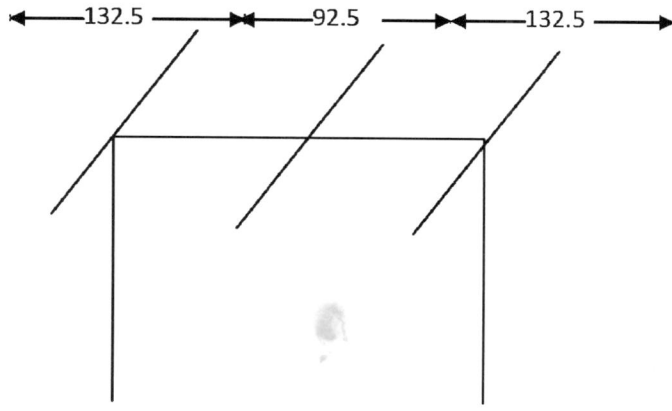

Figure AII-1. Effective Width for Beam Stick Elements

Check S: (based on Meng and Lui, 2002)

For the model:

$$I_{mz} = 2 \times M_s \times (\frac{L^2}{12} + \frac{S^2}{\cos^2 \theta}) + M_s \times (\frac{L^2}{12})$$

$$= 2 \times 27,503.29 \times (\frac{268^2}{12} + \frac{7.71^2}{0.379}) + 27,503.29 \times (\frac{268^2}{12})$$

$$= 502.48\text{E6 (for stick beams)}$$

(AII-1)

For the actual deck of the bridge:

$$I_{mz} = \sum_{i=1}^{2} [\frac{M_i}{12}(L^2 + \frac{W_i^2}{\cos^2 \theta})] + \sum_{j=1}^{n_w} [M_j(\frac{L^2}{12} + \frac{d_j^2}{\cos^2 \theta})]$$

(AII-2)

$$M_i = 357.5 \times 9.5 \times \frac{150}{144} \times 268 \times \frac{1}{32.2} = 29.44\text{E3 } lb.sec^2/ft$$

$$W_i = \frac{357.5}{12} = 29' \ 10\text{" (in the horizontal direction)}$$

$$M_j = 11.75 \times 50 \times \frac{150}{144} \times 268 \times \frac{1}{32.2} = 5.093\text{E3 } lb.sec^2/ft$$

$$I_{mz} = 2\left[\frac{29,444.714}{12}\left(268^2 + \frac{29.83^2}{0.379} \right) \right] + 2\left[5,093.49\left(\frac{268^2}{12} + \frac{5.25^2}{0.379} \right) \right] +$$

$$2\left[5,093.49\left(\frac{268^2}{12} + \frac{14.33^2}{0.379} \right) \right] = 492.19\text{E6 (for actual bridge)}$$

It can be concluded that S=7.71' will lead to reasonable results as mass moment of inertia for both the model and actual bridge are almost the same with difference ratio about 2% .

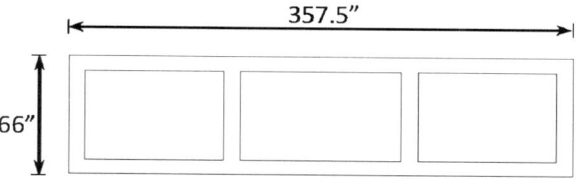

Figure AII.2. Average Cross-Section

$$I_z = 3I_{sz} + 2A_s S^2 \qquad\qquad (\text{AII-3})$$

$$162,136,666.79 = 3I_{sz} + 2 \times 3172.32 \times 92.5^2$$

$$I_{sz} = 39.95E6 \ in.^4 \ (\text{typ.})$$

Appendix III. HAZUS Fragility Curves

Steps discussed in National Bridge Inventory (NBI) document are followed. For different types of regular bridges, median PGA values are reported for each of the damage states (a_2, a_3, a_4, and a_5). The median values for different types of bridges are reported in Tables 17 and 18 in NBI document and Table 7.7 in FEMA (1999). The results of standard bridge are modified by accounting for skew effect (K_{skew}) and the 3D effect (K_{3D}).

Table AIII-1. Modification Rules used to Models 3D Effects (NBI, 1999)

Type	NBI Class	K_{3D}, Conventional Design	K_{3D}, Seismic Design Year > 1990 (>1975 in CA)
Concrete	101-106	$1+0.25/n_p$	$1+0.25/n_p$
Concrete Continuous	201-206	$1+0.33/n$	$1+0.33/n_p$
Steel	301-310	$1+0.09/n_p$; L> 20m $1+0.20/n_p$; L< 20m	$1+0.25/n_p$
Steel Continuous	402-410	$1+0.05/n_p$; L> 20m $1+0.10/n_p$; L< 20m	$1+0.33/n_p$
Prestressed Concrete	501-506	$1+0.25/n_p$	$1+0.25/n_p$
Prestressed Concrete Continuous	601-607	$1+0.33/n$	$1+0.33/n_p$
n= number of spans in bridges; n_p= n-1= number of piers			

$$C_a = S_a \ (T = 0.3 \ sec)$$

$$C_v = S_a \ (T = 1.0 \ sec)$$

$$K_{shape} = \frac{2.5 \cdot C_v}{C_a}$$

(AIII-1)

If K_{shape} is less than unity, short periods govern.

$$K_{skew} = \sqrt{\sin \alpha_{skew}}$$

(AIII-2)

Scaling relations for **damage state 2 (slight damage)**

$$A_2 = \frac{K_{shape} \cdot a_2}{S}$$

(AIII-3)

S is Soil amplification factors (F_a and F_v), in Table 4.10 of the HAZUS Technical Manual (FEMA, 1999); and a_2 is the PGA level in Tables 17 and 18.

Scaling relations for **damage states 3, 4, and 5**

$$A_i = \frac{K_{skew} \cdot K_{3D} \cdot a_i}{S}$$

(AIII-4)

For the continuous bridge Median PGA for each of damage states are 0.91, 0.91, 1.05, and 1.38.

Table AIII-2 Fragility Calculations

	0-Skew	30-Skew	60-Skew
Sa (0.3 Sec.)	0.72g	0.72g	0.72g
Sa (1.0 Sec.)	0.4g	0.4g	0.4g
K_{shape}	1.39	1.39	1.39
n_p	2	2	2
K_{3D}	1.165	1.165	1.165
K_{skew}	N/A[*]	0.931	0.707

[*]not applicable

Table AIII-3 Median PGA for Skewed Bridges using HAZUS

Damage Level	0-Skew	30-Skew	60-Skew
Slight	0.569	0.569	0.569
Moderate	0.663	0.617	0.469
Extensive	0.765	0.711	0.541
Collapse	1.005	0.935	0.710

A dispersion value of 0.4 is recommended by HAZUS technical manual (FEMA, 2003).

Appendix IV. Nomenclature

A: Area

AL: Abutment

A_s: Cross-sectional area of each stick

BP: Bearing pad

D: Column Diameter

D_I: Damage Index

d_j: Perpendicular distance of the jth web with respect to the centerline of the deck

f_c': Concrete compressive strength

f_u: Steel ultimate strength

f_y: Steel yield strength

F_y: Yield force

G: Shear modulus of elasticity

H: Column Height

h: Height

I_e: Effective moment of inertia

I_{ex}: Effective moment of inertia about X-axis

I_{ez}: Effective moment of inertia about Z-axis

IM: Ground motion index

I_{mz}: Mass moment of inertia of the actual bridge deck

$I_{sy(edge)}$: Moment of inertia of the edge sticks about the Y-axis

$I_{sy(middle)}$: Moment of inertia of the middle stick(s) about the Y-axis

I_{sy}: Moment of inertia of each stick about the Y-axis

I_{sz}: Moment of inertia of each stick about the Z-axis

I_y: Moment of inertia of the deck cross-section about the Y-axis

I_y: Weaker moment of inertia

I_z: Moment of inertia of the stick model about the Z-axis

I_z: Stronger moment of inertia

J: Torsional stiffness

$J_{s(edge)}$: Torsional stiffness of the edge stick beams

J_s: Torsional stiffness of stick beams

K_0: initial stiffness

L: Span length

L_p: Plastic length

L_{st}: Limit state

M_i: Mass of top/bottom flange of the bridge deck

M_j: Mass of the jth web

M_w: Moment magnitude

M_y: Yield Moment

M_{yy}: Moment about y-y axis

M_{zz}: Moment about z-z axis

N: Abutment seat width

n: Number of stick beams

p: Pressure

PGA: Peak ground acceleration

PGD: Peak ground displacement

PGV: Peak ground velocity

P_y: Lateral force at first yield

q_y: Shear along the y-y axis

q_z: Shear along the z-z axis

R_{max0}: Response value for bridge without skew

$R_{max\alpha}$: Response value for bridge with skew

R_x: Rotations about X-axis

R_y: Rotations about Y-axis

R_z: Rotations about Z-axis

S: Spacing between stick beams

SK: Shear key

U_-: Deformations away from the soil

U_+: Deformations into the soil

U_x: Displacement in X-direction

U_y: Displacement in Y-direction

U_z: Displacement in Z-direction

W_i: Width of top/bottom flange of the bridge deck

W_m: Tributary width of each stick

W_T: Total width of the deck

α: 90-θ

β: Dispersion value

Δ_y: Yield Deformation

ε_c: Concrete compressive strain

ε_u: Steel ultimate strain

ε_y: Steel yield strain

Θ: skew angle

μ: Ductility

μ_Δ: Rotational ductility

μ_θ: Rotational ductility

ϕ: Curvature

ψ_T: total curvature

ψ_y: Yield curvature

ψ_{yy}: Curvature about y-y axis

ψ_{zz}: Curvature about z-z axis

CPSIA information can be obtained
at www.ICGtesting.com
Printed in the USA
LVIW02n1630181113
361787LV00016B/136

9 781244 055438